Introduction to Modern Algebra and its Applications

Nadiya Gubareni

Associate Professor, Institute of Mathematics
Silesian University of Technology, Gliwice, Poland

CRC Press
Taylor & Francis Group
Boca Raton London New York

CRC Press is an imprint of the
Taylor & Francis Group, an **informa** business

A SCIENCE PUBLISHERS BOOK

CRC Press
Taylor & Francis Group
6000 Broken Sound Parkway NW, Suite 300
Boca Raton, FL 33487-2742

First issued in paperback 2022

© 2021 by Taylor & Francis Group, LLC
CRC Press is an imprint of Taylor & Francis Group, an Informa business

No claim to original U.S. Government works

Version Date: 20200425

ISBN-13: 978-0-367-60908-5 (pbk)
ISBN-13: 978-0-367-82091-6 (hbk)

DOI: 10.1201/9781003015482

Publisher's Note
The publisher has gone to great lengths to ensure the quality of this reprint but points out that some imperfections in the original copies may be apparent.

**Visit the Taylor & Francis Web site at
http://www.taylorandfrancis.com**

**and the CRC Press Web site at
http://www.crcpress.com**

Preface

This book presents various topics of modern algebra and their applications. The purpose of this book is to give a simple and clear presentation of the basic properties and examples of such algebraic structures as groups, rings, fields and finite dimensional algebras and some concrete applications of them in informatics, physics, coding information and cryptography which show the incredible power of algebra.

It should be noted that algebra as a subject has changed significantly over time. If in the Middle Ages the main subject of algebraic research was numbers and operations on them, and in IX–XVI centuries algebra was a science about solutions of algebraic equalities, since the 18th century, algebra has gradually expanded its range and algebraic operations have become the main subjects of algebraic research. The 19th century can be considered as the beginning of the creation of main abstract concepts of algebra, such as a group, a field, a ring and an algebra. However, in the 20th century, modern algebra, or simply algebra, is understood as a general theory of algebraic operations and algebraic structures which are characterized by a high level of abstraction.

The book consists of 11 chapters, Preface and Appendix. In the Appendix, we collect some basic definitions and results from set theory, relations, vector spaces and operations on sets and algebraic operations. The book also contains the index of names, and each chapter ends with a bibliography which was either used in this chapter or can be used for further study of topics considered in it. We apologized to the many authors whose works we have used but possibly not specifically cited.

The first chapter is devoted to elementary number theory. Although, number theory can be considered as a separate part of mathematics, it is very close to algebra and it serves, primarily, as a direct transition from elementary mathematics to modern algebra, and secondly, as a source of numerous elementary examples. This chapter presents only those results and notions from elementary number theory that are actually needed in this book, such as division with remainder, the greatest common divisor and the least common multiple of integers. Two important algorithms, namely, the Euclidean algorithm and the extended Euclidean algorithm, for finding the greatest common divisors of two integers and expressing it as a linear combination of these integers are considered. In this chapter, we consider linear Diophantine equations in two variables and the condition for existence of their solutions, and we also present the general form of these solutions. Congruences are an excellent tool for studying various questions of number theory. In this chapter, we study the basic properties of congruences and the solution of linear congruences in one variable.

The elements of group theory are considered in Chapter 2. Here, we study subgroups, normal subgroups, cosets and quotient groups. In this chapter, we also prove the fundamental result from group theory, Lagrange's theorem, and consider different corollaries from this theorem. At the end of this chapter, we consider group homomorphisms and prove three main theorems of isomorphisms for groups.

Some interesting examples of groups are considered in Chapter 3. In particular, we study in detail groups of permutations and cyclic groups. The concept of a group allows symmetries of geometrical figures to be characterized. In this chapter, we consider symmetry groups of some plane geometrical figures, in particular, symmetries of regular n-polygons for different n and connected with them dihedral groups. At the end of this chapter, we prove the Fundamental Theorem for finite Abelian groups.

Chapter 4 is devoted to the study of elementary ring theory. In this chapter, we introduce a number of basic concepts of ring theory, such as subrings, ideals and quotient rings, and some main properties of them are proved. Also, some special classes of commutative rings, such as integral domains, Euclidean domains and unique factorization domains, are studied in this chapter. At the end of this chapter, we prove the algebraic version of the famous Chinese Remainder Theorem involving rings and ideals, the so-called Generalized Chinese Remainder Theorem.

The concept of a polynomial is one of the central notions in mathematics. Though the ring of polynomials in one variable is a particular example of a ring, it has its own features. Chapter 5 studies the ring of polynomials in one variable together with their basic properties. We consider the theory of divisibility for polynomials presenting the algorithm of division with remainder and the Euclidean algorithm for finding the greatest common divisor of polynomials. The role of prime integers in $\mathbb{Z}$ is played by irreducible polynomials in the polynomial rings. In general, the problem of factorization of polynomials into a product of irreducible polynomials as well as the problem of verifying irreducibility of polynomials is not trivial. For some classes of polynomials over rational numbers, there is an interesting method for verification of irreducibility. This is Eisenstein's criterion, which gives a sufficient condition of irreducibility. A very useful method for rings which leads to the construction of new fields, the method of constructing quotient rings of polynomials by means of irreducible polynomials, is presented in Section 5.7. Another method for construction of new fields, the fields of fractions of integral domains, is considered in Chapter 6.

The elementary introduction to field theory is presented in Chapter 6. In particular, we study the concept of an extension of a field and consider different classes of extensions, such as finite dimensional extensions, algebraic extensions and splitting fields. We also consider algebraic elements and transcendental elements of fields. At the end of the chapter, we study algebraically closed fields and present the factorizations of polynomials over the real and complex numbers.

Some examples of the important applications of rings and fields are considered in Chapter 7. At the beginning of the chapter, we present the numbertheoretic function φ, which is called Euler's φ-function, and its application. We study the main properties of this function and prove two remarkable theorems connected with it, namely Euler's Theorem and Fermat's Little Theorem. We apply this function for solving linear congruences in one variable and present the algorithm of quick modulo exponentiation.

We also consider solving systems of linear congruences with the use of the Chinese Remainder Theorem. In this chapter, we discuss the well known problem of secret sharing, which consists of recovering a secret from a set of shares. The chapter presents tree algorithms for solving this problem, Samir's Threshold Access Algorithm, Mignotte's Threshold Secret Sharing Scheme and Asmuth-Bloom's Threshold Secret Sharing Scheme. At the end of the chapter, we present the cryptographic algorithm RSA, which is the first and most popular asymmetric cryptographic algorithm.

Polynomials in several variables, their properties and applications are considered in Chapter 8. In particular, we study the important class of these polynomials which are symmetric polynomials and prove the Fundamental Theorem on symmetric polynomials. We also study the ideals of the polynomial rings. In this chapter, we consider Noetherian commutative rings, i.e., rings whose ideals are finitely generated, and the main properties of such rings. One of the most important statements in the theory of polynomials is the Hilbert Basis Theorem, proved by David Hilbert (1862–1943) in 1888, which states that every polynomial ring in several variables over a Noetherian ring is also Noetherian. In this chapter, we give the proof of the Hilbert Basis Theorem. The constructive method of finding a finite set of generators of ideals of the ring of polynomials in several variables, which is based on the method of Gröbner bases, is presented in this chapter. The notion of Gröbner basis, main properties and the algorithm for computing these bases introduced and developed by Bruno Buchberger in 1965 are presented in this chapter. The true significance of Gröbner bases shows the facts that many fundamental problems in mathematics can be reduced by simple algorithms to constructing Gröbner bases. Bruno Buchberger also showed many interesting applications of Gröbner bases. Some elementary applications to algebra and elementary algebraic geometry are presented at the end of this chapter.

Finite fields, also called Galois fields, are very important algebraic structures. Galois fields play a significant role in various fields of mathematics: Number theory, group theory, projective geometry and others. They also have many interesting practical applications, particularly in coding theory and cryptography. Chapter 9 considers some basic properties of finite fields and their applications. We study the structure of finite fields and their multiplicative groups. We prove one of the most remarkable properties of finite fields which has many applications, i.e., that the multiplicative group of a finite field is cyclic. In this chapter, we also present one of the most difficult computational problems: The problem of discrete logarithm. For this purpose the notions of primitive roots and indexes are introduced and some their basic properties are given. The discrete logarithm problem is very important and has a significant application in cryptography. The security of many cryptographic algorithms, such as the ElGamal cryptosystem and the Diffie-Hellman scheme considered in this chapter, relies on the presumed hardness of the discrete logarithm problem. Finite fields are also used for construction of error-correcting codes. This chapter presents the main examples of such codes. In particular, we consider linear codes, polynomial codes, cyclic codes and BCH codes.

The theory of finite dimensional algebras is one of the most fundamental fields of modern algebra, which has various applications in other branches of mathematics and theoretical physics. The first examples of non-commutative algebras were given in the works of W.R. Hamilton, A. Cayley and H. Grassman in 1843–1845. Chapter 10 is devoted to the study of hypercomplex numbers, i.e., finite dimensional algebras over the field of real or complex numbers. Here, we study the algebras of quaternions and octonions, and we consider the important construction of Cayley-Dickson, which shows how quaternions, octonions, sedeonions and other algebras are obtained in the natural way from the real numbers. The chapter closes with a brief presentation of dual numbers, double numbers, Clifford and Grassmann algebras.

Quaternions, octonions and other hypercomplex numbers have a lot of applications not only in mathematics, but also in other fields of physics and technics. Chapter 11 presents some their applications in number theory and computer graphics. In particular, at the beginning of this chapter, we prove the identity on representing a product of two n-squares sums of integers as a sum of n squares of some integers using quaternions and octonions. Furthermore, the connection of Fermat's theorem on two-square sum with Gaussian integers and the connection of Lagrange's theorem, is shown which states that each natural number can be written as a sum of four squares, with some kinds of quaternions. At the end of this chapter, we show that each quaternion can be written in the trigonometric form by means of its module and some angle. We close this chapter showing the use of quaternions to write rotations in the three-dimensional real space.

In the first chapters of the book, the material is presented in as detailed a manner as possible, whereas, in the later chapters of the book, some details of the proofs have been omitted because we assume that the Reader already has a sufficient understanding of the subject matter in order to complete the proofs.

It is hard to imagine a Reader who can understand algebra without considering enough examples. This book contains a lot of examples, from simple ones to more complicated ones, which help the Reader to understand the abstract concepts and results discussed in the text and teach how to use them. The author hopes that they will be helpful to the Reader. At the end of each chapter, we present a list of exercises which should not be difficult for the Reader.

This textbook is intended for a wide range of readers, primarily for students and young researchers of applied mathematics, computer science, engineering and other areas of natural and technical sciences. It is written as a self-contained book and can be accessible for independent study on modern algebra and its applications. In particular, we do not assume knowledge of any preliminary information on the theory of groups, rings or finite dimensional algebras.

The text of this book is originated from my lectures, which I have given at the Częstochowa University of Technology and the Silesian University of Technology, Poland, since 2008. The book has also grown from the first version of my textbook "Algebra współczesna i jej zastosowania" published by Publishing House of Częstochowa University of Technology in 2018.

In closing, the author wishes to express her cordial thanks to Aleksander Katkow, who made the pictures for this book. I also thank the CRC Press Science Publishers, in particular Vijay Primlani and Danielle Zarfati, for their very friendly cooperation.

December, 2019   **Nadiya Gubareni**
                                                    Częstochowa, Poland

Contents

Chapter 1

Elements of Number Theory

"God made the integers, all the rest is the work of man."
Leopold Kronecker (1823-1891)

"Mathematics is the queen of the sciences and number theory is the queen of mathematics. She often condescends to render service to astronomy and other natural sciences, but in all relations she is entitled to the first rank."
Carl Friedkich Gauss (1777-1855)

Numbers, in particular natural numbers and integers, have enthralled mathematicians from ancient times to the present. A branch of mathematics which studies the properties of integers is called number theory. The beginnings of this theory can be seen already in the works of Euclid (323-283 BC). The Ancient Greek mathematician, Diophantus (III c. AD), was the first to create number theory as a separate and rigorous mathematical discipline. Unfortunately, after Diophantus, almost nobody developed number theory until the XVIIth century. *"Arithmetics"* of Diophantus, which is so far considered as one of the most mysterious phenomenons in the history of science, had a large influence on the French mathematician Pierre Fermat (1601-1655), who made many important discoveries in number theory. Thanks to his works, the study of integers

Carl Friedrich Gauss
(1777-1855)

again became the center of attention of mathematicians. One of the significant discoveries of P. Fermat was the quite simple theorem, which is now known as Fermat's Little Theorem. Nevertheless it currently has quite significant applications, in particular, in cryptographic algorithms.

The further progress in number theory is connected with such prominent mathematicians as L. Euler (1707-1783), J.L. Lagrange (1736-1813) and C.F.

Gauss (1777-1855). In his famous book *"Disquisitiones arithmeticae"* (1801), C.F. Gauss presented the fundamentals of number theory, systematized the knowledge of predecessors, introduced new methods and made various famous discoveries. In particular, Gauss created a new branch of number theory: Modular arithmetics, i.e., the theory of congruences.

In this chapter, which may be considered as an introduction to elementary number theory, we review some basic concepts of this theory, including divisibility, division with remainder, greatest common divisors, least common multiplies, congruences and their properties. We also describe some important algorithms and their applications in finding the greatest common divisor of two integers, solving linear Diophantine equations in two variables and linear congruences in one variable.

1.1. Divisibility of Integers. Division with Remainder

One of the most important concepts in number theory is the notion of divisibility of two integers.

> ✳ **Definition 1.1.**
> Let $m, n \in \mathbb{Z}$ and $m \neq 0$. We say that a number n is **divisible** by a number m, or equivalently m **divides** n, if there exists an integer k such that $n = mk$. In this case, we say that m is a **divisor** (or a **factor**) of n and n is a **multiple** of m. We write $m|n$ if m divides n. We also write $m \nmid n$ if m does not divide n.

✴ **Examples 1.2.**

- Let $n = 12$, $m = 3$, then $12 = 3 \cdot 4$. So $3|12$, i.e., 3 is a divisor of 12, and 12 is a multiple of 3.

- Let $n = 28$, $m = -7$, then $28 = (-7) \cdot 4$. So $-7|28$, i.e., -7 is a divisor of 28, and 28 is a multiple of -7.

- Let $n = 12$, $m = 5$. Then $5 \nmid 12$, since $12/5$ is not an integer.

The following theorem gives some basic properties of divisibility of integers.

✴ **Theorem 1.3.**
If $a, b, c, d \in \mathbb{Z} \backslash \{0\}$ then:
1. $(a|b) \wedge (a|c) \implies a|(b+c)$
2. $(a|b) \implies a|(bk)$ for all $k \in \mathbb{Z}$
3. $(a|b) \wedge (b|c) \implies a|c$
4. $(a|b) \wedge (c|d) \implies ac|bd$
5. $(a|b) \wedge (b|a) \implies (a = b) \vee (a = -b)$

Proof.

1) $(a|b) \wedge (a|c) \implies \exists (m, n \in \mathbb{Z})$ such that $(b = ma) \wedge (c = na) \implies b + c = (m + n)a \implies a|(b + c)$.

2) $(a|b) \implies \exists (m \in \mathbb{Z})$ such that $b = ma \implies bk = mak \implies a|bk$ for all $k \in \mathbb{Z}$.

3) $(a|b) \wedge (b|c) \implies \exists (m, n \in \mathbb{Z})$ such that $(b = ma) \wedge (c = nb) \implies c = nma \implies a|c$.

4) $(a|b) \wedge (c|d) \implies \exists (m, n \in \mathbb{Z})$ such that $(b = ma) \wedge (d = nc) \implies bd = mnac \implies (ac)|(bd)$

5) $(a|b) \wedge (b|a) \implies \exists (m, n \in \mathbb{Z})$ such that $(b = ma) \wedge (a = nb) \implies ab = mnab \implies mn = 1 \implies m = \pm 1 \implies a = \pm b$. $\quad\square$

Each positive number $n > 1$ has at least two divisors: 1 and n. Depending on number of divisors all positive integers are divided into two classes of numbers: Prime numbers and composite numbers.

> ⁂ **Definition 1.4.**
> A natural number $p \geq 2$ is called a **prime number** (or a **prime**) if it has exactly two positive divisors: 1 and p. A positive integer $n > 1$, which is not a prime, is called a **composite** number.

So a composite number has at least 3 different factors. Note that the integer 1 is not considered to be either a prime number or a composite number.

✳ **Examples 1.5.**

- 2, 3, 5, 7, 11, 13, 17, 23, 29 31 are primes

- $-6, 8, -24, 45$ are composite numbers

The **fundamental theorem of arithmetic** states that every positive integer n can be expressed as a product of primes:

$$n = p_1^{\alpha_1} p_2^{\alpha_2} \cdots p_k^{\alpha_k} \qquad (1.1)$$

where $p_1, p_2, \ldots, p_k$ are distinct primes and $\alpha_1, \alpha_2, \cdots, \alpha_k$ are positive integers. Moreover, this factorization (which is often called the **prime factorization**) is unique, up to a reordering of the prime factors.

We will prove this theorem in a more general form in Chapter 4.

Taking into account the fundamental theorem of arithmetic, we can consider the primes as "building blocks" by which there can be built all integers. Therefore, prime numbers may be considered as atoms of number theory. The primes were introduced by Euclid of Alexandria (323-283 BC) who also proved that there are infinitely many prime numbers.

It is easy to see that m does not always divide n for arbitrary integers m and n. However, there is another kind of division of integers which always gives a quotient and a remainder. This operation is called the **Euclidean division** or **division with remainder** and its existence is based on the following theorem which is often called the **division algorithm**.

❋ **Theorem 1.6 (Division with Remainder).**

Given any integers a, b with $b \neq 0$, there exist unique integers q, r such that

$$a = bq + r, \quad \text{and} \ \ 0 \leq r < |b| \tag{1.2.}$$

Proof.

1) Let $a, b \in \mathbb{Z}$. Suppose that $b > 0$. Consider the set of non-negative integers:

$$S = \{a - bx \ : \ (x \in \mathbb{Z}) \wedge (a - bx \geq 0)\}$$

Note that S is a non-empty set. Indeed, if $a > 0$, then $a \in S$ and if $a \leq 0$, we can always choose a negative integer $x \in \mathbb{Z}$ such that $a - bx > 0$. Therefore, by the Well-Ordering Principle the set S has a least element $r = a - bq \geq 0$. This means that $a - b(q+1) < 0$, otherwise $a - b(q+1) = a - bq - b = r - b < r$. Therefore, $qb \leq a < qb + b$, i.e., $0 \leq r = a - qb < b$. Hence, $a = qb + r$, where $r = a - qb \in \mathbb{Z}$ and satisfies the condition $0 \leq r < b$.

2) Let $b < 0$. Then, $b_1 = -b > 0$. It follows from the previous point, that there exist integer numbers q_1, r_1 such that $a = q_1 b_1 + r_1$, where $r_1 = a - q_1 b_1 \in \mathbb{Z}$ and satisfies the condition $0 \leq r_1 < b_1$. Therefore,

$$a = q_1 b_1 + r_1 = q_1(-b) + r_1 = (-q_1)b + r_1,$$

where $q = -q_1$, $0 \leq r = r_1 < |b| = b_1$.

That proves the existence of q and r satisfying (1.2).

For uniqueness, we suppose that $a = bq + r$ and $a = bq_1 + r_1$, where $0 \leq r < |b|$ and $0 \leq r_1 < |b|$. Then, after subtracting these two equations we obtain $r - r_1 = b(q_1 - q)$ implying that $|r - r_1| = |b| \cdot |q_1 - q|$. However, from two inequalities $0 \leq r < |b|$ and $0 \leq r_1 < |b|$, it follows that $|r - r_1| < |b|$. Hence, $|b| \cdot |q_1 - q| = |r - r_1| < |b|$. Since $b \neq 0$, we have the only possibility $q = q_1$, and so $r_1 = r$. $\square$

In equality (1.2) of the division algorithm, numbers q and r are called, respectively, the **quotient** and the **remainder** of a when a is divided by b. We will write the remainder r as $r_b(a)$ or $a \bmod b$, and the quotient q will be written as $a \operatorname{div} b$.

❋ **Examples 1.7.**

1. If $a = 17$, $b = 5$ then $r = r_5(17) = 17 \bmod 5 = 2$; $q = 17 \operatorname{div} 5 = 3$
2. If $a = 35$, $b = 8$ then $r = r_8(35) = 35 \bmod 8 = 3$; $q = 35 \operatorname{div} 8 = 5$
3. If $a = -5$, $b = 7$ then $q = -1$, $r = 2$.
4. If $a = 5$, $b = -7$ then $q = 0$, $r = 5$.

It is easy to obtain the following useful theorem and its corollary:

✳ **Theorem 1.8.**

For an arbitrary integer b and a positive integer n the number $b^n - 1$ is divided by $b - 1$, and

$$b^n - 1 = (b - 1)(b^{n-1} + b^{n-2} + \cdots + b + 1) \qquad (1.3)$$

✳ **Corollary 1.9.**

For an arbitrary integer b and positive integers n, m the number $b^{nm} - 1$ is divided by $b^m - 1$, and

$$b^{nm} - 1 = (b^m - 1)(b^{m(n-1)} + b^{m(n-2)} + \cdots + b^m + 1) \qquad (1.4)$$

✳ **Example 1.10.**

By Corollary 1.9, $2^{35} - 1 = (2^5 - 1)(2^{30} + 2^{25} + \cdots + 2^5 + 1)$, so $31 | (2^{35} - 1)$. Analogously $127 | (2^{35} - 1)$, since $2^7 - 1 = 127$.

1.2. The Greatest Common Divisor. The Euclidean Algorithm

In this section, we introduce two important notions of number theory: The greatest common divisor and the least common multiple of integers, and consider the Euclidean algorithm, which is one of the oldest algorithms in mathematics. It was known already in ancient times and was used for the computation of the greatest common divisors of two integers. This algorithm was described by the ancient Greek mathematician Euclid, and was included in his famous book "Elements". So far, it is known as the Euclidean Algorithm.

✳ **Definition 1.11.**

Let $a, b \in \mathbb{Z}$ with $(a \neq 0) \vee (b \neq 0)$. A **common divisor** of integers a, b is an integer d such that $d|a$ and $d|b$.

The largest positive integer d such that $d|a$ and $d|b$ is called the **greatest common divisor** of a and b. The greatest common divisor of a and b is denoted by $\gcd(a, b)$. So,

$$\gcd(a, b) = \max\{s \in \mathbb{Z}^+ \ : \ (s|a) \wedge (s|b)\}. \qquad (1.5)$$

In other words, a positive integer d is the greatest common divisor of integers a, b if it satisfies the following conditions:
1) $(d|a) \wedge (d|b)$
2) $(s|a) \wedge (s|b) \implies s \leq d$.

❋ Remarks 1.12.

1. The greatest common divisor always exists, since each integer has only a finite set of divisors.

2. It follows from Definition 1.11, that $\gcd(a,0) = a$ for any integer $a \neq 0$.

3. Note that:

$$\gcd(a,b) = \gcd(-a,b) = \gcd(a,-b) = \gcd(-a,-b) \qquad (1.6)$$

✳ Examples 1.13.

1. $\gcd(15,0) = 15$.
2. $\gcd(14,7) = 7$.
3. $\gcd(-14,7) = \gcd(14,-7) = \gcd(-14,-7) = \gcd(14,7) = 7$.
4. $\gcd(120,500) = 20$.

Analogously, we can introduce the notion of the greatest common divisor of n integers.

❋ Definition 1.14.

The **greatest common divisor** of integers $a_1, a_2, \ldots, a_n$, which are not all equal to zero, is the largest positive integer d that divides each integer a_i for $i = 1, \ldots, n$, that is if it satisfies the following conditions:

1) $d|a_i$ for each $i = 1, \ldots, n$
2) if $(s|a_i)$ for each $i = 1, \ldots, n$ then $s \leq d$.

✳ Examples 1.15.

1. $\gcd(15,0,5) = 5$.
2. $\gcd(14,7,21) = 7$.
3. $\gcd(120,500,240,600) = 20$.

❋ Definition 1.16.

The **least common multiple** of non-zero integers $a_1, a_2, \ldots, a_n$, where $n \geq 2$, is said to be the smallest positive integer M divisible by all a_i for $i = 1, 2, \ldots, n$. It is denoted by $\operatorname{lcm}(a_1, a_2, \ldots, a_n)$. So

$$\operatorname{lcm}(a_1, a_2, \ldots, a_n) = \min\{S \in \mathbb{Z}^+ \; : \; a_i | S, \; i = 1, 2, \ldots, n\}. \qquad (1.7)$$

In other words, a positive integer M is the least common multiple of non-zero integers $a_1, a_2, \ldots, a_n$ if it satisfies the following conditions:

1) $a_i|M$ for each $i = 1, 2, \ldots, n$
2) if $a_i|S$ for each $i = 1, 2, \ldots, n$ then $S \geq M$.

✴ Examples 1.17.

1. $\operatorname{lcm}(5,8) = 40$
2. $\operatorname{lcm}(6,-8) = 24$.

The natural question arises:

- **How do we calculate the greatest common divisor and the least common multiple of integers?**

If we have the factorization of positive integers a and b into products of prime factors:

$$a = p_1^{\alpha_1} p_2^{\alpha_2} \cdots p_k^{\alpha_k}, \quad b = p_1^{\beta_1} p_2^{\beta_2} \cdots p_k^{\beta_k},$$

where $\alpha_i, \beta_i \geq 0$, then the answer is simple. Namely, in this case it easy to show that:

$$\boxed{\gcd(a, b) = p_1^{\gamma_1} p_2^{\gamma_2} \cdots p_k^{\gamma_k}} \tag{1.8}$$

where $\gamma_i = \min(\alpha_i, \beta_i)$, and

$$\boxed{\operatorname{lcm}(a, b) = p_1^{\gamma_1} p_2^{\gamma_2} \cdots p_k^{\gamma_k}} \tag{1.9}$$

where $\gamma_i = \max(\alpha_i, \beta_i)$.

✳ Example 1.18.

Compute $\gcd(240, 560)$ and $\operatorname{lcm}(240, 560)$.

Since the prime factorizations of 240 and 560 are: $240 = 2^4 \cdot 3 \cdot 5 = 2^4 \cdot 3^1 \cdot 5^1 \cdot 7^0$, and $560 = 2^4 \cdot 5 \cdot 7 = 2^4 \cdot 3^0 \cdot 5^1 \cdot 7^1$, the greatest common divisor is: $\gcd(240, 560) = 2^4 \cdot 3^0 \cdot 5^1 \cdot 7^0 = 80$ and the least common multiple is $\operatorname{lcm}(240, 560) = 2^4 \cdot 3^1 \cdot 5^1 \cdot 7^1 = 1680$.

Unfortunately, we often do not know the prime factorizations of integers, and finding such factorizations for large enough integers is often computationally a very difficult problem using existing methods. Luckily, there exists a very effective method for finding the greatest common divisor of two integers without knowing the prime factorizations of these integers. This is the **Euclidean algorithm**, which is based on the following results.

✳ Lemma 1.19.

Suppose $a, b \in \mathbb{Z}$ and $b > 0$. If $b \mid a$, then

$$\gcd(a, b) = b. \tag{1.10}$$

Proof. Since $b \mid a$ and $b \mid b$, b is a common divisor of a and b. If c is a common divisor of a and b, then $(c \mid a) \wedge (c \mid b)$, so $c \leq b$. Since $b > 0$, it follows by Definition 1.11 that $b = \gcd(a, b)$. $\qquad\square$

✳ Lemma 1.20.

If $a, b, q, r \in \mathbb{Z}$ with $b > 0$, $0 \leq r < b$ and $a = bq + r$, then

$$\gcd(a, b) = \gcd(b, r) \tag{1.11}$$

Proof.

Let $d = \gcd(a,b)$ and $d_1 = \gcd(b,r)$, where $a = bq + r$. Then, $(d_1|b) \wedge (d_1|r) \implies d_1|a$. So $(d_1|b) \wedge (d_1|a)$. Hence, it follows by Definition 1.11 that $d_1 \leq d$. On the other hand, from the equality $r = a - bq$, it follows that $d|b$ and $d|r$. Hence, by Definition 1.11, we obtain that $d \leq d_1$. So, $d_1 \leq d \leq d_1$, which implies $d = d_1$. $\qquad\square$

✳ Example 1.21.
$\gcd(45,12) = \gcd(12, r_{12}(45)) = \gcd(12,9) = \gcd(9, r_9(12)) = \gcd(9,3) = \gcd(3, r_3(9)) = \gcd(3,0) = 3$.

If we have large enough numbers, then it is most likely that we do not know the prime factorizations of these numbers. In this case, we can apply the Euclidean Algorithm, which works in the following way.

Since $\gcd(a,b) = \gcd(|a|, |b|)$, we can suppose that $a \geq b > 0$. If $b|a$, then $\gcd(a,b) = b$, by Lemma 1.19. If $b \nmid a$, we start by taking $r_0 = a$, $r_1 = b$ and then by successively applying the division algorithm we get the following sequence of quotients and remainders:

$$
\begin{array}{lclc}
r_0 & = & r_1 q_1 + r_2, & 0 < r_2 < r_1 \\
r_1 & = & r_2 q_2 + r_3, & 0 < r_3 < r_2 \\
r_2 & = & r_3 q_3 + r_4, & 0 < r_4 < r_3 \\
\vdots & \vdots & \vdots & \vdots \\
r_{i-1} & = & r_i q_i + r_{i+1}, & 0 < r_{i+1} < r_i \\
\vdots & \vdots & \vdots & \vdots \\
r_{n-2} & = & r_{n-1} q_{n-1} + r_n, & 0 < r_n < r_{n-1} \\
r_{n-1} & = & r_n q_n + 0 &
\end{array}
\tag{1.12}
$$

where r_n is the last non-zero remainder. This number always exists because we have a decreasing bounded sequence of natural numbers:

$$0 = r_{n+1} < r_n < r_{n-1} < \cdots < r_3 < r_2 < r_1 \leq r_0 = a,$$

which can have only a finite number of elements.

Then, by Lemma 1.20, we get that $r_n = \gcd(a,b)$, since:

$$\gcd(a,b) = \gcd(r_0, r_1) = \gcd(r_1, r_2) = \cdots = \gcd(r_{n-2}, r_{n-1}) =$$

$$= \gcd(r_{n-1}, r_n) = \gcd(r_n, 0) = r_n$$

The algorithm (1.12) described above is called the **Euclidean Algorithm**. So, we obtain the following result:

✳ Theorem 1.22.
The Euclidean Algorithm (1.12) always provides in the result the greatest common divisor of integers a, b in a finite number of steps and $\gcd(a,b) = r_n$, where r_n is the last non-zero remainder in (1.12).

✳ **Example 1.23.**

Find the greatest common divisor of 327 and 234, gcd(327, 234), using the Euclidean algorithm.

Consistently applying the division algorithm, we get:

$327 = 234 \cdot 1 + 93$

$234 = 93 \cdot 2 + 48$

$93 = 48 \cdot 1 + 45$

$48 = 45 \cdot 1 + 3$

$45 = 3 \cdot 15$

Hence, gcd(327, 234) = 3.

Using pseudocode, the Euclidean algorithm can be written in the following form:

The Euclidean Algorithm

procedure $gcd(a, b)$
input: integers $a \geq b > 0$, x, y
   $x := a$, $y := b$
   while $y \neq 0$
     begin
       $r := x \bmod y$
       $x := y$
       $y := r$
     end
   return x
output: $gcd(a, b) := x$

1.3. The Extended Euclidean Algorithm

There is an important and very useful result which shows that the greatest common divisor of two integers can be presented as a linear combination of these integers.

✳ **Theorem 1.24.**

Let $a, b \in \mathbb{Z}$, not both zero, and $d = \gcd(a, b)$. Then, there exist integers x, y such that

$$d = xa + yb \qquad (1.13)$$

Proof.

Let $a, b \in \mathbb{Z}$, not both zero, and $d = \gcd(a, b)$. Consider the set of positive integers:

$$S = \{ua + vb \ : \ (u, v \in \mathbb{Z}) \wedge (ua + vb > 0)\}$$

S is a non-empty subset in the set of natural numbers and so, by the Well-Ordering Principle, it has a least element $s = xa + yb$ for some $x, y \in \mathbb{Z}$. We show that $s|a$ and $s|b$.

Assume that $a = sq + r$, where $0 \leq r < s$. Then, $r = a - sq = a - (xa + yb) = (1 - x)a - yb \in S$. Therefore, $(r \in S) \wedge (0 \leq r < s)$. Since s is a least element in S, we obtain that $r = 0$, i.e., $s|a$. Analogously, we can show that $s|b$. Therefore, by definition of $\gcd(a, b)$ we obtain that $s \leq d$. On the other hand, we have that $d|s$, since $s = xa + yb$ and $(d|a) \wedge (d|b)$. Hence, $d \leq s$. Therefore, $s = d$. $\qquad\square$

✱ **Remarks 1.25.**

1. Note, that numbers x and y in (1.13) are not unique. Moreover, there are an infinite number of ways to express $\gcd(a, b)$ as a linear combination $xa + yb$ (we will show this in section 1.5).

2. The proof of theorem 1.24 does not give any explicate algorithm for finding numbers x, y in (1.13). The **extended Euclidean algorithm** may be used to efficiently compute these numbers. It is based on the following procedure, which is obtained from the Euclidean Algorithm by back substitution.

We rewrite equalities (1.12) in the following form:

$$
\begin{aligned}
r_2 &= r_0 - r_1 q_1 \\
r_3 &= r_1 - r_2 q_2 \\
r_4 &= r_2 - r_3 q_3 \\
\vdots \quad &\vdots \quad \vdots \\
r_{i+1} &= r_{i-1} - r_i q_i \\
\vdots \quad &\vdots \quad \vdots \\
r_{n-1} &= r_{n-3} - r_{n-2} q_{n-2} \\
r_n &= r_{n-2} - r_{n-1} q_{n-1}
\end{aligned}
\tag{1.14}
$$

Using these equalities, we begin from the last equality and by successively substituting the equality above in it we obtain the sequence of equalities, which gives the required numbers x and y:

$$d = r_n = r_{n-2} - r_{n-1} q_{n-1} = r_{n-2} u_1 + r_{n-1} v_1 =$$

$$= r_{n-2} u_1 + (r_{n-3} - r_{n-2} q_{n-2}) v_1 = r_{n-3} v_1 + (u_1 - q_{n-2} v_1) r_{n-2} =$$

$$= r_{n-3} u_2 + r_{n-2} v_2 = \cdots = r_1 u_{n-2} + r_2 v_{n-2} = r_1 u_{n-2} + (r_0 - r_1 q_1) v_{n-2} =$$

$$= r_0 v_{n-2} + r_1 (u_{n-2} - q_1 v_{n-2}) = r_0 u_{n-1} + r_1 v_{n-1} = ax + by$$

✳ Example 1.26.

Find integers x, y such that $\gcd(31, 23) = 31x + 23y$.

Consistently applying the division algorithm, we get:

$$31 = 23 \cdot 1 + 8 \quad \Rightarrow \quad 8 = 31 - 23 \cdot 1$$
$$23 = 8 \cdot 2 + 7 \quad \Rightarrow \quad 7 = 23 - 8 \cdot 2$$
$$8 = 7 \cdot 1 + 1 \quad \Rightarrow \quad 1 = 8 - 7 \cdot 1$$
$$7 = 7 \cdot 1$$

Therefore, $\gcd(31, 23) = 1$ and

$$1 = 8 - 7 \cdot 1 = 8 - (23 - 8 \cdot 2) = 8 \cdot 3 - 23 = (31 - 23 \cdot 1) \cdot 3 - 23 = 31 \cdot 3 - 23 \cdot 4.$$

So $1 = 31x + 23y$, where $x = 3$, $y = -4$.

We can present the algorithm for finding the numbers x, y such that $\gcd(a, b) = ax + by$ in a recursive form. First, from (1.14) we have:

$$d = r_n = r_{n-2} - r_{n-1}q_{n-1} = r_{n-2}u_1 + r_{n-1}v_1 \tag{1.15}$$

where $u_1 = 1$, $v_1 = -q_{n-1}$. We will show that

$$d = r_{n-i-1}u_i + r_{n-i}v_i \tag{1.16}$$

for all $1 \le i \le n - 1$. For $i = 1$ we have (1.15). Suppose that

$$d = r_{n-m-1}u_m + r_{n-m}v_m \tag{1.17}$$

Then, from (1.14) $r_{n-m} = r_{n-m-2} - r_{n-m-1}q_{n-m-1}$ and substituting to (1.17) we obtain:

$$d = r_{n-m-1}u_m + (r_{n-m-2} - r_{n-m-1}q_{n-m-1})v_m =$$

$$= r_{n-m-2}v_m + r_{n-m-1}(u_m - q_{n-m-1}v_m) = r_{n-m-2}u_{m+1} + r_{n-m-1}v_{m+1},$$

where $u_{m+1} = v_m$ and $v_{m+1} = u_m - q_{n-m-1}v_m$. So, equality (1.16) is correct for all $1 \le i \le n - 1$. In particular, if $i = n - 1$, we obtain

$$d = r_0 u_{n-1} + r_1 v_{n-1} = ax + by, \tag{1.18}$$

where $u_{n-1} = x$ and $v_{n-1} = y$.

In this way, using the Euclidean algorithm (1.12), we get the algorithm for finding numbers $x = u_{n-1}$ and $y = v_{n-1}$ such that $\gcd(a, b) = ax + by$ in the following recursive form:

$$\begin{cases} u_0 = 0, \ v_0 = 1 \\ u_1 = 1, \ v_1 = -q_{n-1} \\ u_{i+1} = v_i \\ v_{i+1} = u_i - q_{n-i-1}v_i \end{cases} \tag{1.19}$$

This is a backward recurrence algorithm, which is an implementation of the **Extended Euclidean Algorithm** for expressing the greatest common divisor of two numbers as a linear combination of these numbers. It can be written in the form of the following table:

i	0	1	2	$\cdots$	$n-2$	$n-1$
q_{n-i-1}	q_{n-1}	q_{n-2}	q_{n-3}	$\cdots$	q_1	
u_i	0	1	v_1	$\cdots$	v_{n-3}	v_{n-2}
v_i	1	$-q_{n-1}$	$u_1 - q_{n-2}v_1$	$\cdots$	$u_{n-3} - q_2 v_{n-3}$	$u_{n-2} - q_1 v_{n-2}$

✳ Example 1.27.

Find integers x, y such that $\gcd(31, 23) = 31x + 23y$ using the Extended Euclidean algorithm in the recursive form (1.19).

From example 1.26, we have that $\gcd(31, 23) = 1$ and $q_1 = 1$, $q_2 = 2$, $q_3 = 1$, $n = 4$. We have the following table:

i	0	1	2	3
q_{n-i-1}	1	2	1	
u_i	0	1	-1	3
v_i	1	$-q_3 = -1$	$u_1 - q_2 v_1 = 3$	$u_2 - q_1 v_2 = -4$

So that $x = u_3 = 3$, $y = v_3 = -4$ and $1 = 3 \cdot 31 + (-4) \cdot 23$.

✳ Example 1.28.

Find integers x, y such that $\gcd(327, 234) = 327x + 234y$ using the Extended Euclidean algorithm in the recursive form (1.19).

From example 1.23, we have $\gcd(327, 234) = 3$ and $q_1 = 1$, $q_2 = 2$, $q_3 = 1$, $q_4 = 1$, $n = 5$. We have the following table:

i	0	1	2	3	4
q_{n-i-1}	1	1	2	1	
u_i	0	1	-1	3	-5
v_i	1	-1	2	-5	7

So that $x = u_4 = -5$, $y = v_4 = 7$ and $3 = (-5) \cdot 327 + 7 \cdot 234$.

The Extended Euclidean Algorithm in the recursive form (1.19) is a nice algorithm for hand computation, but it is not good enough for implementation on a computer, because it has to store the array of all the Euclidean Algorithm quotients q_i, and moreover, the dimension of this array is not known beforehand. There are another versions of this algorithm one of which we consider below.

For given integers a, b, there exist numbers x, y such that $\gcd(a, b) = ax + by$, by Theorem 1.24. Suppose that numbers r_i are as in the Euclidean algorithm (1.14), i.e., $r_{i+1} = r_{i-1} - r_i q_i$, where $q_i = r_i \operatorname{div} r_{i-1}$,

$r_{i+1} = r_i \bmod r_{i-1}$, $r_0 = a$ and $r_1 = b$. We will show that for all i there exist sequences of integers x_i, y_i such that

$$r_i = ax_i + by_i. \qquad (1.20)$$

Since for $i = 0$ and for $i = 1$ we have $r_0 = a = a \cdot 1 + b \cdot 0$ and $r_1 = b = a \cdot 0 + b \cdot 1$ we can put $x_0 = 1$, $x_1 = 0$, $y_0 = 0$, $y_1 = 1$. We will prove the equality (1.20) for all $i > 1$:

$$r_{i+1} = r_{i-1} - r_i q_i = (ax_{i-1} + by_{i-1}) - (ax_i + by_i) =$$

$$= a(x_{i-1} - x_i q_i) + b(y_{i-1} - y_i q_i) = ax_{i+1} + by_{i+1}$$

So that the equality (1.20) is correct for all i and r_i, x_i, y_i can be computed using the following recursion rules:

$$\begin{cases} r_{i+1} = r_{i-1} - r_i q_i \\ x_{i+1} = x_{i-1} - x_i q_i \\ y_{i+1} = y_{i-1} - y_i q_i \end{cases} \qquad (1.21)$$

while $r_{n+1} = 0$. Then, $r_n = \gcd(a, b)$, $x = x_n$, $y = y_n$, and $\gcd(a, b) = r_n = ax_n + by_n$. The integers x_n, y_n are called the **Bèzout coefficients** of integers a and b.

The Extended Euclidean Algorithm (1.21) can be written in the form of the following table:

i	0	1	2	$\cdots$	$k+1$	$\cdots$	n	$n+1$
r_i	a	b	$r_0 \bmod r_1$	$\cdots$	$r_{k-1} \bmod r_k$	$\cdots$	$r_{n-2} \bmod r_{n-1}$	0
q_{i-1}			$r_0 \operatorname{div} r_1$	$\cdots$	$r_{k-1} \operatorname{div} r_k$	$\cdots$	$r_{n-2} \operatorname{div} r_{n-1}$	
x_i	1	0	1	$\cdots$	$x_{k-1} - x_k q_k$	$\cdots$	$x_{n-2} - x_{n-1} q_{n-1}$	
y_i	0	1	$-q_1$	$\cdots$	$y_{k-1} - y_k q_k$	$\cdots$	$y_{n-2} - y_{n-1} q_{n-1}$	

where $\gcd(a, b) = r_n$, $x = x_n$, $y = y_n$ and $\gcd(a, b) = ax + by$.

✳ Example 1.29.

Find $\gcd(327, 234)$ and express it as a linear combination of 327 and 234. We have the following table:

i	0	1	2	3	4	5	6
r_i	$a = 327$	$b = 234$	93	48	45	3	0
q_{i-1}			1	2	1	1	
x_i	1	0	1	-1	2	-5	
y_i	0	1	-1	2	-3	7	

Hence, $n = 5$, $\gcd(327, 234) = r_5 = 3$, $x = x_5 = -5$, $y = y_5 = 7$ and $3 = (-5) \cdot 327 + 7 \cdot 234$.

1.4. Relatively Primes

> **✳ Definition 1.30.**
> Two integers a, b are called **relatively prime** or **coprime** if
> $$\gcd(a, b) = 1. \qquad (1.22)$$

✳ Examples 1.31.
1. The numbers $392 = 2^3 \cdot 7^2$ and $2025 = 3^4 \cdot 5^2$ are relatively prime.
2. The numbers 234 and 109 are relatively prime, since $\gcd(109, 234) = 1$.

From Theorem 1.24, we immediately obtain the following result, which gives an equivalent definition of relatively prime integers.

✳ Corollary 1.32.
Integers a, b are relatively prime if and only if there exist integers x, y such that
$$ax + by = 1 \qquad (1.23)$$

The main properties of relatively prime integers are given in the form of the following theorem.

✳ Theorem 1.33 (Euclid).
Let $a, b, c \in \mathbb{Z}$ and $\gcd(a, b) = 1$. Then
1) $b|ac \ \Rightarrow \ b|c$
2) $(a|c) \wedge (b|c) \ \Rightarrow \ ab|c$

Proof.
1) If $\gcd(a, b) = 1$, then there exist integers x, y such that $1 = ax + by$, by Corollary 1.32. Multiplying both sides of this equation by c, we obtain $c = acx + bcy$. Taking into account that $(b|ac) \wedge (b|bc)$, it follows that $b|(acx + bcy)$, and, hence, $b|c$.
2) If $\gcd(a, b) = 1$, then there exist integers x, y such that $1 = ax + by$. Since $(a|c) \wedge (b|c)$, there exist integers k, t such that $c = ak = bt$. Therefore,
$$c = acx + bcy = abtx + baky = ab(tx + ky),$$
and, hence, $ab|c$. $\qquad \square$

✳ Corollary 1.34.
If p is a prime integer and $p|ab$ then $p|a$ or $p|b$.

Proof. Assume that $p \nmid a$, then $\gcd(p, a) = 1$. Therefore, $p|b$, by Theorem 1.33. $\qquad \square$

✳ Corollary 1.35.
If p is a prime integer and $p|a_1a_2\cdots a_n$ then $p|a_i$ for some i.

Proof. We will prove the statement by induction on n. For $n = 1$ the statement is obvious and for $n = 2$ the statement is true, by Corollary 1.34. Suppose that statement is true for $n = k$ and we will verify it for $n = k + 1$. Let $x = a_1a_2\cdots a_k$ and $y = a_{k+1}$. Assume that $p|xy$. If $p|y$, then $p|a_{k+1}$. If $p \nmid y$, then $p|x$, by Corollary 1.34. In this case, $p|a_i$ for some i, by the induction hypothesis. So, in both cases, $p|a_i$ for some i, i.e., the hypothesis holds for $n = k + 1$ as well. Therefore, the statement is true for arbitrary n. $\square$

The definition of the greatest common divisor of integers can be given in the following equivalent form which is useful for applying:

✳ Proposition 1.36.
A positive integer d is the greatest common divisor of two integers a, b, not both zero, if and only if it satisfies the following conditions:
(i) $(d|a) \wedge (d|b)$
(ii) for any $s \in \mathbb{Z}$ such that $(s|a) \wedge (s|b)$ it follows that $s|d$.

Proof.

$\Rightarrow$. Let d be the greatest common divisor of integers a and b. Then, $(d|a) \wedge (d|b)$, by Definition 1.11. It follows from Theorem 1.24 that there exist integers x, y such that $d = ax + by$. Therefore, if $(s|a) \wedge (s|b)$ then $s|d$.

$\Leftarrow$. Suppose that a positive integer d satisfies the condition (i) and (ii). Let $F = \gcd(a, b)$. Then, $(F|a) \wedge (F|b)$ and $d \leq F$, by Definition 1.11. On the other hand, $F|d$ by condition (ii). Therefore, $F \leq d$. In this way, $d \leq F \leq d$. Hence, $d = F = \gcd(a, b)$. $\square$

Similar to Proposition 1.36 for two integers we can prove the following statement which gives the equivalent definition of the greatest common divisor of n integers:

✳ Proposition 1.37.
A positive integer d is the greatest common divisor of n integers $m_1, m_2, \ldots, m_n$, which are not all equal to zero, if and only if it satisfies the following conditions:
(1) $(d|m_i)$ for all $i = 1, \ldots, n$
(2) for any $s \in \mathbb{Z}$ such that $(s|m_i)$ for all $i = 1, \ldots, n$ it follows that $s|d$.

The following lemma allows the least common multiple of two integers to be found by their greatest common divisor without knowing the prime factorizations of these integers.

✳ Lemma 1.38.

For non-zero integers a, b, the following equality holds:

$$\gcd(a, b) \cdot \operatorname{lcm}(a, b) = |ab|. \tag{1.23}$$

Proof.

Let $d = \gcd(a, b)$, $M = \operatorname{lcm}(a, b)$ and $m = |ab|/d$. We show that $M = m$. Without loss of generality, we can suppose that $a, b > 0$. Then $a = a_1 d$, $b = b_1 d$ and $m = a_1 b_1 d = a b_1 = a_1 b$. This implies that $a | m$ and $b | m$, i.e., m is a common multiple of a and b. We will show that m is the least such number, i.e., $(a|n) \wedge (b|n) \implies m \le n$.

Let $(a|n) \wedge (b|n)$, then there exist integers t, s such that $(n = at) \wedge (n = bs)$. Therefore, $n = a_1 dt = b_1 ds$, and, hence, $a_1 t = b_1 s$. Since $\gcd(a_1, b_1) = 1$, it follows from Theorem 1.33(1) that $(a_1|s) \wedge (b_1|t)$, i.e., there exist integers u, v such that $(s = a_1 u) \wedge (t = b_1 v)$. In this way, $n = a_1 dt = a_1 b_1 dv = mv$, hence, $m \le n$, which means that $m = M = \operatorname{lcm}(a, b)$, by Definition 1.16.   □

✳ Corollary 1.39.

If $\gcd(a, b) = 1$ then $\operatorname{lcm}(a, b) = |ab|$.

✳ Examples 1.40.

1. Find $\operatorname{lcm}(327, 234)$.

From Example 1.23 we know that $\gcd(327, 234) = 3$. So we get, by Lemma 1.38, that $\operatorname{lcm}(327, 234) = (327 \cdot 234)/3 = 25506$.

2. Find $\operatorname{lcm}(31, 23)$.

From Example 1.26 we know that $\gcd(31, 23) = 1$. So we get, by Corollary 1.39, that $\operatorname{lcm}(31, 23) = 31 \cdot 23 = 713$.

1.5. Linear Diophantine Equations

In general, a **Diophantine equation** is an equation of the following form

$$P(x_1, x_2, \ldots, x_n) = 0 \tag{1.24}$$

where P is a polynomial in n variables $x_1, x_2, \ldots, x_n$ with integer coefficients, and whose solutions are restricted to integers. This type of equation is named after the ancient mathematician Diophantus, who lived in Alexandria in the 3rd century AD.

In this section, we consider **linear Diophantine equations** in two variables x, y which has the following form:

$$ax + by = c, \tag{1.25}$$

where a, b, c are given integers.

An **integer solution** (we will simply call it a **solution**) of this equation is the ordered pair (u, v), where $u, v \in \mathbb{Z}$, such that $au + bv = c$. Note that a Diophantine equation does not always have a solution.

The following theorem gives the condition for existence of solutions for linear Diophantine equations.

✳ **Theorem 1.41.**
A linear Diophantine equation $ax + by = c$ in two variables x, y with $a, b, c \in \mathbb{Z}$ has solutions if and only if $d | c$, where $d = \gcd(a, b)$.

Proof.
$\Rightarrow$. Suppose that equation (1.25) has a solution, i.e., there exist integers u, v such that $au + bv = c$. If $d = \gcd(a, b)$, then $a = a_1 d$, $b = b_1 d$, where $\gcd(a_1, b_1) = 1$. Therefore, $c = a_1 du + b_1 dv = (a_1 u + b_1 v)d$, which implies that $d | c$.

$\Leftarrow$. Suppose that $d | c$, where $d = \gcd(a, b)$. By Theorem 1.24, there exist integers u, v such that $au + bv = c$, i.e., the pair of integers (u, v) is a solution of Equation (1.25). $\qquad\square$

From the proof of this theorem, we immediately obtain the following corollary which will be used to solve Equation (1.25).

✳ **Corollary 1.42.**
To solve a linear Diophantine equation $ax + by = c$ is equivalent to solving a linear Diophantine equation $a_1 x + b_1 y = c_1$, where $a = a_1 d$, $b = b_1 d$, $c = c_1 d$, and $\gcd(a_1, b_1) = 1$.

✳ **Theorem 1.43.**
A linear Diophantine equation $ax + by = c$, where x, y are variables, $a, b, c \in \mathbb{Z}$ and $\gcd(a, b) = d$ has no solution if d does not divide c. If $d | c$ there are an infinite number of integer solutions to this equation. All general solutions are given by the following form:

$$x = x_0 + b/dt = x_0 + b_1 t, \quad y = y_0 - a/dt = y_0 - a_1 t, \qquad (1.26)$$

where $t \in \mathbb{Z}$, (x_0, y_0) is a particular solution of the equation $ax + by = c$ and $\gcd(a_1, b_1) = 1$.

Proof.
The first part of the theorem is given by Theorem 1.41. We will prove the second part of the theorem.

1) First, we will show that (1.26) is a solution of Equation (1.25):

$$a(x_0 + b_1 t) + b(y_0 - a_1 t) = ax_0 + by_0 + (ab_1 - ba_1)t = ax_0 + by_0 + (a_1 db_1 - b_1 da_1)t = c.$$

2) Now, we will show that each solution of Equation (1.25) has the form (1.26).

Let (x_0, y_0) be a particular solution of the equation $ax + by = c$, i.e., $ax_0 + by_0 = c$. Then, $a(x - x_0) + b(y - y_0) = 0$ or $a(x - x_0) = b(y_0 - y)$. If $\gcd(a, b) = d$ then $a = a_1 d$, $b = b_1 d$, $c = c_1 d$, and $\gcd(a_1, b_1) = 1$. Therefore, $da_1(x - x_0) = db_1(y_0 - y)$ or $a_1(x - x_0) = b_1(y_0 - y)$. Since $\gcd(a_1, b_1) = 1$, we have $b_1 | (x - x_0)$, i.e., $x - x_0 = b_1 t$ or $x = x_0 + b_1 t$. Then, $a_1 b_1 t = b_1(y_0 - y)$ or $a_1 t = y_0 - y$. So $y = y_0 - a_1 t$. $\qquad\square$

✱ Examples 1.44.

1. Solve the equation $13x + 7y = 1$.

   In this case, $\gcd(13, 7) = 1$.

   Using the Extended Euclidean Algorithm, we find a particular solution: $x_0 = -1$, $y_0 = 2$. Then, the general solution:

   $x = -1 + 7t$, $y = 2 - 13t$, where $t \in \mathbb{Z}$.

2. Solve the equation $9x + 15y = 7$.

   In this case, $\gcd(9, 15) = 3$. Since 3 does not divide 7, the equation has no solutions, by Theorem 1.41.

3. Solve the equation $1000x + 73y = 1$.

   In this case, $\gcd(1000, 73) = 1$

   Using the Extended Euclidean Algorithm, we find a particular solution: $x_0 = -10$, $y_0 = 137$. Therefore, the general solution:

   $x = -10 + 73t$, $y = 137 - 1000t$, where $t \in \mathbb{Z}$.

4. Solve the equation $15x + 27y = 3$.

   In this case, $\gcd(15, 27) = 3$

   Since $3|3$, the given equation is solvable and equivalent to the equation: $5x + 9y = 1$, where $\gcd(5, 9) = 1$.

   Using the Extended Euclidean Algorithm, we find a particular solution: $x_0 = 2$, $y_0 = -1$. Therefore,

   $x = 2 + 9t$, $y = -1 - 5t$, where $t \in \mathbb{Z}$.

5. Solve the equation $38x + 30y = 6$.

   In this case, $\gcd(38, 30) = 2$

   Since $2|6$, the given equation is solvable and equivalent to the equation: $19x + 15y = 3$, where $\gcd(19, 15) = 1$.

   Using the Extended Euclidean Algorithm, we find integers $u = 4$, $v = -5$ such that $19u + 15v = 1$. Since $19(3u) + 15(3v) = 3$, the pair of integers (x_0, y_0), where $x_0 = 3u = 12$, $y_0 = 3v = -15$, is a particular solution of the equation $19x + 15y = 3$. Therefore, the general solution of this equation has the following form: $x = 12 + 15t$, $y = -15 - 19t$, where $t \in \mathbb{Z}$.

1.6. Congruences and their Properties

The theory of congruences is one of the basic instruments in number theory.

✱ **Definition 1.45.**
Let m be a positive integer and $x, y \in \mathbb{Z}$. We say that x, y are **congruent modulo** m if their difference $x - y$ is divisible by m, which means that there exists an integer k such that $x - y = km$ or $x = y + km$. We use the notation:

$$x \equiv y \,(\mathrm{mod}\, m) \tag{1.27}$$

which is called a **congruence**, and the number m is called the **modulus** of this congruence.

If $x - y$ is not divisible by m then we say that x and y are incongruent modulo m and we write $x \not\equiv y \,(\mathrm{mod}\, m)$.

✳ **Examples 1.46.**
  1. $33 \equiv 3(\mathrm{mod}\, 5)$ since $33 - 3 = 6 \cdot 5$.
  2. $29 \equiv 5(\mathrm{mod}\, 8)$ since $29 - 5 = 3 \cdot 8$.

We consider the relation:

$$\mathbf{R} = \{(x, y) \in \mathbb{Z} \times \mathbb{Z} \ : \ x \equiv y(\mathrm{mod}\, m)\},$$

which is called the **congruence relation modulo** m.

✳ **Theorem 1.47.**
The congruence relation modulo m

$$\mathbf{R} = \{(x, y) \in \mathbb{Z} \times \mathbb{Z} \ : \ x \equiv y(\mathrm{mod}\, m)\},$$

is an equivalence relation in the set of integers $\mathbb{Z}$.

Proof.
1. Reflexivity:
$\forall (x \in \mathbb{Z})[x \equiv x \,(\mathrm{mod}\, m)]$, since $x - x = 0 = 0 \cdot m$.
2. Symmetry:
$\forall (x, y \in \mathbb{Z})[x \equiv y \,(\mathrm{mod}\, m) \ \Leftrightarrow \ y \equiv x(\mathrm{mod}\, m)]$, since $x \equiv y \,(\mathrm{mod}\, m) \ \Leftrightarrow \ m|(y - x) \ \Leftrightarrow \ m|(x - y) \ \Leftrightarrow \ y \equiv x(\mathrm{mod}\, m)$.
3. Transitivity:
$\forall (x, y, z \in \mathbb{Z})[(x \equiv y(\mathrm{mod}\, m)) \wedge (y \equiv z(\mathrm{mod}\, m)) \ \Rightarrow \ x \equiv z(\mathrm{mod}\, m)]$, since $(x \equiv y(\mathrm{mod}\, m)) \wedge (y \equiv z(\mathrm{mod}\, m)) \ \Rightarrow \ (m|(y - x)) \wedge (m|(z - y)) \ \Rightarrow \ m|((y - x) + (z - y)) \ \Rightarrow \ m|(z - x) \ \Rightarrow \ x \equiv z(\mathrm{mod}\, m)$. $\square$

Therefore, the set of all integers with respect to congruence relation modulo m is decomposed into equivalence classes, which are called **congruence classes** or **residue classes modulo** m.

The number of congruence classes modulo m is equal to m and all numbers belonging to one class have the same remainder when they are divided by m. These m congruence classes modulo m are denoted by:

$$[0], [1], [2], \ldots, [m-1],$$

where

$$[a] = \{a + km \ : \ k \in \mathbb{Z}\} = \{a, a \pm m, a \pm 2m, a \pm 3m, \ldots\}$$

Note that every integer belongs to one and only one congruence class modulo m, and any two integers of different residue classes modulo m are incongruent modulo m. Therefore, all these classes form a partition of the set of integers.

The quotient set of the set of integers $\mathbb{Z}$ with respect to congruent relation modulo m is the set of congruent classes modulo m and it is denoted by $\mathbb{Z}/m\mathbb{Z}$. In this way:

$$\mathbb{Z}/m\mathbb{Z} = \{[0], [1], [2], \ldots, [m-1]\} \tag{1.28}$$

For any integer a and any positive integer m there are integers $q, r \in \mathbb{Z}$ such that $a = mq + r$ and $0 \leq r < m$. The integer r is called the **least residue** of a modulo m. Every integer has the unique least residue modulo m among the integers $0, 1, \ldots, m-1$. It is obvious that all these integers are not pairwise congruent modulo m.

> ✳ **Definition 1.48.**
> The set of integers
> $$\mathbb{Z}_m = \{0, 1, \ldots, m-1\} \tag{1.29}$$
> is called the **least residue system modulo** m.
>
> A subset $S \subset \mathbb{Z}$ is called a **complete residue system modulo** m if S contains exactly m elements and all these elements are not pairwise congruent modulo m.

Note that the least residue system modulo m is a complete residue system modulo m. While the least residue system modulo m is unique, there are infinitely many complete residue systems modulo m.

✳ **Example 1.49.**
For $m = 4$ we have four distinct residue classes:
$[0] = \{\ldots, -8, -4, 0, 4, 8 \ldots\}$
$[1] = \{\ldots, -7, -3, 1, 5, 9 \ldots\}$
$[2] = \{\ldots, -6, -2, 2, 6, 10 \ldots\}$
$[3] = \{\ldots, -5, -1, 3, 7, 11 \ldots\}$
So that $\mathbb{Z}/4\mathbb{Z} = \{[0], [1], [2], [3]\}$.
$\mathbb{Z}_4 = \{0, 1, 2, 3\}$ is the least residue system modulo 4 which is also a complete residue system modulo 4. $S = \{4, -3, 6, 7\}$ is another example of a complete residue system modulo 4.

Congruences have properties which are analogous to properties of algebraic equations. The basic properties of congruences are given in the form of the following theorem:

✳ **Theorem 1.50.**

Let $a, b, c, a_1, b_1, a_2, b_2, m \in \mathbb{Z}$ with $m > 0$.

1) If $a_1 \equiv a_2 \pmod{m}$ and $b_1 \equiv b_2 \pmod{m}$, then $a_1 + b_1 \equiv a_2 + b_2 \pmod{m}$.
2) If $a_1 \equiv a_2 \pmod{m}$ and $b_1 \equiv b_2 \pmod{m}$, then $a_1 \cdot b_1 \equiv a_2 \cdot b_2 \pmod{m}$.
3) If $a_1 \equiv a_2 \pmod{m}$, then $ka_1 \equiv ka_2 \pmod{m}$ for all $k \in \mathbb{Z}$.
4) If $a \equiv b \pmod{m}$, then $a^n \equiv b^n \pmod{m}$ for all $n \in \mathbb{N}$.
5) If $a \equiv b \pmod{m}$, then $P(a) \equiv P(b) \pmod{m}$ for all $P(x) \in \mathbb{Z}[x]$.
6) If $a + b \equiv c \pmod{m}$, then $a \equiv c - b \pmod{m}$.
7) If $a \equiv b \pmod{m}$, then $a + k \equiv b + k \pmod{m}$ for all $k \in \mathbb{Z}$
8) If $a \equiv b \pmod{m}$, then $\gcd(a, m) = \gcd(b, m)$.
9) If $(ak \equiv bk \pmod{m}) \wedge (\gcd(k, m) = 1)$, then $a \equiv b \pmod{m}$

Proof.

1) $(a_1 \equiv a_2 \pmod{m}) \wedge (b_1 \equiv b_2 \pmod{m}) \Rightarrow (a_1 = a_2 + mt) \wedge (b_1 = b_2 + ms) \Rightarrow a_1 + b_1 = a_2 + b_2 + mt + ms = a_2 + b_2 + m(t + s) \Rightarrow a_1 + b_1 \equiv a_2 + b_2 \pmod{m}$.

2) $(a_1 \equiv a_2 \pmod{m}) \wedge (b_1 \equiv b_2 \pmod{m}) \Rightarrow (a_1 = a_2 + mt) \wedge (b_1 = b_2 + ms) \Rightarrow a_1 b_1 = a_2 b_2 + b_2 mt + a_2 ms + mtms = a_2 b_2 + m(tms + b_2 t + a_2 s + s) \Rightarrow a_1 b_1 \equiv a_2 b_2 \pmod{m}$.

3) It follows from part 2 of the theorem when $b_1 = b_2 = c$

4) It follows from part 2 of the theorem applying it n times.

5) It follows immediately from the previous parts of the theorem.

6) $a + b \equiv c \pmod{m} \Rightarrow a + b = c + mk \Rightarrow a = c - b + mk \Rightarrow a \equiv c - b \pmod{m}$

7) It follows from part 1 of the theorem.

8) $a \equiv b \pmod{m} \Rightarrow r_m(a) = r_m(b)$. From Lemma 1.20, we have that $\gcd(a, m) = \gcd(m, r_m(a))$ and $\gcd(b, m) = \gcd(m, r_m(b))$. Therefore, $\gcd(a, m) = \gcd(b, m)$.

9) $(ak \equiv bk \pmod{m}) \Rightarrow m | k(a - b)$. Since $\gcd(k, m) = 1$, it follows from Theorem 1.31(1) that $m | (a - b)$. Therefore, $a \equiv b \pmod{m}$. $\square$

✳ **Example 1.51.**

The congruence $20 \equiv 26 \pmod{6}$ is correct.

However, the congruence $10 \equiv 13 \pmod{6}$ obtained from the given congruence by dividing the both its sides by $\gcd(20, 26, 6) = 2 \neq 1$ is not correct.

Nevertheless, it is easy to prove the following properties of congruences for all $a, b, k \in \mathbb{Z}$ with $k \neq 0$:

1. If $ak \equiv bk \pmod{mk}$, then $a \equiv b \pmod{m}$.
2. If $a \equiv b \pmod{mk}$, then $a \equiv b \pmod{m}$.

✱ Examples 1.52.

1. Find which of the integers $255, 258, -220$ is congruent to $d = 11$ modulo 19.

   Since $255 - 11 = 244$ is not divisible by 19, $255 \not\equiv 11 (\operatorname{mod} 19)$.

   Since $258 - 11 = 247 = 19 \cdot 13$, $258 \equiv 11 (\operatorname{mod} 19)$.

   Since $-220 - 11 = -231$ is not divisible by 19, $-220 \not\equiv 11 (\operatorname{mod} 19)$.

2. Show that $27^{72} \cdot 26^{26} \equiv 14^{14} (\operatorname{mod} 10)$.

   Since $27 \equiv -3 (\operatorname{mod} 10)$, it follows from Theorem 1.50(4) that $27^{72} \equiv (-3)^{72} (\operatorname{mod} 10)$ or $27^{72} \equiv 3^{72} (\operatorname{mod} 10)$. Since $3^{72} = (3^2)^{36} \equiv (-1)^{36} \equiv 1 (\operatorname{mod} 10)$, we obtain $27^{72} \equiv 1 (\operatorname{mod} 10)$.

   Analogously, $26^{26} \equiv (-4)^{26} \equiv ((-4)^2)^{13} \equiv (6)^{13} \equiv 6 \cdot (6^2)^3 \equiv 6^4 \equiv 6 (\operatorname{mod} 10)$.

   Therefore, from Theorem 1.50(2) we have: $27^{72} \cdot 26^{26} \equiv 6 (\operatorname{mod} 10)$.

   On the other hand, $14 \equiv 4 (\operatorname{mod} 10)$. Therefore, $14^{14} \equiv 4^{14} \equiv (4^2)^7 \equiv 6^7 \equiv 6 (\operatorname{mod} 10)$.

   In this way, $27^{72} \cdot 26^{26} \equiv 14^{14} (\operatorname{mod} 10)$.

3. Prove that, for any positive integer n, it holds that $31 | (2^{5n} - 1)$.

   Since $2^5 = 32 \equiv 1 (\operatorname{mod} 31)$, from Theorem 1.50(4) we have $2^{5n} \equiv 1^n \equiv 1 (\operatorname{mod} 31)$, i.e., $2^{5n} - 1 \equiv 0 (\operatorname{mod} 31)$, which means that $31 | (2^{5n} - 1)$.

✱ **Theorem 1.53.**

Let a, b, x, m be positive integers and $\gcd(x, m) = 1$. If $x^a \equiv 1 (\operatorname{mod} m)$ and $x^b \equiv 1 (\operatorname{mod} m)$ then $x^d \equiv 1 (\operatorname{mod} m)$, where $d = \gcd(a, b)$.

Proof.

By Theorem 1.24, there exist integers u, v such that $au + bv = d$. Then, $x^d = x^{au+bv} = (x^a)^u (x^b)^v \equiv 1 (\operatorname{mod} m)$. $\qquad \square$

1.7. Linear Congruences

✱ **Definition 1.54.**

An equation of the form

$$ax \equiv b (\operatorname{mod} m), \tag{1.30}$$

where $a, b, m \in \mathbb{Z}$, with $m > 0$, and x is a variable, is called a **linear congruence** in one variable x.

An integer solution of congruence (1.30) is a residue class $[u]$ modulo m which satisfies this congruence, i.e.,

$$av \equiv b (\operatorname{mod} m)$$

for all $v \in [u]$. If $0 \le u < m$, then a solution of the congruence is written in the form:

$$x \equiv u (\bmod\, m).$$

Since there exist only m distinct residue classes modulo m, each congruence (1.30) possesses at most m incongruent integer solutions[1].

If $\gcd(a, m) = 1$, then a solution of a congruence $ax \equiv 1 (\bmod\, m)$ is called an **inverse of** a **modulo** m.

✳ **Theorem 1.55.**

A linear congruence $ax \equiv b (\bmod\, m)$, where $a, b, m \in \mathbb{Z}$, has an integer solution if and only if a linear Diophantine equation

$$ax + my = b$$

has an integer solution.

Proof.

$\Rightarrow$ Suppose that x_0 is a solution of a linear congruence $ax \equiv b (\bmod\, m)$, i.e., $ax_0 \equiv b (\bmod\, m)$. Then, by definition of congruence, $ax_0 = b + my_0$ for some $y_0 \in \mathbb{Z}$, which means that $(x_0, -y_0)$ is a solution of the linear Diophantine equation $ax + my = b$.

$\Leftarrow$ Suppose that (x_0, y_0) is a solution of a linear Diophantine equation $ax + my = b$, i.e., $ax_0 + my_0 = b$. This means that $ax_0 \equiv b (\bmod\, m)$, i.e., x_0 is a solution of a linear congruence $ax \equiv b (\bmod\, m)$. □

✳ **Theorem 1.56.**

1) A linear congruence $ax \equiv b (\bmod\, m)$, where $a, b, m \in \mathbb{Z}$, with $m > 0$, has integer solutions if and only if $d | b$, where $d = \gcd(a, m)$.

2) If $\gcd(a, m) = 1$ the congruence $ax \equiv b (\bmod\, m)$ has a unique solution which is the residue class modulo m of the form $[sb]$, where $s = x_0 \in \mathbb{Z}$ is a solution of the congruence $ax \equiv 1 (\bmod\, m)$.

3) If $\gcd(a, m) = d$, the congruence $ax \equiv b (\bmod\, m)$ has d incongruent solutions modulo m, which are residue classes modulo m of the form:

$$[x_0], [x_0 + m_1], [x_0 + 2m_1], \ldots, [x_0 + (d-1)m_1],$$

where x_0 is a solution of the congruence $a_1 x \equiv b_1 (\bmod\, m_1)$, and $a_1 = a/d$, $b_1 = b/d$, $m_1 = m/d$.

Proof.

1) This follows from Theorem 1.55 and Theorem 1.41.

2) If $\gcd(a, m) = 1$, then, by Corollary 1.32, there exist integers s, t such that $as + mt = 1$. This implies that $as \equiv 1 (\bmod\, m)$. From Theorem 1.50(3), we obtain that $asb \equiv b (\bmod\, m)$. Consequently, $x \equiv sb (\bmod\, m)$ is an integer

---

[1] When we refer to the number of integer solutions of $ax \equiv b (\bmod\, m)$, we mean the number of incongruent integers modulo m satisfies the congruence.

solution of the given congruence. Note that $s = x_0, t = y_0$ is a particular solution of a linear Diophantine equation $ax + my = 1$.

3) If $\gcd(a, m) = d$ and $d|b$, then $a = a_1 d$, $b = b_1 d$, $m = m_1 d$, where $\gcd(a_1, b_1) = 1$ and we have the equivalent congruence $a_1 x \equiv b_1 \pmod{m_1}$. Using the previous case, we obtain that $x \equiv sb_1 \pmod{m_1}$ is an integer solution of the given congruence. Note that $s = x_0, t = y_0$ is a particular solution of a linear Diophantine equation $a_1 x + m_1 y = 1$. Thus, we have d incongruent integer solutions of the congruence $ax \equiv b \pmod{m}$ which are congruence classes modulo m of the following form:

$$[x_0], [x_0 + m_1], [x_0 + 2m_1], \ldots, [x_0 + (d-1)m_1],$$

because

$$a(x + m_1 k) = ax + am_1 k = ax + a_1 dm_1 k =$$
$$= ax + a_1 dm \equiv ax \pmod{m} \equiv b \pmod{m}.$$

$\square$

✳ Examples 1.57.

1. Solve the congruence $6x \equiv 1 \pmod 7$.

Since $\gcd(6, 7) = 1$, the congruence has one integer solution.

Using the Extended Euclidean algorithm, we find a particular solution of the Diophantine equation $6x + 7y = 1$: $x_0 = 5$. Therefore, $x \equiv 5 \pmod 7$ is the integer solution of the given congruence.

2. Solve the congruence $9x \equiv 7 \pmod{15}$.

Since $\gcd(9, 15) = 3$ is not divisible by 7, the congruence has no integer solutions.

3. Solve the congruence $15x \equiv 6 \pmod{27}$.

Since $\gcd(15, 27) = 3$ divides 6, the congruence has three incongruent integer solutions. In this case, $a_1 = 5$, $b_1 = 2$ and we have the equivalent congruence: $5x \equiv 2 \pmod 9$ with $\gcd(5, 9) = 1$.

Using the Extended Euclidean algorithm we find a particular solution of the Diophantine equation $5x + 9y = 2$: $x_0 = 4$. Therefore, $x \equiv 4 \pmod 9$ is an integer solution of the congruence $5x \equiv 2 \pmod 9$. Consequently, we have three incongruent integer solutions of the given congruence $15x \equiv 6 \pmod{27}$, which are the following residue classes modulo 27:

$$[4], [13], [22]$$

So that all integer solutions of the given congruence can be written in the following form:

$$x \equiv 4 \pmod{27}, x \equiv 13 \pmod{27}, x \equiv 22 \pmod{27}$$

1.8. Exercises

1. Prove that there are infinitely many prime numbers.

2. Find the quotient and the remainder when an integer a is divided by an integer b if:

   (a) $a = 94$, $b = 17$

   (b) $a = 17$, $b = -94$

   (c) $a = -17$, $b = 94$

   (d) $a = -17$, $b = -94$

3. It is known that when an integer a is divided by an integer b the remainder is r. Find the remainder when an integer $-a$ is divided by an integer b.

4. It is known that when an integer a is divided by the integer 19 the remainder is equal to 5. Find the remainder r when an integer $12a$ is divided by the integer 19.

5. Solve the system of equations: $\gcd(a, b) = 10$, $\mathrm{lcm}(a, b) = 100$.

6. Using the Euclidean algorithm, find $\gcd(a, b)$ if:

   (a) $a = 844$, $b = 5442$

   (b) $a = 9744$, $b = -1248$

   (c) $a = -2891$, $b = 1589$

   (d) $a = 321$, $b = 843$

   (e) $a = 2166$, $b = 6099$

   (f) $a = 6787$, $b = 7194$

   (g) $a = 23521$, $b = 75217$

7. Find the integer solutions of the following Diophantine equations:

   (a) $10x + 2y = 5$

   (b) $12x + 20y = 4$

   (c) $27x + 15y = 3$

   (d) $7x + 5y = 1$

   (e) $15x + 24y = 9$

   (f) $16x + 4y = 1830$

   (g) $21x + 19y = 5$

8. Prove that $\gcd(m, n) = \gcd(n, m(\bmod n))$ for all $m, n \in \mathbb{Z}$ and $n \neq 0$.

9. Prove that $\gcd(km, kn) = k \cdot \gcd(m, n)$.

10. Prove that if $\gcd(m, n) = 1$, then $\gcd(mk, n) = \gcd(k, n)$.

11. Prove that if $\gcd(m, k) = 1$ and $\gcd(n, k) = 1$, then $\gcd(mn, k) = 1$.

12. Prove that $\gcd(k^m - 1, k^n - 1) = k^{\gcd(m,n)} - 1$ for all $m, n \in \mathbb{Z}$ and $k \geq 2$.

13. Find $\gcd(3^{288} - 1, 3^{216} - 1)$.

14. Prove that if $\gcd(m, n) = 1$, then $\gcd(2^m - 1, 2^n - 1) = 1$.

15. Prove that when n^2, where n is a positive integer, is divided by 4, the remainder is equal to either to 0 or 1.

16. Prove that, in a Pythagoras triangle (a rectangular triangle which sides are positive integers), the length of the least side is divisible by 3.

17. Solve the linear congruence:

    (a) $29x \equiv 1 \pmod{17}$
    (b) $3x \equiv 4 \pmod 9$
    (c) $6x \equiv 14 \pmod{22}$

18. Using the properties of congruences, prove that:

    (a) $7 | (3^{30} - 2^{30})$,
    (b) $31 | (30^{99} + 61^{100})$

19. Using the properties of congruences, prove that, for any natural number n, it holds that:

    (a) $13 | (1 + 3^{3n+1} + 9^{3n+1})$,
    (b) $19 | (5^{2n+3} + 3^{n+3}2^n)$.

20. Prove that the equation $x^2 - 4x + 12y = 19$ has no integer solutions.

References

[1] Dummit, D.S. and R.M. Foote. 2004. Abstract Algebra (3rd ed.), John Wiley & Sons.

[2] Hardy, G.H. and E.M. Wright. 2008. An Introduction to the Theory of Numbers (6th Ed.), Oxford University Press.

[3] Ireland, K. and M. Rosen. 1990. A Classical Introduction to Modern Number Theory (2nd Ed.) Springer.

[4] Koblitz, N. 1994. A Course in Number Theory and Cryptography, Springer-Verlag New-York.

[5] Niederreiter, H. and A. Winterhof. 2010. Applied Number Theory, Springer.

[6] Niven, I. and H.S. Zuckerman. 1966. Introduction to the Theory of Numbers (2nd Ed.), John Wiley & Sons.

[7] Rademacher, H. 1977. Lectures on Elementary Number Theory, Krieger.

[8] Rosen, K.H. 1993. Elementary Number Theory and Its Applications, Addison-Wesley.

[9] Schroeder, M.R. 1986. Number Theory in Science and Communication, Springer.

[10] Slinko, A. 2015. Algebra for Applications, Springer.

[11] Stillwell, J. 2003. Elements of Number Theory. Undergraduate Texts in Mathematics. Springer.

[12] Yan, Song Y. 2000. Number Theory for Computing, Springer-Verlag, Berlin, Heidelberg.

Chapter 2

Elements of Group Theory

"The Theory of Groups is a branch of mathematics in which one does something to something and then compares the result with the result obtained from doing the same thing to something else, or something else to the same thing."
James Newman

"With me, everything turns into mathematics."
Rene Descartes

In algebra, we generally deal with algebraic structures. Operations on their elements are often assumed to be associative.

The most important examples of algebraic structures with one associative operation are semigroups and groups, the basic properties of which are considered in this chapter.

The theory of semigroups has various applications in biology, theory of automatous, theory of formal languages, where special semigroups are studied.

The notion of a group is one of the most important and deepest concepts in mathematics. Historically, a group was the first example of an abstract algebraic structure. The beginnings of group theory appeared in the study of polynomial equations, in number theory and in geometry. Great mathematicians, J.L. Lagrange, N.H. Abel, E. Galois, L. Euler, C.F. Gauss, A. Möbius and F. Klein, were pioneers in the study of this branch of mathematics. In particular, N.H. Abel (1824) and E. Galois (1830) showed a deep connections between properties of permutation groups and polynomial equations. N.H. Abel proved in 1823 that it is impossible to solve an arbitrary polynomial equation of the fifth degree by radicals, which

Niels Henrik Abel
(1802-1829)

means that there are no formulae for the expression of all roots of a general polynomial of the fifth degree using only the four basic operations: Addition, subtraction, multiplication and division, as well as the taking of radicals, on the arithmetical combinations of its coefficients.

In 1830, E. Galois, who was one of the most tragic yet simultaneously the most romantic figures among all great mathematicians, created a new mathematical theory, which is now known as Galois theory, before reaching the age of 19. This theory decisively solved the problem of roots of arbitrary polynomial equations in one variable, i.e., it determines when a polynomial equation is solvable by radicals and when it is not solvable. The key for solving this problem was using symmetry, namely permutation groups. Galois was the first who, in his notes, written on the night before the duel, introduced such notions as a group (using by him to a permutation group) and a subgroup, an isomorphism and a homomorphism of groups. Unfortunately, the works of Galois stayed practically unknown during his lifetime. Only a few years later after his death in 1846, his notes

Evariste Galois
(1811-1832)

were published by Liouville and became generally available for mathematicians. In 1870, C. Jordan published his extensive work, where he systematically represented group theory, Galois theory and their applications to geometric problems.

Now, group theory is a central part of abstract algebra and has wide applications in different areas of mathematics and elsewhere. Special classes of groups play an important role in crystallography, particle physics, biology, chemistry and other fields of science.

2.1. Semigroups, Monoids and Groups

The most general and simplest examples of algebraic structures with one associative binary operation are semigroups and groups, the most important properties of which will be considered in this section.

The first axiomatic definition of a group was given by English mathematician Arthur Cayley (1821-1895). Unfortunately, in his definition, the main property of a group was missed, namely the existence of an inverse to each element of a group. The modern full axiomatic definition of a group was proposed only in 1902 by E. Hundington and E. Moore.

Though the concept of a group is rather simple, it leads to incredibly rich and complex theory.

> ✳ **Definition 2.1.**
> A **semigroup** is an algebraic structure $(G, \bullet)$ consisting of nonempty set G together with one binary operation $\bullet$ which is associative.

✳ Examples 2.2.

The following sets are semigroups:

1. The set $\{2n \mid n \in \mathbb{N}\}$ of all even natural numbers under addition.

2. The set of all natural numbers $\mathbb{N}$ under addition.

3. The set of all natural numbers $\mathbb{N}$ under multiplication.

4. The set of all integers $\mathbb{Z}$ under addition.

5. The set of all integers $\mathbb{Z}$ under multiplication.

6. The set of all real numbers $\mathbb{R}$ under either addition or multiplication.

7. The set of all negative real numbers under addition.

✳ Definition 2.3.

A **monoid** is an algebraic structure $(G, \bullet)$ consisting of nonempty set G together with one binary operation $\bullet$ which satisfies the following conditions:

1. The operation $\bullet$ is associative.

2. There exists a neutral element with respect to the operation $\bullet$.

From Theorem A.11 it follows that a monoid has the unique neutral element, with respect to the operation defined on it, which is called the **identity** and is usually denoted by e.

The important example of a monoid is a free monoid considered below.

Let A be a set which is called an **alphabet**. Let A^n be a set of all sequences (strings) of n elements of the set A:

$$A^n = \{a_1 a_2 \ldots a_n \mid a_i \in A\}.$$

Elements of the set A^n are called **words** of the length n over A. The empty sequence is considered as a word of the length 0 and denoted by Λ. We denote the set of all words over A by:

$$FM(A) = A^0 \cup A \cup A^2 \cup \ldots = \bigcup_{n=0}^{\infty} A^n,$$

where $A^0 = \{\Lambda\}$.

On the set $FM(A)$, we can define the following operation, which is called **string concatenation** and is denoted by $*$:

$$* : FM(A) \times FM(A) \longrightarrow FM(A)$$

such that

$$a_1 a_2 \ldots a_n * b_1 b_2 \ldots b_m = a_1 a_2 \ldots a_n b_1 b_2 \ldots b_m$$

for all $a_i, b_j \in A$.

The identity element of $FM(A)$ with respect to the operation of string concatenation is the empty string Λ.

✳ Definition 2.4.

A **free monoid** generated by a set A is a monoid $(FM(A), *)$, where $*$ is the operation of string concatenation, with the empty sequence Λ being the identity.

A free monoid $(FM(A), *)$ is also denoted by A^*.

✱ Examples 2.5.

1. If $A = \{0, 1\}$ then

$$FM(A) = \{\Lambda, 0, 1, 00, 01, 10, 11, 000, 001, 010, 011, 100, 101, 110, 111, \ldots\}.$$

$$001 * 101 = 001101, \quad 01 * 001100 = 01001100.$$

2. If $A = \{a, b, c\}$ then

$$FM(A) = \{\Lambda, a, b, c, aa, ab, ac, ba, bb, bc, ca, cb, cc, aaa, aab, aac, \ldots\}$$

$$aabc * cccabba = aabccccabba, \quad bac * bbccccaaa = bacbbccccaaa$$

✳ Definition 2.6.

A **group** is an algebraic structure $(G, \bullet)$ consisting of non-empty set G together with one binary operation $\bullet$ which satisfies the following conditions:

1. The operation $\bullet$ is associative, i.e.,

$$(a \bullet b) \bullet c = a \bullet (b \bullet c) \quad \text{for all } a, b, c \in G.$$

2. There exists the identity element with respect to the operation $\bullet$, i.e., $\exists (e \in G)$ such that

$$e \bullet a = a \bullet e = a \quad \text{for all } a \in G.$$

3. For each element $g \in G$, there exists the inverse element $h \in G$ such that

$$g \bullet h = h \bullet g = e,$$

where e is the identity element of G.

From Theorem A.11, it follows that the identity element of a group G is defined uniquely. It is usually denoted by e. Similarly, from Theorem A.13, it follows that every element of G has only one inverse.

If the group operation is denoted by $+$, then we say that the group G is **additive**. In this case, the identity element of G is written by 0, it is also called the **zero element**, and the inverse of an element $a \in G$ is denoted by $-a$.

If we use multiplicative symbolic, i.e., the group operation is denoted by $\cdot$, then the identity element of G is written by 1, and the inverse of an element $a \in G$ is denoted by a^{-1}. In this case, we say that G is a **multiplicative group**.

✳ **Examples 2.7.**

1. The set of all integers $\mathbb{Z}$ under addition forms a group with 0 being the identity element, and $-a$ being the inverse of $a \in \mathbb{Z}$. However, this set under multiplication does not form a group, since inverses do not exist for any integers other than ± 1.

2. The set of all positive rational numbers $\mathbb{Q}^+$ under multiplication forms a group. The neutral element of this group is 1 and the inverse element to $m/n \in \mathbb{Q}^+$ is n/m.

3. The set of all real numbers $\mathbb{R}$ under addition forms a group. The set of all non-zero real numbers $\mathbb{R}\backslash\{0\}$ under multiplication forms also a group.

4. The set $\{1, -1\}$ under multiplication $\cdot$ defined by the following table:

$\cdot$	1	-1
1	1	-1
-1	-1	1

forms a group.

5. The set of all nonsingular square $n \times n$-matrices

$$GL(n, \mathbb{R}) = \{\mathbf{A} \in M_n(\mathbb{R}) \mid \det(\mathbf{A}) \neq 0\}$$

under multiplication of matrices forms a group which is called the **total linear group**.

Other examples of groups will be given later.

✳ **Definition 2.8.**

A group $(G, \bullet)$ is called **commutative** if the operation $\bullet$ is commutative, i.e.,

$$a \bullet b = b \bullet a \quad \text{for all } a, b \in G.$$

Otherwise a group G is called **non-commutative**.

Commutative groups are often called **Abelian** in honor of the outstanding Norwegian mathematician N.H. Abel (1802–1829) who studied classes of algebraic equations connected with commutative groups. For an Abelian group, we usually use additive notation and call it an additive group.

✱ Example 2.9.
   The groups 1-4, which are given in Example 2.7, are commutative groups, but the group $GL(n\mathbb{R})$ is non-commutative for $n > 1$.

✳ Definition 2.10.
   A group $(G, \bullet)$ is called **finite** if G is a finite set, i.e., has a finite number of elements. Otherwise a group is called **infinite**. The number of elements of a finite group G is called the **order** of G and is denoted by $|G|$. By assumption, the order of an infinite group is equal to ∞.

✱ Example 2.11.
   Let $m > 0$ be a natural number. Consider the quotient set $\mathbb{Z}/m\mathbb{Z} = \{[0], [1], \ldots, [m-1]\}$ of residue classes modulo m. In this set, we can define the operation of addition in the following form:

$$[a] + [b] = [a + b] \tag{2.1}$$

for all $[a], [b] \in \mathbb{Z}/m\mathbb{Z}$. This operation is well-defined, i.e., a residue class $[a] + [b]$ does not depend on the choice of representees in classes $[a], [b]$. Indeed, let $[a] = [a_1]$, $[b] = [b_1]$. Then, from the definition of the congruence relation, we have $a - a_1 = mk$, $b - b_1 = mn$ for some $k, n \in \mathbb{Z}$. Therefore, $(a + b) - (a_1 + b_1) = m(k + n)$, which means that $[a + b] = [a_1 + b_1]$. From properties of addition for integers, we get commutativity and associativity of addition in $\mathbb{Z}/m\mathbb{Z}$:

$$[a] + [b] = [a + b] = [b + a] = [b] + [a]$$
$$([a] + [b]) + [c] = [a + b] + [c] = [(a + b) + c] = [a + (b + c)] =$$
$$= [a] + [b + c] = [a] + ([b] + [c])$$

The neutral element is the class $[0]$, since

$$[a] + [0] = [a + 0] = [a] = [0 + a] = [0] + [a].$$

The inverse to an element $[a]$ is a class $[-a]$, since

$$[a] + [-a] = [a - a] = [0] = [-a] + [a].$$

So that, $\mathbb{Z}/m\mathbb{Z}$ is a finite commutative group of order m under the operation (2.1).

✳ **Example 2.12.**

Let $m > 0$ be a natural number. Consider the least residue system modulo m: $\mathbb{Z}_m = \{0, 1, \ldots, m-1\}$ with operation $\oplus$ defined as follows:

$$a \oplus b = (a + b) \bmod m, \tag{2.2}$$

where $n \bmod m$ means the remainder on division n by m. The element 0 is the neutral element of $\mathbb{Z}_m$, and an element $m - x$ is the inverse to $x \in \mathbb{Z}_m$. It is easy to see that

$$a \oplus b = (a + b) \bmod m = (b + a) \bmod m = b \oplus a$$

$$(a \oplus b) \oplus c = a \oplus (b \oplus c).$$

Therefore, $\mathbb{Z}_m$ is a finite commutative group of order m under operation (2.2).

From the definition of a group we can easy obtain the following important property for groups, called the **cancellation law**.

✳ **Theorem 2.13.**

For a group $(G, \bullet)$ and for all elements $a, x, y \in G$, we have:
1) if $a \bullet x = a \bullet y$ then $x = y$ (left cancellation law);
2) if $x \bullet a = y \bullet a$ then $x = y$ (right cancellation law).

Proof.

1) Let a^{-1} be the inverse element to an element $a \in G$, e the identity element of G and $a \bullet x = a \bullet y$. Then, multiplying this equality on the left side by a^{-1} and using the basic axioms of a group we obtain:

$$a^{-1} \bullet (a \bullet x) = a^{-1} \bullet (a \bullet y) \Rightarrow (a^{-1} \bullet a) \bullet x = (a^{-1} \bullet a) \bullet y \Rightarrow e \bullet x = e \bullet y \Rightarrow x = y.$$

2) Analogously, we can prove the second part of the theorem. □

✳ **Theorem 2.14.**

For a group $(G, \bullet)$ and all elements $a, b \in G$, the equality below holds:

$$(a \bullet b)^{-1} = b^{-1} \bullet a^{-1}. \tag{2.3}$$

Proof.

Let e be the identity element of a group G. Then, using the associative law we have:

$$(b^{-1} \bullet a^{-1}) \bullet (a \bullet b) = b^{-1} \bullet (a^{-1} \bullet a) \bullet b = b^{-1} \bullet e \bullet b = b^{-1} \bullet b = e.$$

On the other hand:

$$(a \bullet b) \bullet (b^{-1} \bullet a^{-1}) = a \bullet (b \bullet b^{-1}) \bullet a^{-1} = a \bullet e \bullet a^{-1} = a \bullet a^{-1} = e.$$

From the definition of the inverse, we get equality (2.3). □

> ✳ **Theorem 2.15.**
> For a group $(G, \bullet)$ and for all elements $a, b, x, y \in G$, the equations
>
> $$a \bullet x = b \text{ and } y \bullet a = b$$
>
> have solutions in G.

Proof.
Let $a \bullet x = b$, then, multiplying on the left side by a^{-1}, we get $x = a^{-1} \bullet b$. Indeed, $a \bullet (a^{-1} \bullet b) = (a \bullet a^{-1}) \bullet b = e \bullet b = b$, i.e., $a^{-1} \bullet b$ is a solution of the equation $a \bullet x = b$. Analogously, we can show that $y = b \bullet a^{-1}$ is a solution of the equation $y \bullet a = b$. □

There are a few ways to represent groups. One of them is a representation of a finite group $(G, \bullet)$ in the form of the Cayley table. If $|G| = n$ and $G = \{g_1, g_2, \ldots, g_n\}$, then the operation $\bullet$ can be defined by giving all possible products $g_i \bullet g_j$ in an $n \times n$-table whose rows and columns are numerated by elements of the group and, in the intersection of the i-th rows and the j-th column, there is the result of the product $g_i \bullet g_j$:

$\bullet$	g_1	g_2	g_3	$\cdots$	g_n
g_1	$g_1 \bullet g_1$	$g_1 \bullet g_2$	$g_1 \bullet g_3$	$\cdots$	$g_1 \bullet g_n$
g_2	$g_2 \bullet g_1$	$g_2 \bullet g_2$	$g_2 \bullet g_3$	$\cdots$	$g_2 \bullet g_n$
g_3	$g_3 \bullet g_1$	$g_3 \bullet g_2$	$g_3 \bullet g_3$	$\cdots$	$g_3 \bullet g_n$
$\cdots$	$\cdots$	$\cdots$	$\cdots$	$\cdots$	$\cdots$
g_n	$g_n \bullet g_1$	$g_n \bullet g_2$	$g_n \bullet g_3$	$\cdots$	$g_n \bullet g_n$

As a corollary from Theorem 2.13, we obtain that all elements in each row (column) of the Cayley table are different elements of the group.

In an arbitrary multiplicative group G, we can recursively define the powers of an element $a \in G$ with the identity element e as follows:

$$a^0 = e, \quad a^{n+1} = a^n a, \quad a^{-n} = (a^n)^{-1}$$

for any $n \in \mathbb{Z}$. It is easy to prove (which we leave to the Reader as an exercise) that in a group G it the following properties hold:

$$a^n a^m = a^{n+m},$$

$$(a^n)^m = a^{nm}$$

$$(ab)^n = b^n a^n$$

for all $a, b \in G$ and for all $n, m \in \mathbb{Z}$.

Remark 2.16.

If a group G is additive, we write na instead of a^n. In this case, we have the following properties:

$$na + ma = (n + m)a,$$

$$m(na) = (mn)a$$

$$n(a + b) = na + nb$$

for all $a, b \in G$ and for all $n, m \in \mathbb{Z}$.

✳ Definition 2.17.

An element a of a group G with neutral element e is said to be of a **finite order** if there is a positive integer n such that $a^n = e$. The least positive integer such that $a^n = e$ is called the **order** of an element $a \in G$ and denoted by $\mathrm{ord}(a)$. If there is no such finite number, we say that the element a has **infinite order**.

If G is a finite group, then each of its elements has the order less than the order of the group. Indeed, if a multiplicative group G has m elements then for an element $a \in G$ $m + 1$ elements $e, a, a^2, \ldots, a^m$ are not distinct, which means that there are at least two numbers m and n such that $n < m$ and $a^n = a^m$. Then, $a^{m-n} = e$, i.e., $\mathrm{ord}(a) \leq m - n < m$.

2.2. Subgroups. Cyclic Groups

✳ Definition 2.18.

A nonempty subset H of a group $(G, \bullet)$ is said to be a **subgroup** of G if H itself is a group under the operation $\bullet$ determining on G, i.e., if the following properties hold:
  1) $a \bullet b \in H$ for all $a, b \in H$,
  2) $a^{-1} \in H$ for all $a \in H$,

From this definition, it follows that $e \in H$, where e is the identity of G. A group which contains only one element e is called **trivial**. It is obvious that $\{e\}$ and G are subgroups of G. A subgroup $H \subseteq G$ is called **proper** if $H \neq \{e\}$ and $H \neq G$. We write $H \leq G$ if H is a subgroup of G, and we write $H < G$ if we want emphasize that H is a proper subgroup of G. Note, that any subgroup of a commutative group is also commutative.

✳ Examples 2.19.

1. The set of all positive real numbers is a subgroup of the group of all non-zero real numbers under multiplication.

2. The set of all odd integers is a subgroup of the group of all integers under addition.

3. If a is an element of a group G then the set of all powers of a is a subgroup of G.

4. The set

$$SL(n, \mathbb{R}) = \{\mathbf{A} \in GL(n, \mathbb{R}) \mid \det(\mathbf{A}) = 1\}$$

is a subgroup of the group $GL(n, \mathbb{R})$ and it is called the **special linear group**.

✳ Example 2.20.
We will show that every non-trivial subgroup of the additive group $\mathbb{Z}$ has the form $m\mathbb{Z}$, where $m \in \mathbb{N}$.

Let H be a non-trivial subgroup of the additive group $\mathbb{Z}$. Note that, if $x \in H$, then $(-x) \in H$ as well. Then, by the Well Ordering Principle, there is the least positive integer $m \in H$. Let n be an element of H. Then, there are $q, r \in \mathbb{Z}$ such that $n = qm + r$ where $0 \le r < m$. Since $r = n - qm \in H$ and m is the least positive integer belonging to H, $r = 0$, i.e., $n = mq$. Therefore, $H = m\mathbb{Z}$.

Important examples of groups are cyclic groups, which are defined as follows:

✳ Definition 2.21.
A group G is called **cyclic** if there is an element $a \in G$ such that each element of G is a power of a. In this case, a is called a **generator** for G.

✳ Example 2.22.
1. The additive group of integers $\mathbb{Z}$ is an infinite cyclic group, with 1 being a generator.
2. The additive group $\mathbb{Z}_m$ is a finite cyclic with 1 being a generator. The order of this group is equal to m.
3. The multiplicative group $G = \{e, a, b, c\}$ with the binary operation given by the following table:

	e	a	b	c
e	e	a	b	c
a	a	b	c	e
b	b	c	e	a
c	c	e	a	b

is a cyclic group of order 4. The generator of this group is the element a, since $e = a^0$, $b = a^2$, $c = ab = a^3$.

A cyclic group of order n is denoted by C_n. If a is a generator for the group C_n, then we write $C_n = \langle a \rangle$ or

$$C_n = \{e, a, a^2, \ldots, a^{n-1} \mid a^n = e\}.$$

Every cyclic group is commutative, since for all integers n, m the following equalities hold:

$$a^n a^m = a^{n+m} = a^{m+n} = a^m a^n.$$

✴ **Example 2.23.**

The multiplicative group $G = \{1, -1, i, -i\}$ with the group operation given in the form of the following table:

	1	−1	i	$-i$
1	1	−1	i	$-i$
−1	−1	1	$-i$	i
i	i	$-i$	−1	1
$-i$	$-i$	i	1	−1

is a cyclic group: $G = \{i, i^2, i^3, i^4 = 1\}$.

✴ **Theorem 2.24.**

The order of an element x of a group G is equal to the order of a cyclic subgroup generated by this element.

Proof. Let G be a multiplicative group with the identity element e. Let $x \in G$ and $\mathrm{ord}(x) = n$. Consider the cyclic subgroup $H = \{x, x^2, \ldots, x^n = 1\}$ of the group G generated by the element x. Then, $|H| = n = \mathrm{ord}(x)$. ☐

✴ **Theorem 2.25.**

Let g be an element of a multiplicative group G with the identity element e and $\mathrm{ord}(g) = m$. Then, $g^k = e$ if and only if $m \mid k$.

Proof.

$\Rightarrow$. Let $k = ms + r$ and $0 \leq r < m$. Then, $g^r = g^{k-ms} = (g^k)(g^m)^{-s} = e$. Since $\mathrm{ord}(g) = m$ and $0 \leq r < m$, we get $r = 0$, i.e., $m \mid k$.

$\Leftarrow$. If $k = ms$, then $g^k = g^{ms} = e$. ☐

✴ **Corollary 2.26.**

Let g be an element of a multiplicative group G and $\mathrm{ord}(g) = m$. Then, $g^k = g^s$ if and only if $k \equiv s \pmod{m}$.

Proof.

Let G be a multiplicative group with the identity element e and $g \in G$. Then,

$$g^k = g^s \iff g^{k-s} = e.$$

By Theorem 2.25, the last equality is equivalent to the condition $m|(k-s)$, which means that $k \equiv s(\mathrm{mod}\, m)$. □

✳ Theorem 2.27.
If an element g of a multiplicative group G has a finite order m and $n \in \mathbb{Z}$, then $\mathrm{ord}(g^n) = m/k$, where $k = \gcd(m, n)$.

Proof.
Let G be a multiplicative group with the identity element e and $g \in G$. Suppose that $\mathrm{ord}(g) = m$. Then, $(g^n)^{m/k} = (g^m)^{n/k} = e$. If $\mathrm{ord}(g^n) = s$ then $g^{ns} = e$. Therefore, by Theorem 2.25, it follows that $m|ns$ and $s|(m/k)$. Let $m = ku$, $n = kv$ and $\gcd(u, v) = 1$. Then $u|vs$ and $s|u$, which implies that $u|s$ and $s|u$. Hence, $s = u = m/k$. □

2.3. Permutation Groups

One of the most known examples of groups are groups of permutations.

✳ Definition 2.28.
A **permutation** is a bijection from a nonempty set X to itself.

If X is a finite set having n elements, then we say that a **permutation** is of degree n. For simplicity in notations, we will consider that a finite set $X = \{1, 2, \ldots, n\}$. Then, any permutation $\sigma : X \to X$ can be written as a matrix:

$$\sigma = \begin{pmatrix} 1 & 2 & \cdots & n \\ a_1 & a_2 & \cdots & a_n \end{pmatrix}, \tag{2.4}$$

where $\sigma(i) = a_i \in X$ for $i = 1, 2, \ldots, n$. Since σ is a bijection, all numbers a_i are distinct and each of them is one of the numbers $1, 2, \ldots, n$.

The set of all permutations of degree n is denoted by S_n.

✳ Examples 2.29.
1. Let $S = \{1, 2, 3, 4\}$ and a bijection $\sigma : S \to S$ is given by:

$$\sigma(1) = 4, \ \sigma(2) = 2, \ \sigma(3) = 1, \ \sigma(4) = 3$$

then the permutation σ can be written as follows:

$$\sigma = \begin{pmatrix} 1 & 2 & 3 & 4 \\ 4 & 2 & 1 & 3 \end{pmatrix}.$$

2. Let $X = \{1, 2\}$. Then, S_2 consists of two elements:

$$e = \begin{pmatrix} 1 & 2 \\ 1 & 2 \end{pmatrix}, \ s = \begin{pmatrix} 1 & 2 \\ 2 & 1 \end{pmatrix}.$$

3. Let $X = \{1, 2, 3\}$. Then, S_3 consists of six elements:

$$e = \begin{pmatrix} 1 & 2 & 3 \\ 1 & 2 & 3 \end{pmatrix}, \quad s_1 = \begin{pmatrix} 1 & 2 & 3 \\ 2 & 3 & 1 \end{pmatrix}, s_2 = \begin{pmatrix} 1 & 2 & 3 \\ 3 & 1 & 2 \end{pmatrix},$$

$$s_3 = \begin{pmatrix} 1 & 2 & 3 \\ 1 & 3 & 2 \end{pmatrix}, \quad s_4 = \begin{pmatrix} 1 & 2 & 3 \\ 3 & 2 & 1 \end{pmatrix}, s_5 = \begin{pmatrix} 1 & 2 & 3 \\ 2 & 1 & 3 \end{pmatrix}.$$

✳ Proposition 2.30.
There are exactly $n!$ permutations of degree n.

Proof.
Each permutation σ of degree n has the form (2.4), where all numbers a_i are distinct and each of them is one of the numbers $1, 2, \ldots, n$. So, the value of a_1 admits n distinct possibilities. When a_1 is chosen, there are only $n - 1$ possibilities for a_2, since $a_2 \neq a_1$. After the choice of a_2 there are only $n - 2$ possibilities for a_3. Continuing this process we have only one possibility for the choice of a_n. Thus, we have

$$n(n-1)(n-2) \cdots 2 \cdot 1 = n!$$

possibilities for σ. □

Since each permutation σ is uniquely determined by given values $\sigma(i) = a_i$ for all i, it is immaterial in what order we write numbers in the first row of a permutation, it is only important that each element a_i be written below the corresponding number i. For example, the permutations

$$\begin{pmatrix} 1 & 2 & 3 & 4 & 5 \\ 2 & 4 & 1 & 5 & 3 \end{pmatrix} \quad \text{and} \quad \begin{pmatrix} 3 & 2 & 5 & 1 & 4 \\ 1 & 4 & 3 & 2 & 5 \end{pmatrix}$$

are two different records of the same permutation.

Since permutations are bijective mappings of a set to itself, there exists their composition, which is also a permutation and is called the product of permutations. Therefore, we can define the product of two permutations σ and τ of a set $X = \{1, 2, \ldots, n\}$ as their composition, i.e., their subsequent execution one after the other beginning from the right permutation. As a result of this operation for two permutations σ and τ we get a permutation which will be written as $\sigma \circ \tau$. For simplicity, the symbol of operation $\circ$ is often omitted. So that

$$(\sigma\tau)(i) = (\sigma \circ \tau)(i) = \sigma(\tau(i)), \quad \text{for all } i \in X.$$

✶ Examples 2.31.
Find the product $s_1 s_5$ of two permutations:

$$s_1 = \begin{pmatrix} 1 & 2 & 3 \\ 2 & 3 & 1 \end{pmatrix}, \quad s_5 = \begin{pmatrix} 1 & 2 & 3 \\ 2 & 1 & 3 \end{pmatrix}.$$

Since $1 \mapsto 2$ by the permutation s_5 and $2 \mapsto 3$ by the permutation s_1, we have that $1 \mapsto 3$ by the product $s_1 s_5$. Furthermore, $2 \mapsto 1$ by the permutation s_5 and $1 \mapsto 2$ by the permutation s_1. Therefore, $2 \mapsto 2$ by the product $s_1 s_5$. Finally $3 \mapsto 3$ by the permutation s_5 and $3 \mapsto 1$ by the permutation s_1. Therefore, $3 \mapsto 1$ by the product $s_1 s_5$. As a result, we get the permutation:

$$s_1 s_5 = \begin{pmatrix} 1 & 2 & 3 \\ 2 & 3 & 1 \end{pmatrix} \begin{pmatrix} 1 & 2 & 3 \\ 2 & 1 & 3 \end{pmatrix} = \begin{pmatrix} 1 & 2 & 3 \\ 3 & 2 & 1 \end{pmatrix} = s_4.$$

Note that the operation of composition of permutations is associative, which follows from associativity of composition of functions. However, for $n > 2$, it is not commutative. For example,

$$s_1 s_4 = \begin{pmatrix} 1 & 2 & 3 \\ 2 & 3 & 1 \end{pmatrix} \begin{pmatrix} 1 & 2 & 3 \\ 3 & 2 & 1 \end{pmatrix} = \begin{pmatrix} 1 & 2 & 3 \\ 1 & 3 & 2 \end{pmatrix} = s_3,$$

on the other hand

$$s_4 s_1 = \begin{pmatrix} 1 & 2 & 3 \\ 3 & 2 & 1 \end{pmatrix} \begin{pmatrix} 1 & 2 & 3 \\ 2 & 3 & 1 \end{pmatrix} = \begin{pmatrix} 1 & 2 & 3 \\ 2 & 1 & 3 \end{pmatrix} = s_5.$$

The set S_n of all permutations of degree n has an identity element with respect to the operation of composition which does not change any element of a set X, i.e., the permutation:

$$e = \begin{pmatrix} 1 & 2 & \cdots & n \\ 1 & 2 & \cdots & n \end{pmatrix},$$

which is called the **identity permutation** (or the **identity**) and

$$\sigma e = e\sigma = \sigma$$

for each permutation $\sigma \in S_n$.

Finally, the inverse to a permutation

$$\sigma = \begin{pmatrix} 1 & 2 & \cdots & n \\ a_1 & a_2 & \cdots & a_n \end{pmatrix},$$

is a permutation:

$$\sigma^{-1} = \begin{pmatrix} a_1 & a_2 & \cdots & a_n \\ 1 & 2 & \cdots & n \end{pmatrix},$$

which is obtained from the permutation σ by changing upper and low rows.

Therefore, the set S_n of all permutations of degree n under the operation of composition forms a group, which is called the **symmetric group of degree** n. This group is finite and its order is equal to $n!$. Each subgroup of the group S_n is called a **permutation group**. Since the operation of composition of permutations is associative and not commutative, the group S_n is associative but not commutative.

2.4. Cosets. Lagrange's Theorem

In this section, we introduce the notion of cosets of a group and show that cosets define a partition of this group with respect to some equivalence relation on it.

> ✳ **Definition 2.32.**
> Let H be a subgroup of a group G and let $a \in G$. The set
>
> $$aH = \{ax \in G \;:\; x \in H\} \tag{2.5}$$
>
> is called the **left coset** of H generated by a. A **right coset** of H generated by a is the set:
> $$Ha = \{xa \in G \;:\; x \in H\} \tag{2.6}$$

Remark 2.33.
Note that, in additive notation, we write $a + H$ and $H + a$ for the right coset and the left coset, respectively. In general, right cosets and left cosets may be different. However, if G is an Abelian group, then the right coset Ha is equal to the left coset aH of any subgroup H and for any element $a \in G$. In this case, we do not need to distinguish left and right cosets, and we call them **cosets**.

✳ **Example 2.34.**
Let $H \leq G$, where $G = \mathbb{Z}_6 = \{0,1,2,3,4,5\}$ is an additive group of the least residue classes modulo 6 under addition modulo 6, and let $H = \{0,3\}$ be a subgroup of G. The cosets of H in G are the sets:

$$0 + H = \{0,3\} \quad 3 + H = \{0,3\} = H$$
$$1 + H = \{1,4\} \quad 4 + H = \{4,1\}$$
$$2 + H = \{2,5\} \quad 5 + H = \{5,2\}$$

So we see that
$$0 + H = 3 + H = H$$
$$1 + H = 4 + H$$
$$2 + H = 5 + H$$
$$(0 + H) \cap (1 + H) = (0 + H) \cap (2 + H) = (1 + H) \cap (2 + H) = \emptyset$$
$$G = H \cup (1 + H) \cup (2 + H) \tag{2.7}$$

✳ **Example 2.35.**
Let $X = \{1,2,3\}$ and $S_3 = \{e, a, b, c, d, f\}$, where

$$e = \begin{pmatrix} 1 & 2 & 3 \\ 1 & 2 & 3 \end{pmatrix}, \quad a = \begin{pmatrix} 1 & 2 & 3 \\ 2 & 3 & 1 \end{pmatrix}, \quad b = \begin{pmatrix} 1 & 2 & 3 \\ 3 & 1 & 2 \end{pmatrix},$$

$$c = \begin{pmatrix} 1 & 2 & 3 \\ 1 & 3 & 2 \end{pmatrix}, \quad d = \begin{pmatrix} 1 & 2 & 3 \\ 3 & 2 & 1 \end{pmatrix}, \quad f = \begin{pmatrix} 1 & 2 & 3 \\ 2 & 1 & 3 \end{pmatrix}.$$

Let $H = \{e, a, b\}$. Since $a^2 = b$, $ab = ba = e$, $b^2 = a$, H is a subgroup of the group S_3. Consider the left cosets of H. For the element $d \in G$, we have: $dH = \{d, da, db\}$. Since:

$$da = \begin{pmatrix} 1 & 2 & 3 \\ 3 & 2 & 1 \end{pmatrix} \begin{pmatrix} 1 & 2 & 3 \\ 2 & 3 & 1 \end{pmatrix} = \begin{pmatrix} 1 & 2 & 3 \\ 2 & 1 & 3 \end{pmatrix} = f,$$

$$db = \begin{pmatrix} 1 & 2 & 3 \\ 3 & 2 & 1 \end{pmatrix} \begin{pmatrix} 1 & 2 & 3 \\ 3 & 1 & 2 \end{pmatrix} = \begin{pmatrix} 1 & 2 & 3 \\ 1 & 3 & 2 \end{pmatrix} = c,$$

we obtain that $dH = \{d, f, c\}$. Therefore:

$$G = H \cup dH, \quad H \cap dH = \emptyset, \quad |H| = |dH| = 3.$$

✳ Example 2.36.

Let $X = \{1, 2, 3\}$ and $S_3 = \{e, a, b, c, d, f\}$, where e, a, b, c, d, f as in the previous example. Since $f^2 = e$, $K = \{e, f\}$ is a subgroup of S_3. Consider left cosets of the subgroup K: $aK = \{a, af\} = \{a, d\}$, $bK = \{b, bf\} = \{b, c\}$. Therefore, we get:

$$G = K \cup aK \cup bK, \quad K \cap aK = K \cap bK = \emptyset, \quad |K| = |aK| = |bK| = 2.$$

Consider the right cosets of K: $Ka = \{a, fa\} = \{a, c\}$, $Kb = \{b, fb\} = \{b, d\}$. So, we get:

$$G = K \cup Ka \cup Kb, \quad K \cap Ka = K \cap Kb = Ka \cap Kb = \emptyset, \quad |K| = |Ka| = |Kb| = 2.$$

Note that $aK = \{a, d\} \neq \{a, c\} = Ka$ and $bK = \{b, c\} \neq \{b, d\} = Kb$, but $G = K \cup aK \cup bK = K \cup Ka \cup Kb$.

✳ Theorem 2.37.

Let H be a subgroup of a group G. Then the left cosets of H form a partition of G. The same is true for right cosets:

1. $G = \bigcup \{Ha \; : \; a \in G\}$.

2. $Ha \cap Hb = \emptyset$ if and only if $Ha \neq Hb$.

3. $G = \bigcup \{aH \; : \; a \in G\}$.

4. $aH \cap bH = \emptyset$ if and only if $aH \neq bH$.

Proof. Since the identity element $e \in H$, we have $a \in aH$ and $a \in Ha$ for any elements $a \in G$.

Suppose that $Ha \cap Hb \neq \emptyset$, then there is an element $x \in G$ such that $x \in Ha \cap Hb$, i.e., $x = ha = gb$ for some $h, g \in H$. Hence, $a = (h^{-1}g)b \in Hb$, which implies that $Ha \subseteq Hb$. Analogously, we get that $Hb \subseteq Ha$. So that $Ha = Hb$.

For left cosets, the proof is similar. $\square$

Since any partition of a set defines an equivalence relation, Theorem 2.37 shows that each subgroup H of G defines two equivalence relations. For all $a, b \in G$, we can define two relations as follows:

1) $a \sim_l b \Leftrightarrow aH = bH$
2) $a \sim_r b \Leftrightarrow Ha = Hb$

which are equivalence relations, by Theorem 2.37.

✳ Theorem 2.38.

Let H be a subgroup of a group G.

1. Two left (right) cosets of H are equinumerous.

2. A left coset of H is equinumerous with a right coset of H.

3. There exists a bijection φ between the set of all right cosets of H and the set of left cosets of H, which is defined as follows: $\varphi(aH) = Ha^{-1}$ for any $a \in G$.

Proof.

1) Consider two left cosets Ha and Hb of H and define the mapping $\varphi : Ha \to Hb$ as follows: $\varphi(ha) = hb$ for each element $h \in H$. Obviously, this mapping is surjective. It is also injective, since

$$\varphi(ha) = \varphi(ga) \Rightarrow hb = gb \Rightarrow h = g \Rightarrow ha = ga.$$

So, φ is a bijection. Hence, $\#(Ha) = \#(Hb)$. Analogously, we can show that right cosets are also equinumerous.

2) This follows from part 1 of the theorem and the fact that the coset H is both a right coset and a left coset.

3) Consider a mapping of the set of all right cosets of H to the set of all left cosets of H, which is given as follows:

$$\varphi(aH) = Ha^{-1}.$$

We show that this mapping is well defined, that is, it does not depend on the representatives of the coset aH. Let $aH = bH$. We show that, in this case, $Ha^{-1} = Hb^{-1}$ as well. Indeed, we have a sequence of implications:

$$aH = bH \Rightarrow b = ah \Rightarrow b^{-1} = h^{-1}a^{-1} \in Ha^{-1} \Rightarrow Hb^{-1} \subseteq Ha^{-1}.$$

Analogously, we obtain that $Ha^{-1} \subseteq Hb^{-1}$. Hence, $Hb^{-1} = Ha^{-1}$.

In a similar way, we show that $Hb^{-1} = Ha^{-1}$ implies that $aH = bH$. So the mapping φ is well defined and it is obviously surjective. It is also injective because by proving the above we have that

$$\varphi(aH) = \varphi(bH) \;\Rightarrow\; Ha^{-1} = Hb^{-1} \;\Rightarrow\; aH = bH.$$

$\square$

✳ Definition 2.39.

The number of distinct left (right) cosets of a subgroup H in a group G is called the **index** of H in G and is denoted by $[G:H]$. This number is either a positive integer or ∞.

✳ Theorem 2.40 (Lagrange's Theorem).

If H is a subgroup of a finite group G, then:

$$|G| = |H| \cdot [G:H]. \tag{2.8}$$

In particular, $|H|$ divides $|G|$ and:

$$[G:H] = |G|/|H|. \tag{2.9}$$

Proof.

From Theorem 2.37, it follows that we have a partition

$$G = \bigcup\{Ha \;:\; a \in G\},$$

Since, by Theorem 2.38, all right (left) cosets are equinumerous, we have $|Hx| = |H|$. Then,

$$|G| = \sum |Ha|,$$

where a runs through a set of representatives of cosets, since $Ha \cap Hb = \emptyset$ if $Ha \neq Hb$. Since the number of disjoint cosets in a partition of G is equal to the index $[G:H]$, we have $|G| = |H| \cdot [G:H]$.

In particular, if G is a finite group of order n and $m = |H|$, then $[G:H] = |G|/|H| = n/m = k \in \mathbb{Z}^{+}$. Hence, $m|n$ and $n = km$. $\square$

Joseph Lagrange
(1736-1813)

✳ Examples 2.41.

- Let $G = \mathbb{Z}_6 = \{0,1,2,3,4,5\}$, $H = \{0,3\}$.
  Then $G = H \cup (1+H) \cup (2+H)$,
  $|H| = 2$, $[G:H] = 3$, $|G| = 6 = |H| \cdot [G:H]$.

- Let $G = S_3 = \{e, a, b, c, d, f\}$, $H = \{e, a, b\}$.

  Then $G = H \cup dH$, $|H| = 3$, $[G : H] = 2$, $|G| = 6 = |H| \cdot [G : H]$.

- Let $G = S_3 = \{e, a, b, c, d, f\}$, $K = \{e, f\}$.

  Then $G = K \cup Ka \cup Kb$, $|H| = 2$, $[G : H] = 3$, $|G| = 6 = |H| \cdot [G : H]$.

✳ Theorem 2.42.

1. Let G be a finite group of order n and $a \in G$. Then, $\mathrm{ord}(a)$ divides n.
2. In particular, if $n = p$ is a prime, then G is a cyclic group.

Proof.

1. This follows from Lagrange's Theorem, because the order of an element $a \in G$ is equal to the order of the cyclic subgroup $H = \langle a \rangle$ generated by this element, i.e., $\mathrm{ord}(a) = |H|$.

2. Let $|G| = p$ be a prime and e the identity element of G. We take $e \neq a \in G$. Then, $\mathrm{ord}(a) | p$. Since p is prime and $a \neq e$, $\mathrm{ord}(a) = p = |G|$. Therefore, the cyclic group generated by a coincides with G. $\qquad\square$

✳ Theorem 2.43.

If G is a finite group with the identity element e then :

$$g^{|G|} = e \tag{2.10}$$

for any element $g \in G$.

Proof.

If $\mathrm{ord}(g) = n$, then $g^n = e$. By Theorem 2.42, $|G| = nk$. Therefore, $g^{|G|} = g^{nk} = (g^n)^k = e$. $\qquad\square$

2.5. Normal Subgroups and Quotient Groups

As shown in the previous section, left cosets are not always equal to right cosets of an arbitrary subgroup. In this section, we will study a special class of subgroups for which these cosets are identical.

✳ Definition 2.44.

A subgroup H of a group G is called a **normal subgroup** of G if:

$$Hg = gH \tag{2.11}$$

for all elements $g \in G$, which is equivalent to the following:

$$H = gHg^{-1} \tag{2.12}$$

We write in this case: $H \trianglelefteq G$, or $H \triangleleft G$ if we want to emphasize that H is a proper normal subgroup of G. It is obvious that $\{e\} \trianglelefteq G$ and $G \trianglelefteq G$. It is also clear that if G is a commutative group then each of its subgroups are normal.

The notion of a normal subgroup can be defined in different ways. We give the equivalent definition of a normal subgroup by the following theorem.

✳ Theorem 2.45.
A subgroup H of a group G is normal if and only if $ghg^{-1} \in H$ for all $g \in G$ and for all $h \in H$.

Proof.
$\Rightarrow$. It follows from Definition 2.44.

$\Leftarrow$. Let $g \in G$, $h \in H$ and $ghg^{-1} \in H$. Then, we have the sequence of implications:

$$ghg^{-1} = a \in H \;\Rightarrow\; h = gag^{-1} \in gHg^{-1} \;\Rightarrow\; H \subseteq gHg^{-1}.$$

On the other hand, from assumption, $gHg^{-1} \subseteq H$. Therefore, $H = gHg^{-1}$ for all $g \in G$, i.e., H is a normal subgroup of G. □

✳ Definition 2.46.
Two elements g, h of a group G are called **conjugate** if there exists an element $x \in G$ such that

$$g = x^{-1}hx \in G \tag{2.13}.$$

From Theorem 2.45, we immediately obtain the following result:

✳ Theorem 2.47.
A subgroup H of a group G is a normal subgroup in G if and only if it is equal to the set of all elements which are conjugate with elements of H.

✳ Theorem 2.48.
Each subgroup H of a group G of index 2 is normal.

Proof.
Let $H \leq G$. Consider the following partition of G:

$$G = H \cup (G \backslash H).$$

If $x \in H$, then $xH = H = Hx$. If $x \notin X$, then $x \in G \backslash H$, and neither $H \neq xH$ nor $H \neq Hx$. So that $G \backslash H = xH = Hx$, since $[G : H] = 2$. Therefore, H is a normal subgroup of a group G. □

✳ Example 2.49.
Let $S_3 = \{e, a, b, c, d, f\}$ and $H = \{e, a, b\}$, as in Example 2.45. Then, $[S_3 : H] = 2$. Therefore, H is a normal subgroup in S_3, by Theorem 2.48.

✴ Example 2.50 (Example of a subgroup which is not normal).

Let $S_3 = \{e, a, b, c, d, f\}$ and $K = \{e, f\}$, as in Example 2.36. Since $aK = \{a, d\} \neq Ka = \{a, c\}$, the subgroup K is not normal in S_3.

✳ Definition 2.51.

A non-trivial group G whose only normal subgroups are the trivial subgroup $\{e\}$ and the group G itself is called a **simple group**.

✴ Theorem 2.52.
If p is a prime and $|G| = p$, then G is a simple group.

Proof.
The statement follows from the fact that if $|G| = p$ is a prime then the only subgroups of G are the trivial subgroup $\{e\}$ and the group G itself, by Lagrange's Theorem. □

✴ Theorem 2.53.
If $H \trianglelefteq G$ then the cosets of H under the operation

$$(aH) \circ (bH) = (ab)H \tag{2.14}$$

form a group which is called the **quotient group** (or the **factor group**) of G by H and denoted by G/H.

Proof.
First, we will show that the operation defined by (2.14) is well defined, i.e., it does not depend on the choice of representatives $a, b \in G$ of cosets aH and bH. Let $aH = cH$ and $bH = dH$. We will show that, in this case, $(ab)H = (cd)H$.

Since $c \in aH$ and $d \in bH$, $c = ah_1$ and $d = bh_2$, where $h_1, h_2 \in H$. Then, $cd = (ah_1)(bh_2)$. Since H is a normal subgroup of G, $bH = Hb$. Hence, $h_1 b = bh_3$. Therefore, $cd = (ab)(h_3 h_2) \in (ab)H$ and so $(cd)H \subseteq (ab)H$. Analogously, $(ab)H \subseteq (cd)H$. Hence, $(ab)H = (cd)H$.

We will now show that the operation (2.14) satisfies all axioms of a group.
1. Associativity:
$((aH) \circ (bH)) \circ (cH) = (aH) \circ ((bH) \circ (cH))$
L$= ((aH) \circ (bH)) \circ (cH) = ((ab)H) \circ (cH) = ((ab)c)H$
R$= (aH) \circ ((bH) \circ (cH)) = (aH) \circ ((bc)H) = (a(bc))H$
Since $(ab)c = a(bc)$ in G, we prove that operation (2.14) is associative.
2. The identity element in G/H is given by the coset $eH = H$, where e is the identity element of G. Really,
$(aH) \circ (eH) = (ae)H = aH$
$(eH) \circ (aH) = (ea)H = aH$.

3. The inverse of a coset aH is a coset $(aH)^{-1} = a^{-1}H$. Really,

$a^{-1}H \circ (aH) = (a^{-1}a)H = eH = H$

$(aH) \circ a^{-1}H = (aa^{-1})H = eH = H$

In this way, G/H is a group under the operation (2.14). $\qquad\square$

✳ Example 2.54.

Consider the additive group $G = \mathbb{Z}$ and the subgroup $H = 3\mathbb{Z}$. Since G is a commutative group, $G/H = \mathbb{Z}/3\mathbb{Z} = \{3\mathbb{Z}, 1 + 3\mathbb{Z}, 2 + 3\mathbb{Z}\}$ is a quotient group under the operation $\oplus$ given in the form of the following table:

$\oplus$	$3\mathbb{Z}$	$1 + 3\mathbb{Z}$	$2 + 3\mathbb{Z}$
$3\mathbb{Z}$	$3\mathbb{Z}$	$1 + 3\mathbb{Z}$	$2 + 3\mathbb{Z}$
$1 + 3\mathbb{Z}$	$1 + 3\mathbb{Z}$	$2 + 3\mathbb{Z}$	$3\mathbb{Z}$
$2 + 3\mathbb{Z}$	$2 + 3\mathbb{Z}$	$3\mathbb{Z}$	$1 + 3\mathbb{Z}$

✳ Example 2.55.

Let $\mathbb{Z}$, $H = m\mathbb{Z}$, then the quotient group $G/H = \mathbb{Z}/m\mathbb{Z} = (\{[0], [1], \ldots, [m-1]\}, +)$ is the set of residue classes modulo m under the operation of addition, given as follows:

$$[a] + [b] = [a + b] \tag{2.15}$$

for all $[a], [b] \in \mathbb{Z}/m\mathbb{Z}$. This operation is well defined, i.e., for any $[a], [b] \in \mathbb{Z}/m\mathbb{Z}$, the class $[a + b]$ does not depend on the choice of representatives in classes $[a], [b]$. Let $a_1, b_1 \in \mathbb{Z}$ and $[a_1] = [a], [b_1] = [b]$. From the definition of congruence relation modulo m, it follows that there are numbers $c, d \in \mathbb{Z}$ such that $a - a_1 = mc$ and $b_1 - b = md$. Adding these equalities, we obtain:

$$m(c + d) = mc + md = (a_1 - a) + (b - b_1) = (a_1 + b_1) - (a + b).$$

Hence, m divides $(a + b_1) - (a + b)$, which means that $[a + b] = [a_1 + b_1]$.

It is also easy to show that

1) $([a] + [b]) + [c] = [a] + ([b] + [c])$

2) $[a] + [0] = [0] + [a] = [a]$

3) $[a] + [-a] = [-a] + [a] = [0]$

This means that the class $[0]$ is the identity element of $\mathbb{Z}/m\mathbb{Z}$ with respect to addition, and the element $[-a]$ is the inverse to an element $[a]$. Therefore, $\mathbb{Z}/m\mathbb{Z}$ is a finite commutative group of order m under the operation of addition (2.15).

If p is a prime, then $\mathbb{Z}/p\mathbb{Z}$ is a simple group of order p.

✳ Example 2.56.

Consider the set $(\mathbb{Z}/p\mathbb{Z})^* = \{[1], [2], \ldots, [p-1]\}$ of all residue classes modulo p with the operation of multiplication given as follows:

$$[a][b] = [ab] \tag{2.16}$$

This operation has the properties:

1) $([a][b])[c] = [a]([b][c])$

2) $[a][1] = [1][a] = [a]$

3) The equation $[a][x] = [1]$ has a solution for any $[a] \in (\mathbb{Z}/p\mathbb{Z})^*$. Really, since $\gcd(a, p) = 1$, by Corollary 1.32 there exist integers x, y such that $ax + py = 1$. Then, $[a]^{-1} = [a^{-1}] = [x]$ is a solution of this equation. So, $(\mathbb{Z}/p\mathbb{Z})^*$ is a finite group called the **multiplicative group of residue classes modulo** p. It is obvious that $(\mathbb{Z}/p\mathbb{Z})^*$ is a commutative group of order $p - 1$.

⁕ **Definition 2.57.**

The number of positive integers not exceeding a positive number m and relatively prime to m is called **Euler's totient function** (or **Euler's φ-function**). It is denoted by $\varphi(m)$.

Note that 1 is considered to be a relatively prime to m.

⁕ **Examples 2.58.**

1. Since we have only 2 positive integer not exceeding to 4 which are relatively prime to 4, namely 1 and 3, then $\varphi(4) = 2$.

2. $\varphi(7) = 6$, since numbers 1,2,3,4,5,6 are relatively prime to 7.

⁕ **Example 2.59.**

Consider the set $(\mathbb{Z}/m\mathbb{Z})^* = \{[k] : k \in \mathbb{N} \text{ and } \gcd(k, m) = 1\}$ of residue classes modulo m under the operation of multiplication of the following form:

$$[a][b] = [ab] \tag{2.17}$$

This operation has the following properties:

1) $([a][b])[c] = [a]([b][c])$

2) $[a][1] = [1][a] = [a]$

3) The equation $[a][x] = [1]$ has a solution for any $[a] \in (\mathbb{Z}/m\mathbb{Z})^*$. Really, since $\gcd(a, m) = 1$, there are integers x, y such that $ax + py = 1$. Then, $[a]^{-1} = [a^{-1}] = [x]$ is a solution of this equation. Hence, $(\mathbb{Z}/m\mathbb{Z})^*$ is a finite group called the **multiplicative group of residue classes modulo** m. Since the number of integers $0 < k < m$ for which $\gcd(a, m) = 1$ is equal to $\varphi(m)$, $(\mathbb{Z}/p\mathbb{Z})^*$ is a commutative group of order $\varphi(m)$.

2.6. Group Homomorphisms

⁕ **Definition 2.60.**

Given two groups $(G, \bullet)$ and $(H, \circ)$, a **group homomorphism** is a map $\varphi : G \to H$ such that

$$\varphi(a \bullet b) = \varphi(a) \circ \varphi(b) \tag{2.18}$$

for all $a, b \in G$.

✳ Example 2.61.

The mapping $\pi : \mathbb{Z} \to \mathbb{Z}/m\mathbb{Z}$ defined as follows: $\pi(a) = [a]$ for all $a \in \mathbb{Z}$ is a homomorphism of the additive groups that follows from the definition of the operation (2.15) in $\mathbb{Z}/m\mathbb{Z}$.

✳ Definition 2.62.

A group homomorphism $\varphi : G \to H$ is called a **monomorphism** if φ is injective, i.e., $\varphi(a_1) = \varphi(a_2)$ if and only if $a_1 = a_2$. A group homomorphism $\varphi : G \to H$ is called an **epimorphism** if φ is surjective, i.e., for any element $y \in H$ there exists an element $x \in G$ such that $\varphi(x) = y$.

✳ Theorem 2.63.

If φ is a homomorphism of a group $(G, \bullet)$ to a group $(H, \circ)$ with identity elements e_1 and e_2, then

1) $\varphi(e_1) = e_2$;
2) $\varphi(a^{-1}) = (\varphi(a))^{-1}$ for all $a \in G$.

Proof.

1) From Definition 2.60 it follows that $\varphi(a) = \varphi(a \bullet e_1) = \varphi(a) \circ \varphi(e_1)$ and $\varphi(a) = \varphi(e_1 \bullet a) = \varphi(e_1) \circ \varphi(a)$, which means that $\varphi(e_1) = e_2$.

2) Analogously, by Definition 2.60, we obtain that $e_2 = \varphi(e_1) = \varphi(a \bullet a^{-1}) = \varphi(a) \circ \varphi(a^{-1})$ and $e_2 = \varphi(e_1) = \varphi(a^{-1} \bullet a) = \varphi(a^{-1}) \circ \varphi(a)$ for an arbitrary element $a \in G$. Therefore, $\varphi(a^{-1}) = (\varphi(a))^{-1}$. □

✳ Definition 2.64.

If $\varphi : G \to H$ is a group homomorphism and e_2 is the identity element of H, then the set

$$\mathrm{Ker}(\varphi) = \{a \in G \ : \ \varphi(a) = e_2\} \tag{2.19}$$

is called the **kernel** of the homomorphism φ

The set of elements of H of the form $\varphi(a)$, where $a \in G$ is called the **image** of the homomorphism φ and is denoted by $\mathrm{Im}(\varphi)$, i.e.,

$$\mathrm{Im}(\varphi) = \{h \in H \ : \ \exists (a \in G)[h = \varphi(a)]\} \tag{2.20}$$

If K is a subgroup of H, then the set

$$\varphi^{-1}(K) = \{g \in G \ : \ \varphi(g) \in K\} \tag{2.21}$$

is called the **preimage** of K.

✳ Theorem 2.65.

If φ is a homomorphism of a group $(G, \bullet)$ to a group $(H, \circ)$, then $\mathrm{Ker}(\varphi)$ is a normal subgroup of G.

Proof.

Suppose that e_1, e_2 are the identity elements of G and H, respectively. Let $a_1, a_2 \in \text{Ker}(\varphi)$, then $\varphi(a_1) = \varphi(a_2) = e_2$. Since φ is a group homomorphism, $\varphi(a_1 \bullet a_2) = \varphi(a_1) \circ \varphi(a_2) = e_2 \circ e_2 = e_2$. Therefore, $a_1 \bullet a_2 \in \text{Ker}(\varphi)$. If $a \in \text{Ker}(\varphi)$, then, by Theorem 2.63(2), $\varphi(a^{-1}) = (\varphi(a))^{-1} = e_2^{-1} = e_2$, i.e., $a^{-1} \in \text{Ker}(\varphi)$. By Theorem 2.63(1), $e_1 \in \text{Ker}(\varphi)$. So, $\text{Ker}(\varphi)$ is a subgroup of G.

We will show that $\text{Ker}(\varphi)$ is a normal subgroup of G. Let $s \in \text{Ker}(\varphi)$, then $\varphi(s) = e_2$. For any $g \in G$ we have:

$$\varphi(g \bullet s \bullet g^{-1}) = \varphi(g) \circ \varphi(s) \circ \varphi(g^{-1}) = \varphi(g) \circ e_2 \circ \varphi(g^{-1}) = \varphi(g) \circ (\varphi(g))^{-1} = e_2,$$

which means that $g \bullet s \bullet g^{-1} \in \text{Ker}(\varphi)$. Therefore, $\text{Ker}(\varphi)$ is a normal subgroup, by Theorem 2.45. □

✱ Example 2.66.

Let $G = GL(n, \mathbb{R})$ be a group of nonsingular square matrices of degree n and let $H = \mathbb{R}^*$ be the multiplicative group of nonzero real numbers under multiplication. Consider the mapping $\det : G \to H$ given by $\mathbf{A} \mapsto \det(\mathbf{A})$. Then, det is a homomorphism of these groups because $\det(\mathbf{AB}) = \det(\mathbf{A})\det(\mathbf{B})$ and $\text{Ker}(\det) = \{\mathbf{A} \in GL(n, \mathbb{R}) : \det(\mathbf{A}) = 1\} = SL(n, \mathbb{R})$.

✱ Theorem 2.67.

The image of a homomorphism φ of a group $(G, \bullet)$ to a group $(H, \circ)$ is a subgroup of H.

Proof.

Let $b_1, b_2 \in \text{Im}(\varphi)$, then there are $a_1, a_2 \in G$ such that $b_1 = \varphi(a_1)$, $b_2 = \varphi(a_2)$. Therefore, $b_1 \circ b_2 = \varphi(a_1) \circ \varphi(a_2) = \varphi(a_1 \bullet a_2) \in \text{Im}(\varphi)$. If $b \in \text{Im}(\varphi)$, then there is an element $a \in G$ such that $b = \varphi(a)$. Then, by Theorem 2.63(2), $b^{-1} = \varphi(a)^{-1} = \varphi(a^{-1}) \in \text{Im}(\varphi)$. Since $e_2 = \varphi(e_1) \in \text{Im}(\varphi)$, $\text{Im}(\varphi)$ is a subgroup of H. □

✱ Theorem 2.68.

If φ is a homomorphism of a group $(G, \bullet)$ to a group $(H, \circ)$ with identity elements e_1 and e_2, then:
1. $\text{Ker}(\varphi) = \{e_1\}$ if and only if φ is a monomorphism.
2. $\text{Im}(\varphi) = H$ if and only if φ is an epimorphism.

Proof.

1. Suppose that $\text{Ker}(\varphi) = \{e_1\}$. If $\varphi(a_1) = \varphi(a_2)$ for some $a_1, a_2 \in G$, then $\varphi(a_1 \bullet a_2^{-1}) = \varphi(a_1) \circ [\varphi(a_2)]^{-1} = \varphi(a_2) \circ [\varphi(a_2)]^{-1} = e_2$, implying that $a_1 \bullet a_2^{-1} \in \text{Ker}(\varphi) = \{e_1\}$, hence, $a_1 = a_2$. So φ is a monomorphism. Conversely, let φ be a monomorphism and $a \in \text{Ker}(\varphi)$. Then, $\varphi(a) = e_2 = \varphi(e_1)$, because $e_2 = \varphi(e_1)$, by Theorem 2.63(1). Therefore, $a = e_1$, and so $\text{Ker}(\varphi) = \{e_1\}$.

2. This property follows from the definition of $\text{Im}(\varphi)$. □

✳ Theorem 2.69.
If N is a normal subgroup of a multiplicative group G, then a mapping

$$\varphi : G \to G/N$$

defined by $\varphi(a) = aN$ for all $a \in G$ is a group homomorphism, which is an epimorphism, i.e., $\text{Ker}(\varphi) = N$ and $\text{Im}(\varphi) = G/N$.

Proof.
By the definition of φ and G/N with operation $\circ$, we have:

$$\varphi(ab) = (ab)N = (aN) \circ (bN) = \varphi(a) \circ \varphi(b),$$

which shows that φ is a group homomorphism.

Let $x \in \text{Ker}(\varphi)$, then $\varphi(x) = e_2$ is the identity element of the group G/N. Since $e_2 = N$, we have a sequence of implications:

$$\varphi(x) = \varphi(e_2) \implies xN = N \implies x \in N \implies \text{Ker}(\varphi) \subseteq N.$$

Conversely, if $x \in N$, then $\varphi(x) = xN = N = e_2$, so $N \subseteq \text{Ker}(\varphi)$. Hence, $N = \text{Ker}(\varphi)$. □

The homomorphism $\varphi : G \to G/N$, defined in Theorem 2.69, is called the **canonical** (or **natural**) **map** or **projection**.

The concept of an isomorphism is one of the most central concepts in abstract algebra. Isomorphic groups have the same properties. Therefore, groups are considered up to isomorphism, i.e., we consider equivalence classes with respect to isomorphism.

✳ Definition 2.70.
  A group homomorphism $\varphi : G \to H$ is called an **isomorphism** if it is a bijective map. If there is a group isomorphism $\varphi : G \to H$, then the groups G and H are called **isomorphic** and we write $G \cong H$.

We will show that a group isomorphism is an equivalence relation on the set of all groups. We verify properties of isomorphism $\cong$.
   1. Reflexivity: $G \cong G$
   2. Symmetry: If $G \cong H$, then there is a group isomorphism $\varphi : G \to H$. Then, there is an inverse mapping $\varphi^{-1} : H \to G$ which is also a group isomorphism. Really, since φ is an isomorphism, there exist $g_1, g_2 \in G$ such that $\varphi(g_1) = h_1$ and $\varphi(g_2) = h_2$. Then $\varphi^{-1}(h_1 \circ h_2) = \varphi^{-1}(\varphi(g_1) \circ \varphi(g_2)) = \varphi^{-1}(\varphi(g_1 \bullet g_2)) = g_1 \bullet g_2 = \varphi^{-1}(h_1) \bullet \varphi^{-1}(h_2))$. Thus $H \cong G$.
   3. Transitivity: If $G \cong H$ and $H \cong F$ then there are group isomorphisms $\varphi : G \to H$ and $\psi : H \to F$. Then, there is also their composition $\psi \circ \varphi : G \to$

F. Since a composition of bijections is also a bijection, $\psi \circ \varphi$ is also a group isomorphism, i.e., $G \cong F$.

In this way, the isomorphism relation $\cong$ satisfies all properties of equivalence relation and is, therefore, an equivalence relation on the set of all groups.

From Theorem 2.68, we immediately obtain the following theorem.

✳ Theorem 2.71.

If $\varphi : G \to H$ is a group homomorphism, then φ is a group isomorphism if and only if $\mathrm{Ker}(\varphi) = \{e_1\}$ and $\mathrm{Im}(\varphi) = H$, where e_1 is the identity element of G.

✳ Examples 2.72.

1. The mapping $\varphi : \mathbb{Z}_m \to \mathbb{Z}/m\mathbb{Z}$ defined by $\varphi(a) = [a]$ is a group isomorphism that follows immediately from the definition of the operation of addition on these groups.

2. Let $G = \{\mathbb{R}, +\}$ be the additive group of real numbers, $H = \{\mathbb{R}^+, \cdot\}$ the multiplicative group of positive real numbers, where $\mathbb{R}^+ = \{x \in \mathbb{R} : x > 0\}$. Consider the mapping $\exp : G \to H$ given by $\exp(x) = e^x$. Then, $\exp(x + y) = e^{x+y} = e^x e^y = \exp(x)\exp(y)$, for all $x, y \in G$, implying that exp is a group homomorphism.

If $\exp(x) = 1 = e^x$, then $x = 0$. Therefore, $\mathrm{Ker}(\exp) = \{0\}$, i.e., exp is a monomorphism of groups. Since for each $y \in H$ there exists a real number $x = \ln(y)$ such that $y = e^x = \exp(x)$, exp is an epimorphism of groups. Therefore, exp is a group isomorphism, and groups G and H are isomorphic. The inverse mapping $\ln : H \to G$ given by the formula $x \mapsto \ln(x)$ is also a group isomorphism and there hold the equalities: $\exp(\ln(x)) = x, \ln(\exp(x)) = x$.

3. Let $G = \{\mathbb{R}^+, \cdot\}$ be the multiplicative group of positive real numbers, $H = \mathbb{R}^* = \{\mathbb{R}\backslash\{0\}, \cdot\}$ the multiplicative group of nonzero real numbers. Consider the mapping $| \ | : H \to G$ given by the formula:

$$|x| = \begin{cases} x & \text{if } x > 0 \\ -x & \text{if } x < 0 \end{cases}$$

Since $|xy| = |x| \cdot |y|$, the mapping $| \ |$, which is called **modulo**, is a group homomorphism. It is obviously an epimorphism, but it is not a monomorphism.

4. Let $H = \mathbb{R}^* = \{\mathbb{R}\backslash\{0\}, \cdot\}$ be the multiplicative group of nonzero real numbers, $F = \{1, -1\}$ the group under multiplication as in Example 2.7(4). Consider the mapping $\mathrm{sgn} : H \to F$ given by the formula:

$$\mathrm{sgn}(x) = \begin{cases} 1 & \text{if } x > 0 \\ -1 & \text{if } x < 0 \end{cases}$$

Since $\mathrm{sgn}(xy) = \mathrm{sgn}(x)\mathrm{sgn}(y)$, the mapping sgn is a group homomorphism. It is easy to see that sgn is an epimorphism, but it is not a monomorphism.

5. Let $G = \mathbb{C}^* = \{\mathbb{R}\backslash\{0\}, \cdot\}$ be the multiplicative group of nonzero complex numbers, $H = \{\mathbb{R}^+, \cdot\}$ the multiplicative group of positive real numbers. Then, the mapping $|\ | : G \to H$ given by the formula $|x + iy| = \sqrt{x^2 + y^2}$ is a group homomorphism, since $|xy| = |x| \cdot |y|$. This homomorphism is an epimorphism, but it is not a monomorphism.

6. Let G be a group, $H = \{\mathbb{Z}, +\}$ the additive group of integers. For an element $g \in G$, consider the mapping $f_g : H \to G$ given by the formula: $f_g(n) = g^n$. Since $f_g(n + m) = g^{n+m} = g^n g^m = f_g(n)f_g(m)$, f_g is a group homomorphism. If g is an element of an infinite order, then f_g is a monomorphism, otherwise f_g is not a monomorphism.

> ※ **Definition 2.73.**
> A homomorphism of a group G to itself is called an **endomorphism** of G.
> An isomorphism of a group G to itself is called an **automorphism** of G. The set of all automorphisms of a group G is denoted by $\mathrm{Aut}(G)$.

※ **Theorem 2.74.**
The set of all automorphisms $\mathrm{Aut}(G)$ of a group G under the operation $\circ$, which is a composition of mappings, is a group.

Proof.

1) If $\varphi, \psi \in \mathrm{Aut}(G)$, then $\varphi \circ \psi \in \mathrm{Aut}(G)$ as well;

2) The operation $\circ$ is associative as a composition of mappings;

3) The identical mapping Id_G of a group G to itself is the identity element in the set $\mathrm{Aut}(G)$ with respect to operation $\circ$;

4) If $\varphi \in \mathrm{Aut}(G)$, then $\varphi^{-1} \in \mathrm{Aut}(G)$ as well, because φ is a bijection. $\square$

※ **Example 2.75.**
Consider a cyclic group of order 8: $G = C_8 = \langle x \rangle$ and a mapping $\varphi : G \to G$ given by the formula: $\varphi(g) = g^3$ for all $g \in G$. Since $\varphi(gh) = (gh)^3 = g^3 h^3 = \varphi(g)\varphi(h)$, $\varphi \in \mathrm{Aut}(G)$. Automorphism φ can be written in the form of the following table:

x	$e = x^0$	x	x^2	x^3	x^4	x^5	x^6	x^7
$\varphi(x) = x^3$	$e = x^0$	x^3	x^6	x	x^4	x^7	x^2	x^5

2.7. The Isomorphism Theorems

In this section, we consider three significant theorems concerning group isomorphisms.

✳ **Theorem 2.76 (First Isomorphism Theorem).**
If $\varphi : G \to H$ is a group homomorphism, then:

$$\text{Im}(\varphi) \cong G/\text{Ker}(\varphi) \tag{2.22}$$

Proof.

Let $\varphi : G \to H$ be a group homomorphism of multiplication groups. Since $\text{Ker}(\varphi)$ is a normal subgroup of G, $G/\text{Ker}(\varphi)$ is a quotient group with the operation $\circ$ defined by (2.14). Denote $N = \text{Ker}(\varphi)$ and consider a mapping $f : G/N \to \text{Im}(\varphi)$ given as follows:

$$f(aN) = \varphi(a) \tag{2.23}$$

for all $a \in G$. Since φ is a group homomorphism,

$$f(aN \circ bN) = f((ab)N) = \varphi(ab) = \varphi(a)\varphi(b) = f(aN)f(bN)$$

implying that f is a group homomorphism.

We will show that f is an isomorphism. It follows from (2.23) that f is an epimorphism. Let $f(aN) = e = \varphi(a)$, where e is the identity element of the group G. Then $a \in \text{Ker}(\varphi) = N$, so $aN = N$ is the identity element of the quotient group $G/\text{Ker}(\varphi)$. This means that f is a monomorphism. Thus, f is a group isomorphism. ☐

✳ **Corollary 2.77.**
If $\varphi : G \to H$ is an epimorphism of groups then:

$$H \cong G/\text{Ker}(\varphi) \tag{2.24}$$

✳ **Examples 2.78.**

1. The groups $G = (\mathbb{R}, +)$ and $H = (\mathbb{R}^+, \cdot)$ are isomorphic. An isomorphism of groups $\varphi : G \to H$ can be given, for example, as follows: $\varphi(x) = 2^x$. Really, $\text{Ker}(\varphi) = \{x \in \mathbb{R} : 2^x = 1\} = \{0\}$. Since for any $y \in H$ there is an element $x = \log_2(y)$, we have $2^x = y$, which means that $\text{Im}(\varphi) = H$. So the inverse mapping $\varphi^{-1} : H \to G$ is given by $\varphi^{-1}(x) = \log_2(x)$. Hence, $G \cong H$.

2. Consider the groups $G = (\mathbb{R}, +)$, $H = (\mathbb{Z}, +)$, $T = (\{z \in \mathbb{C} : |z| = 1\}, \cdot)$. The mapping $\varphi : G \to T$ given as follows: $\varphi(x) = e^{2\pi i x}$ is a group homomorphism because $\varphi(x + y) = e^{2\pi i(x+y)} = e^{2\pi i x}e^{2\pi i y} = \varphi(x)\varphi(y)$. This homomorphism is an epimorphism. It follows from the fact that any $z \in \mathbb{C}$ with $|z| = 1$ has the form $z = e^{2\pi i x}$ for some $x \in \mathbb{R}$. We now find the kernel of homomorphism φ. Since $e^{2\pi i x} = 1$ if and only if $x = n \in \mathbb{Z}$, $\text{Ker}(\varphi) = H$. Therefore, by Corollary 2.77 we obtain that $G/H \cong T$.

> **❋ Lemma 2.79.**
> If H is a subgroup of a group G and $N \trianglelefteq G$ then
> 1) $K = HN = NH$ is a subgroup of G, where
>
> $$HN = \{hn \mid h \in H, n \in N\}, \quad NH = \{nh \mid h \in H, n \in N\}$$
>
> 2) $N \trianglelefteq HN$,
> 3) $(H \cap N) \trianglelefteq H$.

Proof.

1) Since $N \trianglelefteq G$, $h^{-1}nh \in N$ for all $h \in H$ and for all $n \in N$, by Theorem 2.45. So $nh \in HN$, i.e., $NH \subseteq HN$. Analogously, we can show that $HN \subseteq NH$. In this way, $HN = NH$. Furthermore, $(hn)(h_1 n_1) = h(nh_1)n_1 = (hh_1)n_2 n_1 \in HN$. If e is the identity element of the group G, then $e = ee \in HN$. The inverse element to hn is $(hn)^{-1} = n^{-1}h^{-1} \in NH = HN$. Hence, $HN = NH$ is a subgroup of G.

2) Since N is a subgroup of HN and N is a normal subgroup of G, the statement follows from Theorem 2.45.

3) Analogously, since $H \cap N$ is a subgroup of H and N is a normal subgroup of G, the statement follows from Theorem 2.45. $\square$

> **❋ Theorem 2.80 (Second Isomorphism Theorem).**
> If H is a subgroup of a group G and $N \trianglelefteq G$, then:
>
> $$(HN)/N \cong H/(N \cap H). \qquad (2.25)$$

Proof.

Consider a mapping $F : H \to (HN)/N$ given as follows:

$$F(h) = hN = (hN) \circ N \in (HN)/N$$

for all $h \in H$.

1) Since $F(h_1 h_2) = (h_1 h_2)N = (h_1 N) \circ (h_2 N) = F(h_1) \circ F(h_2)$, F is a group homomorphism.

2) $\mathrm{Ker}(F) = \{h \in H : hN = N\} = \{h \in H : h \in N\} = H \cap N$.

3) Let $x \in (HN)/N$, then $x = (hn)N$ for some $h \in H$ and $n \in N$. Therefore, $F(h) = hN = (hN) \circ N = (hN) \circ (nN) = (hn)N = x$ implying that F is an epimorphism.

Since by Lemma 2.79(2) $N \trianglelefteq HN$, $\mathrm{Im}(F) = (HN)/N$. Moreover, $\mathrm{Ker}(F) = H \cap N \trianglelefteq H$. Therefore, by Theorem 2.76 we have a group isomorphism: $H/(H \cap N) \cong (HN)/N$. $\square$

> **❋ Theorem 2.81 (Third Isomorphism Theorem).**
> If M, N are normal subgroups of a group G and $M \subset N$ then:
> 1) $M \trianglelefteq N$;
> 2) $(N/M) \trianglelefteq (G/M)$;
> 3) $G/N \cong (G/M)/(N/M)$.

Proof.

The first and the second parts of the theorem are obvious.

Consider a mapping $F : G/M \to G/N$ given by $F(gM) = gN$ for all $g \in G$.

i) Let $a, b \in G$ and assume that $aM = bM$. Then, we have a sequence of implications:

$$ab^{-1} \in M \subset N \;\Rightarrow\; ab^{-1} \in N \;\Rightarrow\; a \in Nb = bN \;\Rightarrow\; aN \subseteq bN$$

Analogously, $bN \subseteq aN$, hence, $aN = bN$.

ii) Since $F[(aM) \circ (bM)] = F[(ab)M] = (ab)N = (aN) \circ (bN) = F(aM) \circ F(bM)$, F is a group homomorphism.

iii) From the definition of F, it follows that F is an epimorphism, i.e., $\text{Im}(F) = G/N$.

iv) $\text{Ker}(F) = \{aM \in G/M \;:\; F(aM) = N\} = \{aM \in G/M \;:\; aN = N\} = \{aM \in G/M \;:\; a \in N\} = N/M$.

Thus, we have that $\text{Im}(F) = G/N$ and $\text{Ker}(F) = N/M$. Then, by Theorem 2.76 we have a group isomorphism:

$$G/N \cong (G/M)/(N/M).$$

$\square$

✳ Example 2.82.

Consider the mapping $\varphi : GL(n, \mathbb{R}) \to \mathbb{R}^*$ given as follows: $\varphi(\mathbf{A}) = \det(\mathbf{A})$ for all $\mathbf{A} \in GL(n, \mathbb{R})$. Then,

$$\text{Ker}(\det) = \{\mathbf{A} \in GL(n, \mathbb{R}) \;:\; \det(\mathbf{A}) = 1\} = SL(n, \mathbb{R})$$

and $\text{Im}(\varphi) = \mathbb{R}^*$ because for each $x \in \mathbb{R}^*$ there is the matrix

$$\begin{pmatrix} x & 0 & \cdots & 0 \\ 0 & 1 & \cdots & 0 \\ \vdots & \vdots & \ddots & \vdots \\ 0 & 0 & \cdots & 1 \end{pmatrix} \in GL(n, \mathbb{R})$$

with $\det(\mathbf{A}) = x$. So, by Theorem 2.76, we have an isomorphism:

$$GL(n, \mathbb{R})/SL(n, \mathbb{R}) \cong \mathbb{R}^*.$$

2.8. Exercises

1. Write the Cayley tables for the set $\{0, 1, 2, 3, 4\}$ under operations: i) addition modulo 5; ii) multiplication modulo 5.

2. Which of the following sets are groups under multiplication:

   (a) the set of all positive real numbers,

   (b) the set of all non-zero real numbers with condition $|x| \leq 1$,

   (c) the set of all non-zero complex numbers,

   (d) the set of all non-zero complex numbers with property $|z| = 1$.

3. Determine which of given algebraic structures is a group and which of them are Abelian groups:

   (a) $(\{1, 2, 3, 4, 6, 12\}, \gcd)$,

   (b) $(\{a + b\sqrt{2} : a, b \in \mathbb{Q}\}, +)$,

   (c) $(\{z \in \mathbb{C} : |z| = 1\}, +)$,

   (d) $(\{z \in \mathbb{C} : |z| = 1\}, \cdot)$,

   (e) The set of matrices: $\left\{ \begin{pmatrix} 1 & 0 \\ 0 & 1 \end{pmatrix}, \begin{pmatrix} -1 & 0 \\ 0 & 1 \end{pmatrix}, \begin{pmatrix} 1 & 0 \\ 0 & -1 \end{pmatrix}, \begin{pmatrix} -1 & 0 \\ 0 & -1 \end{pmatrix} \right\}$
   under multiplication of matrices.

4. Show that the only element of a group satisfying the property $a^2 = a$ is the identity element.

5. Show that if in a group G every element satisfies the property $a^2 = e$, where e is the identity element of G, then G is Abelian.

6. Find $\sigma\tau$, $\tau^{-1}\sigma$, $\sigma\tau^{-3}$, where

$$\sigma = \begin{pmatrix} 1 & 2 & 3 & 4 & 5 & 6 & 7 \\ 6 & 3 & 5 & 4 & 2 & 7 & 1 \end{pmatrix}, \quad \tau = \begin{pmatrix} 1 & 2 & 3 & 4 & 5 & 6 & 7 \\ 3 & 2 & 7 & 4 & 1 & 5 & 6 \end{pmatrix}.$$

7. Find $\tau\sigma$, $\sigma\tau^2$, σ^{-2}, where

$$\sigma = \begin{pmatrix} 1 & 2 & 3 & 4 & 5 & 6 & 7 \\ 6 & 3 & 4 & 5 & 2 & 7 & 1 \end{pmatrix}, \quad \tau = \begin{pmatrix} 1 & 2 & 3 & 4 & 5 & 6 & 7 \\ 3 & 2 & 7 & 6 & 1 & 5 & 4 \end{pmatrix}.$$

8. Find a permutation μ satisfying the equation $\tau\mu\sigma^{-1} = \rho$, where

$$\sigma = \begin{pmatrix} 1 & 2 & 3 & 4 & 5 \\ 3 & 5 & 1 & 4 & 2 \end{pmatrix}, \quad \tau = \begin{pmatrix} 1 & 2 & 3 & 4 & 5 \\ 5 & 4 & 3 & 2 & 1 \end{pmatrix}, \quad \rho = \begin{pmatrix} 1 & 2 & 3 & 4 & 5 \\ 2 & 4 & 5 & 3 & 1 \end{pmatrix}.$$

9. Solve the equation $f^2 x g^{-1} = h$, where

$$f = \begin{pmatrix} 1 & 2 & 3 & 4 & 5 & 6 \\ 3 & 1 & 4 & 2 & 5 & 6 \end{pmatrix}, \quad g = \begin{pmatrix} 1 & 2 & 3 & 4 & 5 & 6 \\ 2 & 5 & 1 & 4 & 6 & 3 \end{pmatrix},$$

$$h = \begin{pmatrix} 1 & 2 & 3 & 4 & 5 & 6 \\ 3 & 2 & 5 & 6 & 1 & 4 \end{pmatrix}.$$

10. Show that the set of matrices

$$H = \left\{ \begin{pmatrix} a & b \\ 0 & 1/a \end{pmatrix} \mid a, b \in \mathbb{R},\ a \neq 0 \right\}$$

under the operation of multiplication is a subgroup of $GL(2, \mathbb{R})$.

11. Find all subgroups of the group S_3.

12. Show that if A, B are subgroups of an Abelian group G, then the set

$$AB = \{ab\ :\ a \in A, b \in B\}$$

is also a subgroup of G.

13. Show that, in the multiplicative groups $(\mathbb{Z}/m\mathbb{Z})^*$ for $m = 3, 5, 16$, all elements satisfy the condition $x^4 = 1$.

14. Find orders of all elements in the multiplicative groups $(\mathbb{Z}/8\mathbb{Z})^*$ and $(\mathbb{Z}/20\mathbb{Z})^*$. Are these groups cyclic?

15. Show that, if $G = \{g_1, g_2, \ldots, g_n\}$ is an Abelian group of order n, then the element $x = g_1 g_2 \cdots g_n$ satisfies the equation $x^2 = e$, where e is the identity element of G.

16. Let $G = \mathbb{Z}_{12}$ be the additive group. Find all cosets of the subgroup $H = \{0, 3, 6, 9\}$.

17. Let $G = \langle x \rangle$ be a cyclic group of order 10. Find all cosets of the subgroup $H = \langle x^2 \rangle$.

18. Find possible orders of subgroups of the additive groups $\mathbb{Z}_6$, $\mathbb{Z}_7$, $\mathbb{Z}_{12}$, $\mathbb{Z}_{12}$, $\mathbb{Z}_{13}$, $\mathbb{Z}_{23}$, $\mathbb{Z}_{24}$.

19. Find the quotient group $3\mathbb{Z}/15\mathbb{Z}$.

20. Find all subgroups of the group $\mathbb{Z}_6$.

21. Find all subgroups of the group $\mathbb{Z}_{12}$.

22. Find all generators of the cyclic group $G = \langle x \rangle$ of order 10.

23. The set $Z(G) = \{x \in G\ :\ \forall (g \in G)(xg = gx)\}$ is called the **center** of a group G. Show that $Z(G)$ is a normal subgroup of G.

24. Find the center of the group of matrices (under multiplication) of the

form: $\begin{pmatrix} 1 & a & b \\ 0 & 1 & c \\ 0 & 0 & 1 \end{pmatrix}$, where $a, b, c \in \mathbb{R}$.

References

[1] Ash, R.B. 2007. Basic Abstract Algebra, Dover Publications.

[2] Dummit, D.S. and R.M. Foote. 2004. Abstract Algebra (3rd Ed.), John Wiley & Sons.

[3] Hall, M.(Jr.) 1959. The Theory of Groups, New York, Macmillan.

[4] Hazewinkel, M., N. Gubareni and V.V. Kirichenko. 2007. Algebra, Rings and Modules, Vol. 2, Springer.

[5] Humphreys, J.F. 1996. A Course in Group Theory, Oxford University Press.

[6] Jacobson, N. 1985. Basic Algebra I, Freeman, San Francisco, H. Freeman & Co.

[7] Judson, Th.W. 1994. Abstract Algebra. Theory and Applications, PWS Publishing Company, Michigan.

[8] Lang, S. 1984. Algebra, Addison-Wesley.

[9] Oggier, F. 2011. Algebraic Methods.
http://www1.spms.ntu.edu.sg/~frederique/AA11.pdf.

[10] Oggier, F. and A.M. Bruckstein. 2013. Groups and Symmetries.
http://www1.spms.ntu.edu.sg/~frederique/groupsymmetryws.pdf.

[11] Rotman, J.J. 1994. An Introduction to the Theory of Groups, Springer-Verlag.

Chapter 3

Examples of Groups

"The most painful thing about mathematics is how far away you are from being able to use it after you have learned it."

James Newman

"Imagination is more important than knowledge."

Albert Einstein

"Where there is matter, there is geometry."

Johannes Kepler

One of the most important examples of groups are groups of transformations, in particular, permutation groups. Permutations were first studied by L.J. Lagrange (1771) and A. Vandermonde (1771). The deep connections between the properties of permutation groups and those of algebraic equations were pointed out by N.H. Abel (1824) and E. Galois (1830). In particular, E. Galois discovered the role played by normal subgroups in problems of solvability of equations by radicals, he also proved that the alternating groups of order ≥ 5 are simple, etc. Also, the treatise of C. Jordan (1870) on permutation groups played an important role in the systematization and development of this branch of algebra.

Groups are a powerful tool for studying symmetric objects. The concept of a group allows symmetries of geometrical figures to be characterized. An important example of groups are symmetry groups whose elements are symmetries of geometric figures. Namely, we can associate to any geometrical figure F all its isometries which transform F to itself. If this group is not trivial, the figure F is said to be symmetric, or to have symmetry. It was, in fact, the approach of J.S. Fiedorow for the problem of classification of all possible structures of crystals, which is one of the basic problems in crystallography. The study of crystallographic groups was started by J.S. Fiedorow and continued by A. Schönflies at the end of the 19-th century. They showed that there are only 17 plane crystallographic groups and there are exactly 230 different three-dimensional crystallographic groups. It was the first example of the application of group theory to natural science. It is also interesting to note that the three-dimensional crystallographic groups were found mathematically before these 230 different types of crystals were actually discovered in nature.

In 1870, while studying the theory of groups, C. Jordan showed its close connections with geometry. He studied the classification of main classes of moving a finite rigid solid in the Euclidean space. C. Jordan listed the basic finite groups of rotations and reflections: Cyclic and dihedral groups, groups of rotations and reflections for a tetrahedron, an octahedron and an icosahedron.

In Chapter 2, we introduced the notion of a permutation and showed that permutations form a group under composition. In the first three sections of this chapter, we study the further properties of permutations. The next two sections are devoted to studying the properties of cyclic groups and the groups of symmetries of some plane geometrical figures. In the last section of this chapter, we consider the main properties and the structure theorems of finite Abelian groups.

3.1. Cycle Notation and Cycle Decomposition of Permutations

In this section, we introduce the cycle notation for permutations. As was shown in the previous chapter permutations can be written as matrices with 2 rows. We will show that they can be also written in the form of product of cycles. For example, the permutation $\sigma = \begin{pmatrix} 1 & 2 & 3 \\ 2 & 3 & 1 \end{pmatrix}$ maps the elements of the set $X = \{1, 2, 3\}$ cyclicly. It is a cycle of degree 3 and, in cycle notation, it is written as $(1\ 2\ 3)$. More precisely, we introduce the following definition.

> ✳ **Definition 3.1.**
> A permutation $\sigma \in S_n = S(X)$ is called a **cyclic permutation** or a **cycle of length** k (or k-**cycle**) if there is a subset $Y = \{a_1, a_2, \ldots, a_k\} \subseteq X$ such that
>
> $$\sigma(a_1) = a_2,\ \sigma(a_2) = a_3, \ldots, \sigma(a_{k-1}) = a_k,\ \sigma(a_k) = a_1,$$
>
> and $\sigma(a_i) = a_i$ for all $a_i \notin Y$.
> In this case, a cycle of length k is denoted by $(a_1\ a_2\ \cdots\ a_k)$.
> By assumption, the identity permutation is a cycle of length 0.

✱ **Examples 3.2.**

1. The permutation $\begin{pmatrix} 1 & 2 & 3 & 4 \\ 3 & 1 & 4 & 2 \end{pmatrix} = (1\ 3\ 4\ 2)$ is a cycle of length 4 in S_4.

2. The permutation $\begin{pmatrix} 1 & 2 & 3 & 4 & 5 & 6 \\ 3 & 1 & 4 & 2 & 5 & 6 \end{pmatrix} = (1\ 3\ 4\ 2)$ is a cycle of length 4 in S_6 because we have $1 \mapsto 3 \mapsto 4 \mapsto 2 \mapsto 1$ and other elements 5 and 6 stay unchanged.

3. The permutation $\begin{pmatrix} 1 & 2 & 3 & 4 & 5 & 6 & 7 \\ 1 & 5 & 3 & 2 & 4 & 6 & 7 \end{pmatrix} = (2\ 5\ 4)$ is a cycle of length 3 in S_7 because we have $2 \mapsto 5 \mapsto 4 \mapsto 2$ and other elements of the set $\{1, 2, 3, 4, 5, 6, 7\}$, namely 1, 3, 6, 7, stay unchanged.

It is easy to show that the inverse of a cycle is also a cycle. More exactly, we have the following proposition.

> **✳ Proposition 3.3.**
> The inverse to a cycle $\sigma = (a_1\ a_2\ \ldots a_{k-1}\ a_k)$ is a cycle $\sigma^{-1} = (a_k\ a_{k-1} \ldots a_2\ a_1)$.

✳ Examples 3.4.
If $\sigma = \begin{pmatrix} 1 & 2 & 3 & 4 & 5 & 6 & 7 \\ 1 & 5 & 3 & 2 & 4 & 6 & 7 \end{pmatrix} = (2\ 5\ 4)$ then $\sigma^{-1} = (4\ 5\ 2) = \begin{pmatrix} 1 & 2 & 3 & 4 & 5 & 6 & 7 \\ 1 & 4 & 3 & 5 & 2 & 6 & 7 \end{pmatrix}.$

> **✳ Theorem 3.5.**
> Any cyclic permutation σ of length k can be decomposed into a product of $k - 1$ cycles of length 2.

Proof.
Let $\sigma = (a_1, a_2, \ldots, a_k)$ be a cycle of length k, then it can be written, for example, as a product of $k - 1$ cycles of length 2, as follows:

$$\sigma = (a_1, a_2, \ldots, a_k) = (a_1\ a_2)(a_2\ a_3) \ldots (a_{k-1}\ a_k)$$

☐

✳ Examples 3.6.
1. $(1\ 3\ 4\ 2) = (1\ 3)(3\ 4)(4\ 1)$
2. $\begin{pmatrix} 1 & 2 & 3 & 4 & 5 \\ 2 & 3 & 4 & 5 & 1 \end{pmatrix} = (1\ 2)(2\ 3)(3\ 4)(4\ 5).$

✳ Definition 3.7.
A cycle of length 2 is called a **transposition**.

Any transposition has a form $\rho = (i\ j)$ and swaps exactly two elements i, j and does not change others:

$$\rho = \begin{pmatrix} 1 & \cdots & i-1 & i & i+1 & \cdots & j-1 & j & j+1 \cdots & n \\ 1 & \cdots & i-1 & j & i+1 & \cdots & j-1 & i & j+1 \cdots & n \end{pmatrix} = (i\ j) \quad (3.1)$$

If $j = i + 1$, then a transposition

$$(i \ i+1) = \begin{pmatrix} 1 & \cdots & i & i+1 & \cdots & n \\ 1 & \cdots & i+1 & i & \cdots & n \end{pmatrix} \tag{3.2}$$

is called an **adjacent transposition** and denoted by τ_i.

It is obvious that $\tau^2 = e$ for any transposition τ.

Using the notion of a transposition, Theorem 3.5 can be rewritten as follows:

❋ **Theorem 3.8.**

Any cyclic permutation σ of length k can be decomposed into a product of $k - 1$ transpositions:

$$\sigma = \rho_1 \rho_2 \cdots \rho_{k-1} \tag{3.3}$$

where each ρ_i is a transposition.

Note that decomposition of a cyclic permutation in a product of transpositions is not unique. For example,

$(1\ 2\ 3) = (1\ 3)(1\ 2) = (2\ 3)(1\ 3) = (1\ 3)(2\ 4)(1\ 2)(1\ 4)$

❋ **Lemma 3.9.**

Any transposition can be decomposed into a product of adjacent transpositions.

Proof. Let $\tau = (i\ j)$ and $i < j$. Then we can write the transposition τ as the following product of adjacent transpositions:

$$(i\ j) = \tau_i \tau_{i+1} \cdots \tau_{j-2} \tau_{j-1} \tau_{j-2} \cdots \tau_{i+1} \tau_i \tag{3.4}$$

□

✳ **Definition 3.10.**

Two cycles $(a_1\ a_2 \ldots a_k)$ and $(b_1\ b_2 \ldots b_m)$ of S_n which have no common elements are called **disjoint**.

❋ **Examples 3.11.**

1. The cycles $(1\ 3)$ and $(2\ 4)$ of S_4 are disjoint.
2. The cycles $(1\ 3\ 4\ 2)$ and $(1\ 4)$ of S_4 are not disjoint.

❋ **Proposition 3.12.**

If cycles $\sigma, \tau \in S_n$ are disjoint, then their composition is commutative, i.e., $\sigma\tau = \tau\sigma$.

Proof.

Let $\sigma = (a_1\ a_2\ \ldots a_{k-1}\ a_k)$ and $\tau = (b_1\ b_2\ \ldots b_{m-1}\ b_m)$ be disjoint cycles in $S_n = S(X)$. We denote $Y = \{a_1\ a_2\ \ldots a_{k-1}\ a_k\} \subseteq X$ and $U =$

$\{b_1 b_2 \ldots b_{m-1} b_m\} \subseteq X$. Then, $Y \cap U = \emptyset$. We must show that $\sigma\tau(x) = \tau\sigma(x)$ for all $x \in X$. We consider three different cases.

1) Suppose $x \notin Y$ and $x \notin U$. Then

$$\sigma\tau(x) = \sigma(x) = x = \tau(x) = \tau\sigma(x)$$

2) Assume that $x \in Y$ but $x \notin U$. Then, $\sigma(x) \in Y$ and so $\sigma(x) \notin U$. Therefore, $\tau\sigma(x) = \sigma(x)$. Since $x \notin U$, $\tau(x) = x$. Therefore, $\sigma\tau(x) = \sigma(x)$. Hence,

$$\tau\sigma(x) = \sigma(x) = \sigma\tau(x)$$

3) Analogously we obtain that

$$\tau\sigma(x) = \tau(x) = \sigma\tau(x)$$

if $x \notin Y$ but $x \in U$.   □

✴ Theorem 3.13.
Each permutation can be written as a product of disjoint cycles. Moreover, any such representation is unique up to the order of the factors.

Proof.

Let $S_n = S(x)$, where $X = \{1, 2, \ldots, n\}$, and $\sigma \in S_n$.

We will prove by using the mathematical induction on the number of elements of the set X.

Consider the set of different elements $X_1 = \{1, \sigma(1), \sigma^2(1), \ldots\}$. This set is finite because it is a subset of a finite set X. Let k be a minimal number such that $\sigma^k(1) \in \{1, \sigma(1), \sigma^2(1), \ldots, \sigma^{k-1}(1)\}$. We will show that $\sigma^k(1) = 1$. Otherwise, if $\sigma^k(1) = \sigma^m(1)$ and $0 < m < k$, then $\sigma^{k-m}(1) = 1$, i.e., $\sigma^{k-m}(1) \in X_1$ and $k - m < k$, which contradicts the choice of the number k. Hence, $\sigma^k(1) = 1$. Therefore, $X_1 = \{1, \sigma(1), \sigma^2(1), \ldots, \sigma^{k-1}(1)\}$ and the permutation σ_1 defined as follows

$$\sigma_1(x) = \begin{cases} \sigma(x) & x \in X_1 \\ x & x \notin X_1 \end{cases}$$

is a cycle.

Now, let $Y = X \backslash X_1$. Then, $X_1 \cap Y = \emptyset$ and we can define the permutation:

$$\tau(x) = \begin{cases} x & x \in X_1 \\ \sigma(x) & x \in Y \end{cases}$$

and $\sigma = \sigma_1 \tau$. Moreover, the permutations σ_1 and τ are disjoint because the sets X_1 and Y are disjoint. Since the permutation τ acts as σ on the set Y which contains less elements than the set X, by the inductive assumption it decomposed into a product of cycles, i.e., $\tau = \sigma_2 \ldots \sigma_r$ where each σ_i is a cycle and all permutations $\sigma_2, \ldots, \sigma_r$ are disjoint. Thus,

$$\sigma = \sigma_1 \sigma_2 \ldots \sigma_r, \tag{3.5}$$

where each σ_i is a cycle and all these cycles are disjoint for $i \in \{1, 2, \ldots, r\}$. By Proposition 3.12, decomposition (3.5) does not depend on the order of factors.   □

✱ **Examples 3.14.**

Write the following permutations in the form of disjoint cycles.

1. $\begin{pmatrix} 1 & 2 & 3 & 4 \\ 2 & 1 & 4 & 3 \end{pmatrix} = (1\ 2)(3\ 4).$

2. $\begin{pmatrix} 1 & 2 & 3 & 4 & 5 & 6 & 7 & 8 & 9 & 10 \\ 1 & 5 & 7 & 2 & 4 & 8 & 6 & 3 & 10 & 9 \end{pmatrix} = (2\ 5\ 4)(3\ 7\ 6\ 8)(9\ 10).$

✳ **Theorem 3.15.**

Any permutation σ of degree n can be written in the form of a product of transpositions:

$$\sigma = \rho_1 \rho_2 \cdots \rho_k. \tag{3.6}$$

Proof.

This theorem follows directly from Theorem 3.8 and Theorem 3.13.   □

Taking into account Lemma 3.9 and this theorem, we obtain the following important result:

✳ **Theorem 3.16.**

Any permutation σ of degree n can be written in the form of a product of adjacent transpositions:

$$\sigma = \tau_{i_1} \tau_{i_2} \cdots \tau_{i_k}. \tag{3.7}$$

3.2. Inversion, Parity and Order of a Permutation

As seen in the previous section, the composition of a permutation into a product of transpositions is far from unique. In this section, we consider some invariants for a permutation which are unique for each permutation.

The identity permutation e is to-nothing permutation and it is characterized by the special property which states that if $i < j$ then $e(i) < e(j)$ for all i, j. It is easy to see that for each other permutation σ there always exists a pair of numbers i, j such that if $i < j$ then $\sigma(i) > \sigma(j)$.

✳ **Definition 3.17.**

Let

$$\sigma = \begin{pmatrix} 1 & 2 & \cdots & n \\ a_1 & a_2 & \cdots & a_n \end{pmatrix}$$

be a permutation of degree n. A pair of numbers (i, j) is an **inversion** of the permutation σ if $i < j$ and $a_i > a_j$. The number of inversions in a permutation σ is denoted by $\text{inv}(\sigma)$.

✳ **Example 3.18.**

Let $\sigma = \begin{pmatrix} 1 & 2 & 3 \\ 3 & 2 & 1 \end{pmatrix}$. Then, pairs $(1, 2)$ and $(1, 3)$ are inversions in the permutation σ. So $\text{inv}(\sigma) = 2$.

❋ **Definition 3.19.**

A permutation of degree n is called **even** if the number of its inversions is even, and it is called **odd** if the number of its inversions is odd.

✳ **Examples 3.20.**

1. The identity permutation e is even because $\text{inv}(\sigma) = 0$.

2. Since $\text{inv}(\sigma) = 2$ in the permutation $\sigma = \begin{pmatrix} 1 & 2 & 3 \\ 3 & 2 & 1 \end{pmatrix}$, this permutation is even.

3. The permutation $\sigma = \begin{pmatrix} 1 & 2 & 3 \\ 2 & 1 & 3 \end{pmatrix}$ is odd because $\text{inv}(\sigma) = 1$.

4. The permutation $\sigma = \begin{pmatrix} 1 & 2 & 3 & 4 \\ 4 & 3 & 1 & 2 \end{pmatrix}$ has five inversions: $(1, 2)$, $(1, 3)$, $(1, 4)$, $(2, 3)$, $(2, 4)$, so $\text{inv}(\sigma) = 5$. Therefore, σ is an odd permutation.

Determining parity of a given permutation by means of calculating the number of its inversions is not an easy problem for a permutation of large enough degree. Another method for determining whether a given permutation is even or odd is connected with decomposition of the permutation into a product of disjoint cycles.

On the set of all permutations S_n we define the following function:

$$\text{sgn}(\sigma) = \begin{cases} 1, & \text{if } \sigma \text{ is even} \\ -1, & \text{if } \sigma \text{ is odd} \end{cases} \tag{3.8}$$

which is called the **sign** of a permutation σ. This function is also can be written as follows:

$$\text{sgn}(\sigma) = (-1)^{\text{inv}(\sigma)}. \tag{3.9}$$

Any adjacent transposition $\tau_i = (i \; i + 1)$ is an odd permutation and $\text{sgn}(\tau_i) = -1$, since it has exactly one inversion.

❋ **Lemma 3.21.**

For any permutation σ and an adjacent transposition $\tau_i = (i \; i + 1)$, the equalities below hold:

$$\text{sgn}(\tau_i \sigma) = -\text{sgn}(\sigma) = \text{sgn}(\sigma \tau_i) \tag{3.10}$$

Proof.

Consider a permutation $\sigma = \begin{pmatrix} 1 & 2 & \cdots & n \\ a_1 & a_2 & \cdots & a_n \end{pmatrix}$ and an adjacent transposition $\tau_i = (i \ i+1)$. Then,

$$\tau_i \sigma = \begin{pmatrix} 1 & \cdots & i & i+1 & \cdots & n \\ a_1 & \cdots & a_{i+1} & a_i & \cdots & a_n \end{pmatrix},$$

where in the second row the elements a_i and a_{i+1} are interchanged. Therefore, if the numbers i and $i+1$ form an inversion in σ, then they will not form an inversion in $\tau_i \sigma$ and otherwise. Since all other inversions in σ and $\tau_i \sigma$ are the same, permutations σ and $\tau_i \sigma$ have the different parity. □

✳ Theorem 3.22.
For any permutations σ and τ, the equality below holds:

$$\mathrm{sgn}(\sigma\tau) = \mathrm{sgn}(\sigma)\mathrm{sgn}(\tau) \tag{3.11}$$

Proof.

By Theorem 3.16, a permutation σ can be decomposed into a product of adjacent transpositions: $\sigma = \tau_{i_1} \tau_{i_2} \cdots \tau_{i_m}$. Therefore, using Lemma 3.21, we obtain:

$$\mathrm{sgn}(\sigma) = \mathrm{sgn}(\tau_{i_1} \tau_{i_2} \cdots \tau_{i_m}) = \mathrm{sgn}(\tau_{i_1}(\tau_{i_2} \cdots \tau_{i_m})) = \mathrm{sgn}(\tau_{i_1})\mathrm{sgn}(\tau_{i_2} \cdots \tau_{i_m}) =$$

$$= \cdots = \mathrm{sgn}(\tau_{i_1})\mathrm{sgn}(\tau_{i_2}) \cdots \mathrm{sgn}(\tau_{i_m})$$

Using the same Lemma 3.21, we have:

$$\mathrm{sgn}(\sigma\tau) = \mathrm{sgn}(\tau_{i_1} \tau_{i_2} \cdots \tau_{i_m} \tau) = \mathrm{sgn}(\tau_{i_1})\mathrm{sgn}(\tau_{i_2} \cdots \tau_{i_m} \tau) =$$

$$= \cdots = \mathrm{sgn}(\tau_{i_1})\mathrm{sgn}(\tau_{i_2}) \cdots \mathrm{sgn}(\tau_{i_m})\mathrm{sgn}(\tau) = \mathrm{sgn}(\sigma)\mathrm{sgn}(\tau).$$

□

✳ Corollary 3.23.
For any permutation σ, the equality below holds:

$$\mathrm{sgn}(\sigma^{-1}) = \mathrm{sgn}(\sigma) \tag{3.12}$$

Proof.
By Theorem 3.22,

$$\mathrm{sgn}(\sigma)\mathrm{sgn}(\sigma^{-1}) = \mathrm{sgn}(\sigma\sigma^{-1}) = \mathrm{sgn}(e) = 1.$$

Therefore, $\mathrm{sgn}(\sigma) = \mathrm{sgn}(\sigma^{-1})$. □

✳ **Lemma 3.24.**

Any transformation τ is an odd permutation, i.e.,

$$\operatorname{sgn}(\tau) = -1 \qquad (3.13)$$

Proof. Since, by Lemma 3.9, a transposition $\tau = (i\ j)$, where $i < j$, can be decomposed into $2(j - i) - 1 = 2m - 1$ of adjacent transpositions and each adjacent transformation is odd, from Theorem 3.22, we have $\operatorname{sgn}(\tau) = (-1)^{2m-1} = -1$. □

✳ **Corollary 3.25.**

For any permutation σ and any transposition τ, the equality below holds:

$$\operatorname{sgn}(\sigma\tau) = -\operatorname{sgn}(\sigma) \qquad (3.14)$$

✳ **Corollary 3.26.**

A cyclic permutation of length k is odd if and only if k is even and it is even if and only if k is odd.

Proof.

By Theorem 3.8, a cyclic permutation σ of length k can be decomposed into a product of $k - 1$ transpositions:

$$\sigma = \rho_1 \rho_2 \cdots \rho_{k-1},$$

where each ρ_i is a transposition. Since $\operatorname{sgn}(\rho_i) = -1$ for all i, from Theorem 3.22, it follows that

$$\operatorname{sgn}(\sigma) = (-1)^{k-1} = \begin{cases} 1 & \text{if } k \text{ is odd} \\ -1 & \text{if } k \text{ is even} \end{cases} \qquad (3.15)$$

□

✳ **Examples 3.27.**

1. The cycle (2 4 5 3 1 6) of length 6 is an odd permutation.
2. The cycle (1 8 9 3 5 6 2 4 7) of length 9 is an even permutation.

✳ **Corollary 3.28.**

If a permutation $\sigma \in S_n$ is decomposed into a product of disjoint cycles $\sigma = \rho_1 \rho_2 \ldots \rho_p$, then

$$\operatorname{sgn}(\sigma) = (-1)^m, \qquad (3.16)$$

where $m = \sum_{i=1}^{p}(k_i - 1)$ and k_i is the length of the cycle ρ_i.

Proof. By Theorem 3.13, each permutation can be written as a product of disjoint cycles:

$$\sigma = \rho_1 \rho_2 \cdots \rho_p$$

and this representation is unique up to the order of the factors. If ρ_i is a cycle of the length k_i, then $\text{sgn}(\rho_i) = (-1)^{k_i - 1}$ by (3.15). Then, from Theorem 3.22, we obtain:

$$\text{sgn}(\sigma) = (-1)^{k_1 - 1} \cdot (-1)^{k_2 - 1} \cdots (-1)^{k_p - 1} = (-1)^m,$$

where $m = \sum_{i=1}^{p}(k_i - 1)$.

So that if m is odd, then the permutation σ is odd, and if m is even, then the permutation σ is even. $\square$

✳ Example 3.29.

Determine the parity of the permutation $\sigma = \begin{pmatrix} 1 & 2 & 3 & 4 & 5 & 6 & 7 & 8 \\ 3 & 5 & 8 & 7 & 1 & 4 & 6 & 2 \end{pmatrix}$

Since $\sigma = (1\ 3\ 8\ 2\ 5)(4\ 7\ 6)$, $\text{sgn}(\sigma) = (-1)^{4+2} = 1$. Hence, σ is an even permutation.

Recall that the order of a permutation $\sigma \in S_n$ is the least positive integers m such that $\sigma^m = e$, where e is the identity permutation in S_n.

✳ Proposition 3.30.
The order of a cycle is equal to the length of this cycle.

Proof.

Let $\sigma = (a_1, a_2, \ldots, a_k)$ be a cycle of length k of $S_n = S(X)$, and $Y = \{a_1, a_2, \ldots, a_k\} \subseteq X$. Then,

$$\sigma(a_1) = a_2, \ \sigma^2(a_1) = \sigma(a_2) = a_3, \ \ldots, \ \sigma^k(a_1) = \sigma(a_k) = a_1.$$

Analogously, $\sigma^k(a_i) = a_i$ for each $a_i \in X$. If $x \in X \backslash Y$, then $\sigma^k(x) = x = \sigma(x)$, i.e., $\sigma^k = e$ is the identity element of the group S_n. So, $\text{ord}(\sigma) \leq k$. Suppose that $\text{ord}(\sigma) = m < k$. Then,

$$\sigma(a_1) = a_2, \ \sigma^2(a_1) = \sigma(a_2) = a_3, \ \ldots, \ \sigma^m(a_1) = a_{m+1} \neq a_1.$$

Therefore, $\sigma^m \neq e$, and so $\text{ord}(\sigma) = k$. $\square$

✳ Examples 3.31.
1. Let $\sigma = (2\ 5\ 1\ 4\ 3)$, then $\text{ord}(\sigma) = 5$.
2. Let $\sigma = (3\ 4\ 7\ 2\ 5\ 1\ 6)$, then $\text{ord}(\sigma) = 7$.

❋ **Theorem 3.32.**

Let $\sigma = \rho_1\rho_2\ldots\rho_p$ be a decomposition of a permutation $\sigma \in S_n$ into a product of disjoint cycles. Then,

$$\text{ord}(\sigma) = \text{lcm}(k_1, k_2, \ldots, k_p), \tag{3.17}$$

where k_i is the length of the cycle τ_i for all i.

Proof.

Let $\sigma = \rho_1\rho_2\ldots\rho_p$ be a decomposition of a permutation $\sigma \in S_n = S(X)$ into a product of disjoint cycles $\rho_1, \ldots, \rho_p$ and each permutation ρ_i acts on the set X_i, where $X = X_1 \cup \ldots \cup X_p$ and $X_i \cap X_j = \emptyset$ for all $i \neq j$. Suppose that $\text{ord}(\sigma) = m$ and $\text{lcm}(k_1, k_2, \ldots, k_m) = k$.

First we will show that $m \leq k$. Since all permutations ρ_i are disjoint cycles, by Proposition 3.12, they commute and so $\sigma^k = \rho_1^k\rho_2^k\ldots\rho_p^k$. Since $k_i | k$ for all i, we obtain that $\rho_i^k = e$ for all i and so $\sigma^k = e$. Therefore, $m \leq k$.

Now, we will show that $m \geq k$. Since $\sigma^m = e$, from Proposition 3.12, it follows that $\sigma^m = \rho_1^m\rho_2^m\ldots\rho_p^m = e$. Hence, $\rho_i^m = e$ for all i. Otherwise, there is i such that $\rho_i^m \neq e$. In this case, there are $r, s \in X_i$ such that $r \neq s$ and $\rho_i^m(r) = s$. Since ρ_i are disjoint permutations, $\rho_j^m(r) = r$ for all $j \neq i$, and, in this case, $\sigma^m(r) = s$, i.e., $\sigma^m \neq e$. This contradiction shows that $\rho_i^m = e$ for all i. Therefore, from Theorem 2.25, it follows that $k_i | m$ for all i. Then, by Definition 1.16, $m \geq k$. Thus, $m \leq k \leq m$, implying that $k = m$.   □

❋ **Example 3.33.**

Find the order of the permutation $\sigma = \begin{pmatrix} 1 & 2 & 3 & 4 & 5 & 6 & 7 & 8 \\ 3 & 5 & 8 & 7 & 1 & 4 & 6 & 2 \end{pmatrix}$

Since $\sigma = (1\ 3\ 8\ 2\ 5)(4\ 7\ 6)$, $\text{ord}(\sigma) = \text{lcm}(5, 3) = 15$.

3.3. Alternating Group

It follows, from Theorem 3.22, that a product of two permutations of the same parity is an even permutation, but a product of two permutations of different parity is an odd permutation. By Corollary 3.23, a permutation has the same parity as the inverse permutation.

So, the product of two even permutations is an even permutation, the identity permutation is even, and the permutation which is the inverse to an even permutation is also even. In this way, the set of all even permutations of degree n under composition is a subgroup of the group S_n.

❋ **Definition 3.34.** The group of all even permutations of degree n is called the **alternating group of degree** n and is denoted by A_n.

Denote by G_2 the group of elements $\{1, -1\}$ with multiplication given by the following table:

·	1	−1
1	1	−1
−1	−1	1

It follows, from Theorem 3.22, that the mapping sgn : $S_n \rightarrow G_2$ given by the formula $\mathrm{sgn}(\sigma) = (-1)^{\mathrm{inv}(\sigma)}$ for all $\sigma \in S_n$ is a group homomorphism. From the above reasoning, we have the following theorem:

> ✳ **Theorem 3.35.**
> Let sgn : $S_n \rightarrow G_2$ be a group homomorphism given by the formula: $\mathrm{sgn}(\sigma) = (-1)^{\mathrm{inv}(\sigma)}$ for any permutation σ. Then $\mathrm{Ker}(\mathrm{sgn}) = A_n$.

The next theorem shows that the number of even and odd permutations of the group S_n is the same.

> ✳ **Theorem 3.36.**
> The order of the alternating group of degree n is equal to $n!/2$.

Proof. Let A_n be the set of all even permutations and B_n be the set of all odd permutations of the group S_n. First, we will show that the sets A_n and B_n have the same number of elements. For this purpose it suffices to determine the bijective mapping between these sets. Let $\tau \in S_n$ be a transposition. Then, for each even permutation $\sigma \in A_n$ the permutation $\tau\sigma$ is odd and so $\tau\sigma \in B_n$. Therefore, we can consider a mapping $\varphi : A_n \rightarrow B_n$ such that $\varphi(\sigma) = \tau\sigma$ for any even permutation $\sigma \in A_n$ and a transposition τ. This mapping is one-to-one, since $\tau\sigma = \tau\pi$ implies that $\sigma = \pi$. Since $\tau^2 = e$, the identity permutation, and for any odd permutation $\rho \in B$ the permutation $\tau\rho$ is even, we have that $\varphi(\tau\rho) = \tau(\tau\rho) = \rho \in B_n$, that shows that φ is onto. So, φ is a bijective mapping, which implies that the number of elements of the sets A_n and B_n is the same. Since the number of elements of the set S_n is equal to $n!$, $|A_n| = n!/2$. □

Since $|S_n| = |A_n| \cdot [S_n : A_n] = 1/2|S_n| \cdot [S_n : A_n]$, we obtain that $[S_n : A_n] = 2$. Hence, taking into account Theorem 2.48 we get:

> ✳ **Corollary 3.37.**
> 1. The alternating group A_n has index 2 in S_n.
> 2. The alternating group A_n is a normal subgroup in S_n.

✳ **Example 3.38.**
The subgroup A_3 of the symmetric group S_3 contains 3 elements:

$$e = \begin{pmatrix} 1 & 2 & 3 \\ 1 & 2 & 3 \end{pmatrix}, \quad s_1 = \begin{pmatrix} 1 & 2 & 3 \\ 2 & 3 & 1 \end{pmatrix}, \quad s_2 = \begin{pmatrix} 1 & 2 & 3 \\ 3 & 1 & 2 \end{pmatrix}.$$

The Cayley table for A_3 is given as follows:

	e	s_1	s_2
e	e	s_1	s_2
s_1	s_1	s_2	e
s_2	s_2	e	s_1

The element e is the identity element of A_3, $s_1^{-1} = s_2$, and $s_2^{-1} = s_1$.

✳ Example 3.39.

Let S_n be the symmetric group of degree n and A_n alternating group. Then, $S_n = A_n \cup B_n$, where B_n is the set of all odd permutations of degree n. Since A_n is a normal subgroup of index 2 in S_n, S_n/A_n is a quotient group of order 2. Let $H = (\{1, -1\}, \cdot)$ be a group with the following Cayley table:

$\cdot$	1	-1
1	1	-1
-1	-1	1

We consider the mapping sgn : $S_n \to H$ given as follows:

$$\text{sgn}(\sigma) = \begin{cases} 1 & \text{if } \sigma \in A_n \\ -1 & \text{if } \sigma \in B_n \end{cases}$$

Since $\text{sgn}(\sigma\tau) = \text{sgn}(\sigma)\text{sgn}(\tau)$ and $\text{sgn}(e) = 1$, sgn is a group homomorphism. Then, from Theorem 7.86, it follows that:

$$S_n/A_n \cong H$$

because $\text{Ker}(\text{sgn}) = \{\sigma \in S_n \ : \ \text{sgn}(\sigma) = 1\} = A_n$ and $\text{Im}(\text{sgn}) = H$.

Recall that a group G whose only normal subgroups are $\{e\}$ and G is called a **simple group**. The following famous theorem, first proved by E. Galois, we give without proof.

✳ Theorem 3.40 (E. Galois).
The alternating group A_n is simple for all $n > 4$.

Recall that any permutation of degree n is a bijection of a set X consisting of n elements to itself. If X is an infinity set, then we also can consider bijections of a set X to itself and all bijections on the set X under composition $\circ$ form a group which is denoted by $S(X)$ and is called the **symmetric group**. The group S_n is a particular example of the symmetric group $S(X)$ when a set X is finite and contains n elements.

The symmetric group is very important when taking into account the following theorem:

✳ Theorem 3.41 (A. Cayley).
Every group G is isomorphic to a subgroup of the symmetric group $(S(G), \circ)$.

Proof.
For each element a of a group G, we define a mapping: $f_a : G \to G$ given by the formula: $f_a(x) = ax$. We will show that f_a is a bijection.

For each $y \in G$, we have: $f_a(a^{-1}y) = a(a^{-1}y) = y$, which shows that f_a is a surjection. Since the equality $ax = ay$ implies that $x = y$, by the cancellation law in the group G, f_a is an injection. So, f_a is a bijection, i.e., $f_a \in S(G)$.

We will show that $(H, \circ)$, where $H = \{f_a \in S(G) : a \in G\}$ is a subgroup of the symmetric group $(S(G), \circ)$, is isomorphic to the group G. We define a mapping $\varphi : G \to S(G)$ given as follows $\varphi(a) = f_a$. Then, $\varphi(ab)(x) = f_{ab}(x) = (ab)x = a(bx) = f_a(bx) = f_a \circ f_b(x) = (\varphi(a) \circ \varphi(b))(x)$, i.e., $\varphi(ab) = f_{ab} = f_a \circ f_b = \varphi(a) \circ \varphi(b)$, which means that φ is a group homomorphism.

We will show that φ is an isomorphism. From the definition, it follows that φ is a surjection. Since the equality $\varphi(a) = \varphi(b)$ implies that $ax = bx$, for any $x \in G$, it follows that $a = b$, by the cancellation law in the group G. So, φ is an injection. Thus, φ is an isomorphism and so $G \cong \operatorname{Im}(\varphi) = H \subseteq S(G)$.
□

✳ Corollary 3.42.
If G is a finite group of order n, then G is isomorphic to a subgroup of the symmetric group S_n.

3.4. Cyclic Groups

In Chapter 2 we introduced the notion of a cyclic group which is generated by one element. In this section, we study the properties of such groups more thoroughly.

✳ Theorem 3.43.
A finite group G of order n is cyclic if and only if G contains an element of order n.

Proof.
$\Rightarrow$. If $G = \langle g \rangle$ is a cyclic group and $|G| = n$, then, from definition of G, we have $\operatorname{ord}(g) = n$.

$\Leftarrow$. Suppose there is an element $g \in G$ such that $|G| = \operatorname{ord}(g) = n$. Consider a cyclic subgroup $H = \langle g \rangle \subseteq G$. Then, $|H| = \operatorname{ord}(g) = n$. Therefore, $H = G = \langle g \rangle$.
□

⊛ **Theorem 3.44.**
Each subgroup of a cyclic group is cyclic.

Proof.
Let $G = \langle g \rangle$ be a cyclic group generated by an element g. Let $H \leq G$ be a non-trivial subgroup of G. We choose an element $h = g^k \in H$ such that k is the least positive integer among all powers of elements in H. Then, we state that $H = \langle h \rangle$. Really, let $x = g^n \in H$. If $n = km + r$, where $0 \leq r < k$, then $x = g^n = g^{km+r} = (g^k)^m g^r = h^m g^r$, so $g^r = xh^{-m} \in H$. Then from the choice of h we obtain that $r = 0$. So $x = h^m$ for an arbitrary element $x \in H$, i.e., H is a cyclic subgroup. $\qquad\square$

⊛ **Theorem 3.45.**
1. If g is a generator of a finite cyclic group G of order n, then g^k is also a generator of G if and only if $\gcd(k, n) = 1$.
2. If G is a finite cyclic group of order n, then G has exactly $\varphi(n)$ distinct generators, where φ is Euler's totient function.

Proof.
1. Let G be a finite cyclic group and $|G| = n$. If g is a generator of G, then $G = \{e, g, g^2, \ldots, g^{n-1} \; : \; g^n = e; n \in \mathbb{N}\}$ and $\operatorname{ord}(g) = n$. If an element g^k is also a generator of G, then $\operatorname{ord}(g^k) = n$. Then, from Theorem 2.27, it follows that $\gcd(k, n) = 1$. Conversely, if $\gcd(k, n) = 1$ then $\operatorname{ord}(g^k) = n$, by Theorem 2.27. This means that the element g^k is a generator of G, by Theorem 2.25.
2. Since there are exactly $\varphi(n)$ integers $0 < k < n$ for which $\gcd(k, n) = 1$, from the first part of this theorem, it follows that the cyclic group G of order n also possesses exactly $\varphi(n)$ different generators. $\qquad\square$

⊛ **Theorem 3.46.**
All finite cyclic groups of order m are isomorphic to the additive group $\mathbb{Z}_m$.

Proof.
Let $G = \langle g \rangle$ be a finite cyclic group and $|G| = m$, i.e., $G = \{e, g, g^2, \ldots, g^{m-1} \; : \; g^m = e; m \in \mathbb{N}\}$, and $\mathbb{Z}_m = \{[0], [1], [2], \ldots, [m-1]\}$. Consider a mapping $f : G \to \mathbb{Z}_m$ given as follows: $f(g^k) = [k]$. Since

$$f(g^k g^n) = f(g^{k+n}) = [k + n] = [k] + [n] = f(g^k) + f(g^n),$$

f is a group homomorphism. It is obvious that $\operatorname{Im}(f) = \mathbb{Z}_m$. $\operatorname{Ker}(f) = \{g^k \in G \; : \; f(g^k) = [0]\}$, i.e., $[k] = [0]$, where $0 \leq k < m$, so $k = 0$. Therefore, $\operatorname{Ker}(f) = \{e\}$, which means that f is an injection. Hence, f is a group homomorphism. $\qquad\square$

⊛ **Theorem 3.47.**
All infinite cyclic groups are isomorphic to the additive group $\mathbb{Z}$.

Proof.

Let $G = \langle g \rangle$ be an infinite cyclic group and $\#(G) = \infty$, i.e., $G = \{\ldots, g^{-2}, g^{-1}, e, g, g^2, \ldots, g^m, \ldots\}$. Consider a mapping $f : G \to \mathbb{Z}$ given as follows: $f(g^k) = k$. Since

$$f(g^k g^n) = f(g^{k+n}) = k + n = f(g^k) + f(g^n),$$

f is a group homomorphism. It is obvious that $\text{Im}(f) = \mathbb{Z}$. Let $\text{Ker}(f) = \{g^k \in G : f(g^k) = 0\}$. Then $k = 0$, i.e., $\text{Ker}(f) = \{e\}$, which means that f is an monomorphism. Thus, f is a group isomorphism. □

✳ Example 3.48.

Consider a group G whose elements are the roots of the n-th degree from 1:

$$z_k = \cos(2\pi k/n) + i\sin(2\pi k/n) = e^{2\pi k/n},$$

where $i^2 = -1$. If $z_m = \cos(2\pi m/n) + i\sin(2\pi m/n) = e^{2\pi m/n}$, then the multiplication of the roots z_k, z_m is defined as follows: $z_k z_m = z_r$, where

$$r = \begin{cases} k + m & \text{if } k + m < n, \\ k + m - n & \text{if } k + m \geq n, \end{cases}$$

i.e., $z_r = \cos(2\pi r/n) + i\sin(2\pi r/n) = e^{2\pi r/n}$.

The element $z_1 = \cos(2\pi/n) + i\sin(2\pi/n) = e^{2\pi/n}$ is a generator of this group, since for each $z_k \in G$ we have: $z_k = (z_1)^k$. Thus, $G = \{z_0 = 1, z_1, z_2, \ldots, z_{n-1}\}$ is a finite cyclic group of order n and, by Theorem 3.46, $G \cong C_n \cong \mathbb{Z}_n$.

3.5. Groups of Symmetries. The Dihedral Groups

Important examples of groups are groups of symmetries, which are subgroups of groups of isometries in the Euclidean space $\mathbb{R}^n$. Recall that an **isometry** in $\mathbb{R}^n$ is a transformation of $\mathbb{R}^n$ to $\mathbb{R}^n$ which preserves a distance between points. In this section, we are mostly interested in symmetries of geometric figures in $\mathbb{R}^2$, which can be described by a finite set of points S and some segments connected these points. We can define a symmetry of an object as a transformation of that object to itself which preserves its essential structure. Then, the set of all symmetries of the object forms a group.

✳ Definition 3.49. A **symmetry** of a set of points S on the plane $\mathbb{R}^2$ is an isometry of $\mathbb{R}^2$ to $\mathbb{R}^2$ which maps the set S to itself.

There are the following special symmetries of a set $S \subset \mathbb{R}^2$ which has a center at the origin $O = (0, 0)$:

- the identity transformation Id: $(x, y) \mapsto (x, y)$;

- the mirror reflection m_x about the vertical axis $x = 0$: $(x, y) \mapsto (-x, y)$;

- the mirror reflection m_y about the horizontal axis $y = 0$: $(x, y) \mapsto (x, -y)$;

- the mirror reflection m_v about the line v which runs through the center of the set S;

- the rotation r_φ about the point O counterclockwise by an angle φ:

$$r_\varphi : (x, y) \mapsto (x \cos \varphi - y \sin \varphi, x \sin \varphi + y cos \varphi)$$

- combinations of the above transformations.

✳ Examples 3.50.

Consider symmetries of the line segment $[-c, c]$ which can be given by the set consisting of two vertices $S = \{(c, 0), (-c, 0) \ : \ c \in \mathbb{R}\}$. It is obvious that the identity transformation $Id(S) = S$ is the trivial symmetry of the set S. The mirror reflection m_x about the vertical axis $x = 0$ is also a symmetry of the set S since:

$$m_x(c, 0) = (-c, 0) \in S \quad \text{and} \quad m_x(-c, 0) = (c, 0) \in S.$$

Moreover, $m_x(m_x(c, 0)) = (c, 0)$ and $m_x(m_x(-c, 0)) = (-c, 0)$, i.e., $m_x^2 = Id$. So we have two symmetries of the set S whose compositions can be written by the following table:

	Id	m_x
Id	Id	m_x
m_x	m_x	Id

It shows that the set of symmetries $\{Id, m_x\}$ of the set S under composition forms a cyclic group of order 2.

Various geometric figures can possess mirror reflections, for example the figures:

Figure 3.1: Geometric figures with mirror reflections.

The first two figures have mirror reflections about the vertical axis, but the last one has the mirror reflection about the horizontal axis.

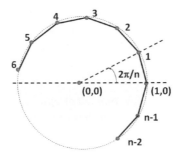

Figure 3.2: Rotations of an n-sided regular polygon.

✳ Examples 3.51.

An important example of subgroups of group of isometrics on the Euclidean space $\mathbb{R}^2$ is the rotation group of an n-sided regular polygon F_n in $\mathbb{R}^2$.

Rotations of an n-sided regular polygon F_n are determined by axes of rotations. We will suppose that the center O of a regular n-polygon F_n is at the origin $(0,0)$, and a rotation axis runs through the point O. We can consider F_n as a finite set which contains n vertices on the unit circle and is labeled as $0, 1, 2, \ldots, n-1$ starting at the point $(1,0)$, and further proceeding counterclockwise.

Let g be a rotation of F_n about the point O counterclockwise by the angle $2\pi/n$. Then, each proper rotation of the regular n-polygon F_n has the form g^k, that is a rotation by the angle $2\pi k/n$. Since g^n is the rotation by the angle 2π, i.e., the identity transformation $e = Id$, the set of all proper rotations of F_n forms a cyclic group $G = \{g^0 = e, g, g^2, \ldots, g^{n-1}\}$ of order n and so

$$G \cong C_n \cong \mathbb{Z}_n.$$

In particular, if $n = 3$ then the group of rotations of an equilateral triangular is the cyclic group C_3, and the group of rotations of a square is the cyclic group C_4.

Many various geometrical figures which are not regular n-polygons can also have rotation symmetry, for example the following figures:

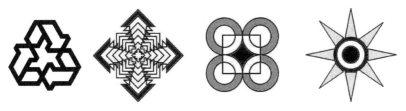

Figure 3.3: Geometric figures with rotation symmetry.

The first figure can be rotated by the angles 120°, 240° and 360°, the second and third figures can be rotated a multiple 90°, but the fourth figure can be rotated a multiple 45° about their centers.

✳ **Definition 3.52.**

The **symmetry group** of a geometrical figure on the Euclidean space $\mathbb{R}^2$ is a group of all symmetries of the figure under composition of transformations.

Figures on the plane (or in the space with more dimension) can be defined by symmetry groups, which are different groups of isometries of whole plane (or space with more dimension).

✳ **Examples 3.53.**

Consider the group of all symmetries of a rectangle which is not a square.

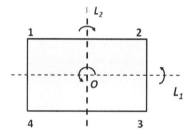

Figure 3.4: Symmetries of a rectangle.

A rectangle can be considered as a set containing 4 vertices labelled as 1, 2, 3, 4, as shown in Figure 3.4. A rectangle has two axes of symmetry which pass through the center of a rectangle: The horizontal axis L_1 and the vertical axis L_2. Denote by a the symmetry obtained by the mirror reflection of the rectangle about the horizontal axis and by b the symmetry obtained by the mirror reflection of the rectangle about the vertical axis. Then

$$a(1) = 4, a(2) = 3, a(3) = 2, a(4) = 1$$

and

$$b(1) = 2, b(2) = 1, b(3) = 4, b(4) = 3.$$

Since any symmetry maps a finite set to itself, symmetries can be written in the form of permutations:

$$a = \begin{pmatrix} 1 & 2 & 3 & 4 \\ 4 & 3 & 2 & 1 \end{pmatrix}, \quad b = \begin{pmatrix} 1 & 2 & 3 & 4 \\ 2 & 1 & 4 & 3 \end{pmatrix}.$$

There are also two another symmetries of a rectangle: c is the rotation of the rectangle about the center O by the angle π, which can be written in the form of the permutation:

$$c = \begin{pmatrix} 1 & 2 & 3 & 4 \\ 3 & 4 & 1 & 2 \end{pmatrix}$$

and $e = Id$ is the identity mapping:

$$e = \begin{pmatrix} 1 & 2 & 3 & 4 \\ 1 & 2 & 3 & 4 \end{pmatrix}.$$

Then, $a^2 = b^2 = c^2 = e$ and so $a^{-1} = a$, $b^{-1} = b$, $c^{-1} = c$. Since $ab = ba = c$, $ac = ca = b$, $bc = cb = a$, all compositions of symmetries e, a, b, c can be written by the following table:

	e	a	b	c
e	e	a	b	c
a	a	e	c	b
b	b	c	e	a
c	c	b	a	e

Table 3.1.

that shows that all symmetries of the rectangle form a group G_4 called the **Klein four-group** in honor of the famous German mathematician F. Klein (1849-1925). The group G_4 is Abelian but not cyclic, since there is no element of order 4.

The group G_4 have 5 subgroups, two of which are $\{e\}$ and G_4, and 3 proper subgroups: $H_1 = \{e, a\}$, $H_2 = \{e, b\}$, $H_3 = \{e, c\}$, which are cyclic groups of order 2. All of them are, of course, normal, since G_4 is an Abelian group.

✳ Examples 3.54.

Consider the group of all symmetries of an equilateral triangle T.

Symmetries of an equilateral triangle T are the following six transformations:

- three counterclockwise rotations r_1, r_2, r_3 about the center O of the triangular T by the angles 0, $2\pi/3$ and $4\pi/3$;

- three axial mirror reflections : s_1, s_2, s_3 about the lines L_1, L_2, L_3 containing triangle perpendiculars (running through each vertex and the center of the opposite sides). (You can imagine them as the rotation by the angle $\pi/2$ about the axis of symmetry in the three-dimensional space).

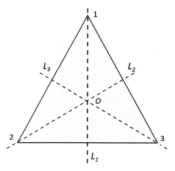

Figure 3.5: Symmetries of an equilateral triangle.

Note that the rotation $r_1 = e = Id$ is the identity transformation. Denote $r_2 = x$. Then, $r_3 = x^2$ and $x^3 = e$. Denote $s_1 = y$. Then, $y^2 = e$, $s_2 = xy$, $s_3 = xy^2$ and $yx = x^2y$, or $xy = yx^{-1}$.

Labelling the vertices of the triangle counterclockwise as 1, 2, 3, we obtain six symmetries corresponding to the following permutations of the vertices:

$$e = \begin{pmatrix} 1 & 2 & 3 \\ 1 & 2 & 3 \end{pmatrix}, \quad x = \begin{pmatrix} 1 & 2 & 3 \\ 2 & 3 & 1 \end{pmatrix}, \quad x^2 = \begin{pmatrix} 1 & 2 & 3 \\ 3 & 1 & 2 \end{pmatrix},$$

$$y = \begin{pmatrix} 1 & 2 & 3 \\ 1 & 3 & 2 \end{pmatrix}, \quad xy = \begin{pmatrix} 1 & 2 & 3 \\ 3 & 2 & 1 \end{pmatrix}, \quad x^2y = \begin{pmatrix} 1 & 2 & 3 \\ 2 & 1 & 3 \end{pmatrix}.$$

The table of compositions of all symmetries of the equilateral triangle T can be written as follows:

	e	x	x^2	y	xy	x^2y
e	e	x	x^2	y	xy	x^2y
x	x	x^2	e	xy	x^2y	y
x^2	x^2	e	x	x^2y	y	xy
y	y	x^2y	xy	e	x^2	x
xy	xy	y	x^2y	x	e	x^2
x^2y	x^2y	xy	y	x^2	x	e

Table 3.2.

In this way, the symmetries of the equilateral triangle T form a group of compositions which satisfy the following properties:

- a composition of symmetries is associative;

- there is the identity element which is the identity transposition e;

- each symmetry has the inverse element.

Thus, the set of symmetries $G = \{e, x, x^2, y, xy, x^2y\}$ forms a group in which the following equalities hold: $x^3 = y^2 = e$, $yx = x^2y$. This group is not commutative and it is isomorphic to the group S_3.

✳ Examples 3.55.

Consider the group of all symmetries of a square.

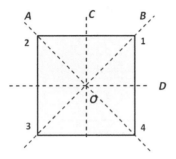

Figure 3.6: Symmetries of a square.

Symmetries of a square are the following eight transformations:

- The counterclockwise rotations r_1, r_2, r_3, r_4 about the center O of the triangular by the angles 0, $\pi/2$, π and $3\pi/2$;

- The axial symmetries: s_1, s_2 about the lines A, B, running through opposite vertexes, and two axial symmetries: s_3, s_4 about the lines C, D, running through the centers of the opposite sides of the square.

Note that the rotation $r_1 = e = Id$ is the identity transformation. Denote $r_2 = x$. Then, $r_3 = x^2$, $r_4 = x^3$ and $x^4 = e$.

If we label the vertices of the square counterclockwise as 1, 2, 3, 4, then symmetries r_1, r_2, r_3, r_4 correspond to the permutations:

$$e = \begin{pmatrix} 1 & 2 & 3 & 4 \\ 1 & 2 & 3 & 4 \end{pmatrix}, \quad x = \begin{pmatrix} 1 & 2 & 3 & 4 \\ 2 & 3 & 4 & 1 \end{pmatrix},$$

$$x^2 = \begin{pmatrix} 1 & 2 & 3 & 4 \\ 3 & 4 & 1 & 2 \end{pmatrix}, \quad x^3 = \begin{pmatrix} 1 & 2 & 3 & 4 \\ 4 & 3 & 2 & 1 \end{pmatrix}.$$

Denote $s_1 = y$. Then $s_2 = x^2y$, $s_3 = xy$, $s_4 = x^3y$ correspond to the permutations:

$$s_1 = \begin{pmatrix} 1 & 2 & 3 & 4 \\ 1 & 4 & 3 & 2 \end{pmatrix} = (2\ 4), \quad s_2 = \begin{pmatrix} 1 & 2 & 3 & 4 \\ 3 & 2 & 1 & 4 \end{pmatrix} = (1\ 3),$$

$$s_3 = \begin{pmatrix} 1 & 2 & 3 & 4 \\ 2 & 1 & 4 & 3 \end{pmatrix} = (1\ 2)(3\ 4), \quad s_4 = \begin{pmatrix} 1 & 2 & 3 & 4 \\ 4 & 3 & 2 & 1 \end{pmatrix} = (1\ 4)(2\ 3).$$

The table of compositions of symmetries of a square has the following form:

	e	x	x^2	x^3	y	xy	x^2y	x^3y
e	e	x	x^2	x^3	y	xy	x^2y	x^3y
x	x	x^2	x^3	e	xy	x^2y	x^3y	y
x^2	x^2	x^3	e	x	x^2y	x^3y	y	xy
x^3	x^3	e	x	x^2	x^3y	y	xy	x^2y
y	y	x^3y	x^2y	xy	e	x^3	x^2	x
xy	xy	y	x^3y	x^2y	x	e	x^3	x^2
x^2y	x^2y	xy	y	x^3y	x^2	x	e	x^3
x^3y	x^3y	x^2y	xy	y	x^3	x^2	x	e

Table 3.3.

Therefore, the set of symmetries $G = \{e, x, x^2, x^3, y, xy, x^2y, x^3y\}$ forms a group in which the following equalities hold: $x^4 = e$, $y^2 = e$, $yx = x^{-1}y$. This group is not commutative and has a subgroup of proper rotations of a square $C_4 = \{e, x, x^2, x^3\}$.

> ✳ **Definition 3.56.**
> A group of all symmetries of an n-sided regular polygon is called the **dihedral group** of order $2n$ and is denoted by D_{2n}. (Sometimes this group is denoted by D_n.)

For an arbitrary n, the dihedral group D_{2n} of order $2n$ is generated by two elements x, y, subject to the relations:

$$x^n = e, \ y^2 = e, \ yx = x^{-1}y \tag{3.18}$$

where e is the identity element of the group D_{2n}, x is the rotation of a regular n-polygon F_n over its center by the angle $2\pi/n$, and y is the mirror reflection over some axis passing through the center of F_n.

The group D_{2n} can be also written in the following form:

$$D_{2n} = \{x, y \ : \ x^n = y^2 = e, yx = x^{n-1}y\} \tag{3.19}$$

For $n \geq 3$ the group D_{2n} is not Abelian since $yx = x^{n-1}y \neq xy$.

✳ **Examples 3.57.**
   1. The dihedral group D_6 is the group of symmetries of an equilateral triangular and can be written in the following form:

$$D_6 = \{x, y \ : \ x^3 = y^2 = e, \ yx = x^2y\}$$

2. The dihedral group D_8 is the group of symmetries of a square and can be written in the following form:

$$D_8 = \{x, y \mid x^4 = y^2 = e, yx = x^3 y\}.$$

3.6. Direct Product of Groups

One of the general constructions in mathematics is the direct product. In this section, we consider the direct product of groups.

Let $(H, \bullet)$ and $(N, \circ)$ be arbitrary multiplicative groups, and let $G = H \times N$ be the Cartesian product, i.e., the set of ordered pairs (h, n) such that $h \in H$ and $n \in N$. We define a binary operation $\cdot$ on G componentwise:

$$(h_1, n_1) \cdot (h_2, n_2) = (h_1 \bullet h_2, n_1 \circ n_2) \tag{3.20}$$

Since $h_1 \bullet h_2 \in H$ and $n_1 \circ n_2 \in N$, G is closed under multiplication $\cdot$. This operation is associative, since operations in H and N are associative. If e_1 is the identity element of H and e_2 is the identity element of N, then $e = (e_1, e_2)$ is the identity element of G. The inverse element to (h, n) is (h^{-1}, n^{-1}). Thus, G is a group, which is called the **external** or **outer direct product** of two groups H and N and is denoted by $H \times N$. From this definition, it immediately follows that $|H \times N| = |H| \cdot |N|$.

For simplicity, we will omit the signs $\bullet$, $\circ$ and $\cdot$ of operations in groups H, N and $H \times N$. Also, we often say the **direct product** of groups instead of the external product of groups.

If we have the direct product of groups $G = H \times N$, then there are two natural embeddings $\mu_1 : H \to G$ and $\mu_2 : N \to G$ given by $\mu_1(h) = (h, e_2)$ and $\mu_2(n) = (e_1, n)$. We denote $\overline{H} = H \times \{e_2\} = \{(h, e_2) : h \in H\}$ and $\overline{N} = \{e_1\} \times N = \{(e_1, n) : n \in N\}$. It is clear that $\mu_1(H) = \overline{H}$ and $\mu_2(N) = \overline{N}$ are subgroups of $G \times N$ and $\overline{H} \cong H$, $\overline{N} \cong N$.

Moreover, $\overline{H}$ and $\overline{N}$ are normal subgroups of $G \times N$ because $(h, n)(h_1, e_2)(h^{-1}, n^{-1}) = (hh_1h^{-1}, ne_2n^{-1}) = (hh_1h^{-1}, e_2) \in \overline{H}$ and $(h, n)(e_1, n_1)(h^{-1}, n^{-1}) = (he_1h^{-1}, nn_1n^{-1}) = (e_1, nn_1n^{-1}) \in \overline{N}$. Since $(h, e_2)(e_1, n) = (h, n) = (e_1, n)(h, e_2)$, the subgroups $\overline{H}$ and $\overline{N}$ commute.

Also, from the definition of $\overline{H}$ and $\overline{N}$, we have that $G \times H = \overline{H} \cdot \overline{N} = \overline{N} \cdot \overline{H}$ and $\overline{H} \cap \overline{N} = \{e\}$, where $\overline{H} \cdot \overline{N} = \{(h, e_2)(e_1, n) : h \in H, n \in N\}$.

Conversely, if we have two normal subgroups H, N of a group G with identity element e and
1) $G = HN = \{hn : h \in H, n \in N\}$
2) $H \cap N = \{e\}$,
then we said that G is an **internal** or **inner direct product** of subgroups H and N.

Note, that if H and N are normal subgroups of a group G with identity element e, and $H \cap N = \{e\}$, then they commute, i.e., $hn = nh$ for all $h \in H$ and for all $n \in N$. Really, for any $h \in H$ and any $n \in N$ we have that

$nhn^{-1}h^{-1} = (nhn^{-1})h^{-1} \in H$, and $nhn^{-1}h^{-1} = n(hn^{-1}h^{-1}) = nn_1 \in N$, i.e., $nhn^{-1}h^{-1} \in H \cap N = \{e\}$, implying that $nh = hn$.

Also it should be noted that, in general, if H and N are arbitrary subgroups of a group G, then NH is not a subgroup of G. However, if either H or N is a normal subgroup of G, then NH is a subgroup of G.

The definitions of the external and internal direct products of groups are closely related to each other due to the following theorem.

✳ **Theorem 3.58.**

If G is the internal direct product of two normal subgroups H and N of G, then they commute and G is isomorphic to the external direct product $H \times N$.

Proof.

1. Let H and N be normal subgroups of a group G with identity element e. We will show that they commute, i.e., $hn = nh$ for all $h \in H$ and for all $n \in N$. Really, for each $h \in H$ and each $n \in N$, we have that $nhn^{-1}h^{-1} = (nhn^{-1})h^{-1} \in H$, and $nhn^{-1}h^{-1} = n(hn^{-1}h^{-1}) \in N$, i.e., $nhn^{-1}h^{-1} \in H \cap N$. Since G is the internal direct product of H and N, $H \cap N = \{e\}$. Therefore, $nhn^{-1}h^{-1} = e$, which implies that $nh = hn$.

2. Since $G = HN$, we can consider a map $f : G \to H \times N$ defined as follows: $f(hn) = (h, n)$. Let $g_1 = h_1 n_1 \in G$ and $g_2 = h_2 n_2 \in G$. Then, $f(g_1 g_2) = f((h_1 n_1)(h_2 n_2)) = f((h_1 h_2)(n_1 n_2)) = (h_1 h_2, n_1 n_2) = (h_1, n_1)(h_2, n_2) = f(g_1)f(g_2)$ because $hn = nh$ for all $h \in H$ and for all $n \in N$. Therefore, f is a homomorphism of groups.

Since $f(hn) = (h, n)$ for any $(h, n) \in H \times N$, f is surjective. Suppose $f(g) = f(hn) = (e, e)$, then $(e, e) = (h, n)$, i.e., $h = n = e$ and so $g = e$, which means that f is injective. Thus, f is an isomorphism of groups G and $H \times N$. □

The notions of the external and internal direct products can be generalized for the case of an arbitrary number of factors.

✳ **Definition 3.59.**

Let $H_1, H_2, \ldots, H_n$ be arbitrary multiplicative groups. The cartesian product $G = H_1 \times H_2 \times \ldots \times H_n$ with componentwise multiplication

$$(h_1, h_2, \ldots, h_n)(h_1', h_2', \ldots, h_n') = (h_1 h_1', h_2 h_2', \ldots, h_n h_n')$$

is called the **external direct product** of groups $H_1, H_2, \ldots, H_n$ and we write $G = \prod_i^n H_i = H_1 \times H_2 \times \ldots \times H_n$. If $H = H_1 = H_2 = \ldots = H_n$, then we often write H^n instead of $H \times H \times \ldots \times H$.

Note that also in this case each group

$$\overline{H_i} = \{(e_1, \ldots, e_{i-1}, h_i, e_{i+1}, \ldots, e_n) \ : \ h_i \in H_i, \ i = 1, 2, \ldots, n\}$$

is a normal subgroup of G, all these subgroups commute and $G = \overline{H_1} \cdot \overline{H_2} \cdots \overline{H_n}$.

> ✳ **Definition 3.60.**
>   Let $H_1, H_2, \ldots, H_n$ be normal subgroups of a multiplicative group G which satisfy the following conditions:
>   1) $G = H_1 H_2 \ldots H_n$
>   2) $H_j \cap \prod_{\substack{i=1 \\ i \neq j}}^{n} H_i = \{e\}$, for all $i, j = 1, 2, \ldots, n$, where e is the identity element of G.
>   Then G is called the **internal direct product** of subgroups $H_1, H_2, \ldots, H_n$.

✱ **Example 3.61.**
   1. Let $H = \mathbb{Z}_2$. Consider the direct product of n copies of this group: $G = H^n = \mathbb{Z}_2^n$. Then G is a set of all binary n-tuples:

$$G = \{(a_1 a_2 \ldots a_n) \ : \ a_i \in \{0, 1\}\}$$

and the binary operation on G is addition modulo 2. For example, in $\mathbb{Z}_2^{11}$ we have the addition of two elements:

$$(00101001110) + (10011110001) = (101101111111).$$

   2. Let G_4 be the Klein four-group considered in Example 3.53, which is the group of all symmetries of a rectangle, with binary operation given by Table 3.1. It is an Abelian group and it has 3 proper subgroups $H_1 = \{e, a\}$, $H_2 = \{e, b\}$ and $H_3 = \{e, c\}$, each of which is isomorphic to the cyclic group C_2. All these subgroups are normal, since G_4 is an Abelian group. The group G_4 is of order 4 and all its elements can be written, for example, as follows: e, a, b, ab, since $c = ab = ba$. Therefore, $G_4 = H_1 H_2$ and $H_1 \cap H_2 = \{e\}$. In this way, G_4 is the internal direct product of subgroups H_1 and H_2. Since $H_1 \cong H_2 \cong C_2$, we obtain $G_4 \cong C_2 \times C_2$, by Theorem 3.58.

Similarly to the case of two groups, we can prove the following result.

> ✱ **Theorem 3.62.**
> If G is the internal direct product of normal subgroups $H_1, H_2, \ldots, H_n$ of G, then they commute and G is isomorphic to the external direct product $\prod_{i}^{n} H_i = H_1 \times H_2 \times \ldots \times H_n$.

Remarks 3.63.
   1. If we have a finite number of Abelian groups $H_1, H_2, \ldots, H_n$, then the direct product of groups is sometimes called the **direct sum** of groups. Moreover, if additive notation is used then the direct sum of these groups is

denoted by $H_1 \oplus H_2 \oplus \ldots \oplus H_n$ and the operation on $H_1 \oplus H_2 \oplus \ldots \oplus H_n$ is written as:

$$(h_1, h_2, \ldots, h_n) + (h'_1, h'_2, \ldots, h'_n) = (h_1 + h'_1, h_2 + h'_2, \ldots, h_n + h'_n)$$

2. If we have an infinite number of Abelian groups then we also can define the direct product of groups and the direct sum of groups. In this case, however, we obtain the different groups. Every element of the direct product has one arbitrary element from each group, while elements of the direct sum have one extra condition: Only a finite number of elements of given groups must be different from the identity elements of groups.

3.7. Finite Abelian Groups

In this section, we consider the structure of finite Abelian groups, one of the most interesting examples of groups. First we consider the particular case of additive groups $\mathbb{Z}_m$ of least residue classes modulo m.

❊ **Proposition 3.64.**

Let $G = G_1 \times G_2 \times \cdots \times G_n$ be the direct product of multiplicative groups. If $g = (g_1, g_2, \ldots, g_n) \in G$ and $\mathrm{ord}(g_i) = k_i$ for all $i = 1, \ldots, n$, then $\mathrm{ord}(g) = \mathrm{lcm}(k_1, k_2, \ldots, k_n)$.

Proof.

Let G_i be a group with the identity element e_i, then $e = (e_1, \ldots, e_n)$ is the identity element of G. Let $\mathrm{ord}(g) = m$ and $\mathrm{lcm}(k_1, k_2, \ldots, k_n) = k$. Since $g^k = (g_1^k, g_2^k, \ldots, g_n^k) = e$, from Theorem 2.25, it follows that $m|k$. On the other hand, $e = g^m = (g_1^m, g_2^m, \ldots, g_n^m) = (e_1, e_2, \ldots, e_n)$ which means that $k_i|m$ for all i. Hence, $k|m$. Thus, $m|k$ and $k|m$, implying that $k = m$.   □

❊ **Theorem 3.65.**

The additive group $\mathbb{Z}_{mn}$ is isomorphic to the direct product of groups $\mathbb{Z}_n \times \mathbb{Z}_m$ if and only if $\gcd(m, n) = 1$.

Proof.

1. Suppose that $\mathbb{Z}_{mn} \cong \mathbb{Z}_n \times \mathbb{Z}_m$ and $\gcd(m, n) = d > 1$. Then, $m = m_1 d$, $n = n_1 d$ and $k = mn/d = m_1 n_1 d = mn_1 = m_1 n < mn$. Hence, $k(a, b) = (e, e)$ is the identity element of $\mathbb{Z}_n \times \mathbb{Z}_m$ for all elements $(a, b) \in \mathbb{Z}_n \times \mathbb{Z}_m$. This means that $\mathrm{ord}(a, b) \leq k < mn$. Therefore, $\mathbb{Z}_{mn}$ is not a cyclic group, by Theorem 3.43. This contradiction shows that $\gcd(m, n) = 1$.

2. Suppose that $\gcd(m, n) = 1$, then $\mathrm{lcm}(m, n) = mn$. Let a be the generator of $\mathbb{Z}_n$ and let b be the generator of $\mathbb{Z}_m$. Then, $\mathrm{ord}(a) = n$ and $\mathrm{ord}(b) = m$. Therefore, $\mathrm{ord}(a, b) = mn$, by Theorem 3.64. Hence, $\mathrm{ord}(a, b) = mn = |\mathbb{Z}_n \times \mathbb{Z}_m|$. Therefore, $\mathbb{Z}_n \times \mathbb{Z}_m$ is a cyclic group of order mn, by Theorem 3.43. In this way, $\mathbb{Z}_n \times \mathbb{Z}_m \cong \mathbb{Z}_{mn}$.   □

This theorem can be generalized to the case of k factors:

✳ Theorem 3.66.

The group $\mathbb{Z}_m$, where $m = n_1 n_2 \cdots n_k$, is isomorphic to the direct product of groups $\mathbb{Z}_{n_1} \times \mathbb{Z}_{n_2} \times \cdots \times \mathbb{Z}_{n_k}$ if and only if $\gcd(n_i, n_j) = 1$ for all i, j with $i \neq j$.

✳ Corollary 3.67.

The group $\mathbb{Z}_m$, where $m = p_1^{\alpha_1} p_2^{\alpha_2} \cdots p_k^{\alpha_k}$ and $p_1, p_2, \ldots, p_k$ are distinct primes, is isomorphic to the direct product of groups $\mathbb{Z}_{p_1^{\alpha_1}} \times \mathbb{Z}_{p_2^{\alpha_2}} \times \cdots \times \mathbb{Z}_{p_k^{\alpha_k}}$.

✳ Definition 3.68.

The smallest positive integer n (if it exists) such that $g^n = e$ for all elements of a group G is called the **exponent** of G. If such an integer does not exist, the exponent is assumed to be equal to infinity.

In other words, the exponent of a group is the least common multiple of the orders of all elements of the group.

If G is a finite group of order n, then, by Lagrange's theorem, the order of all elements divides n. Hence, the exponent of G always exists and divides n.

✳ Proposition 3.69.

The exponent of a finite cyclic group G is equal to its order.

Proof.

Let $|G| = n$. By Lagrange's Theorem, $h^n = e$ for any $h \in G$. Therefore, $\mathrm{ord}(h)|n$, by Theorem 2.25. Hence, $\mathrm{ord}(h) = n/k \leq n$ for some $k \in \mathbb{Z}^+$. On the other hand, if g is a generator of a cyclic group G, then $\mathrm{ord}(g) = n$. Therefore, the exponent of G, which is the least common multiple of orders of all elements of G, is equal to $n = |G|$. $\square$

✳ Proposition 3.70.

If $G = G_1 \times G_2$, where G_1, G_2 are Abelian groups with exponents n_1 and n_2, then the exponent of G is equal to $\mathrm{lcm}(n_1, n_2)$.

Proof.

This follows directly from Definition 3.68 and Proposition 3.64. $\square$

✳ Examples 3.71.

1. The exponent of the additive group $\mathbb{Z}_n$ is equal to n.

2. The exponent of the direct product of groups $\mathbb{Z}_3 \times \mathbb{Z}_3$ is equal to $3 = \mathrm{lcm}(3, 3)$.

3. The exponent of the direct product of groups $\mathbb{Z}_2 \times \mathbb{Z}_5$ is equal to $10 = \mathrm{lcm}(2, 5)$.

✴ Lemma 3.72.

Let G be an Abelian group with the identity element e. If $g \in G$ satisfies the properties $g^{p^k} = e$ and $g^{p^{k-1}} \neq e$ for some a prime p and an integer k, then $\mathrm{ord}(g) = p^k$.

Proof.

    Let $\mathrm{ord}(g) = m$. Since $g^{p^k} = e$, $m | p^k$, by Theorem 2.25. Hence, $m \leq p^k$. Since $g^{p^{k-1}} \neq e$, $p^{k-1} < m \leq p^k$. In this way, $m = p^k$.   □

✴ Lemma 3.73.

Let g_1, g_2 be elements of an Abelian group G such that $\mathrm{ord}(g_1) = m$, $\mathrm{ord}(g_2) = n$ and $\gcd(m, n) = 1$. Then, $\mathrm{ord}(g_1 g_2) = mn$.

Proof.

    Consider cyclic subgroups in an Abelian group G: $G_1 = \langle g_1 \rangle$ and $G_2 = \langle g_2 \rangle$. Then, we can construct the external direct product $H = G_1 \times G_2$ of these subgroups. From condition $\gcd(m, n) = 1$, it follows that $G_1 \cap G_2 = \emptyset$. Since G is Abelian group, $G_1 G_2 = G_2 G_1$. Therefore, by Theorem 3.58, the external direct product of groups G_1 and G_2 is isomorphic to the inner direct product of these subgroups, i.e., $G_1 G_2 \simeq G_1 \times G_2$, which is given as follows: $g_1 g_2 \mapsto (g_1, g_2)$. Using Proposition 3.64, we obtain that $\mathrm{ord}(g_1 g_2) = \mathrm{lcm}(m, n) = mn$, taking into account that $\gcd(m, n) = 1$.   □

✴ Corollary 3.74.

Let $g_1, g_2, \ldots, g_k$ be elements of an Abelian group G such that $\mathrm{ord}(g_i) = m_i$, and $\gcd(m_i, m_j) = 1$ for all $i \neq j$, then $\mathrm{ord}(g_1 g_2 \cdots g_k) = m_1 m_2 \cdots m_k$.

✴ Theorem 3.75.

Let G be an Abelian group with exponent n. Then, G contains an element of order n.

Proof.

    Let $n = p_1^{\alpha_1} p_2^{\alpha_2} \cdots p_k^{\alpha_k}$, where the p_i's are distinct primes. Since n is the least common multiple of orders of all elements of a group G, there is an element $g_i \in G$, for all i, such that $g_i^{\beta_i} \neq e$, where $\beta_i = n/p_i$. Otherwise, there is an i such that $g^{\beta_i} = e$ for all $g \in G$, and so $\beta_i < n$ would be the exponent of G. So, we can choose the elements $g_1, g_2, \ldots, g_k \in G$ such that $g_i^{\beta_i} \neq e$ but $g_i^n = e$. Then, by Lemma 3.72, $\mathrm{ord}(g_i) = p_i^{\alpha_i}$ for all i.

    Consider the element $g = g_1 g_2 \cdots g_k$. Then, by Corollary 3.74, we obtain $\mathrm{ord}(g) = \mathrm{lcm}(\mathrm{ord}(g_1), \mathrm{ord}(g_2), \ldots, \mathrm{ord}(g_k)) = \mathrm{lcm}(p_1^{\alpha_1}, p_2^{\alpha_2}, \cdots, p_k^{\alpha_k}) = n$.   □

✴ Corollary 3.76.

A finite Abelian group G is cyclic if and only if its order is equal to its exponent.

✱ Lemma 3.77.
Let G be a finite Abelian group G of order n. If p is a prime which divides n, then G has an element of order p.

Proof.
Let $n = pm$. We prove by induction on m. If $m = 1$ then $|G| = p$ and G is a cyclic group, and so each non-identity element of G is of order p. Now, we assume that $m > 1$ and the theorem holds for any Abelian group with order $< m$. Let H be a proper subgroup of G. If p divides $|H|$, then, by induction hypothesis, H (and hence G) has an element of order p, and we are done. Assume that p does not divide $|H|$. In this case, we can consider the quotient group G/H whose order is less than m and p divides its order. Then, by induction hypothesis, G/H has an element $\bar{a} = aH \in G/H$ of order p. So, $a^p \in H$. Suppose $\operatorname{ord}(a) = k$. Then, $a^k = e \in H$. Consider the projection map $\pi : G \to G/H$ given by $\pi(g) = \bar{g} = gH$ for all $g \in G$. Then, $(\bar{a})^p = \pi(a)^p = \pi(a^p) = \pi(e) = H$ and $(\bar{a})^k = \pi(a)^k = \pi(a^k) = \pi(e) = H$. Hence, by theorem 2.25, $p|k$. Let $k = ps$, then the element $a^s \in G$ is of order p, since $(a^s)^p = a^k = e$. □

✱ Theorem 3.78.
Let G be a finite Abelian group G with exponent n. Then G has a cyclic subgroup H of order n and a subgroup N of order m such that $G \cong H \times N$ and $n|m$.

Proof.
Let G be a finite Abelian group with exponent n. Then, by Theorem 3.75, G contains an element h of order n. Consider a cyclic subgroup $H = \langle h \rangle \cong \mathbb{Z}_n$. Let N be a subgroup of G of the largest order subject to the condition $H \cap N = \{e\}$. We will show that $G = HN$. Suppose that it is not the case, i.e., $G \neq HN$. Since HN is a normal subgroup of G, we can consider the quotient group $G/(HN)$, which is non-trivial. Therefore, there is a prime p which divides $G/(HN)$ and so it divides n. Therefore, by Lemma 3.77, $G/(HK)$ contains an element of order p. So, there is an element $g \in G$ such that $g \notin HN$ and $g^p \in HK$. Let $g^p = h^k x$, where $h \in H$ and $x \in N$. Consider two cases.

1) $p|k$, i.e., $k = ps$. Consider an element $a = h^{-s}g$. If $a \in N$, then $g = h^s a \in HN$, which is a contradiction. So, $a \notin N$. Consider a subgroup of G generated by the subgroup N and the element a: $N_1 = \langle N, a \rangle$. Suppose $N_1 \cap H \neq \{e\}$, i.e., there is an element $b \neq e$ such that $b \in H$ and $b \in N_1$. Then, there is an element $y \in N$ such that $b = ay = h^r$, implying that $g = h^s a = h^s h^r y^{-1} \in HN$, which is a contradiction. Thus, we have a subgroup N_1 that has a larger order than $|N|$ and $H \cap N_1 = \{e\}$, which contradicts to the choice of the subgroup N.

2) Assume that p does not divide k. Since $g^p = h^k x$, we obtain that $\operatorname{ord}(g) > \operatorname{ord}(h) = n$, this contradicts the fact that n is the exponent of G.

Hence, both cases reduce to contradictions. Therefore, $G = HN$ and $H \cap N = \{e\}$. Since G is an Abelian group, these conditions imply that $G = H \times N$.

If $|N| = m$, then $n|m$. Really, since n is the exponent of G, $h^n = e$ for all $h \in H$. Let $\mathrm{ord}(h) = k$. Then, by Theorem 2.24, $n|k$. On the other hand, by Lagrange's Theorem, $k|m$. Hence, $n|m$, as required. $\square$

❋ Theorem 3.79.

A non-trivial finite Abelian group G is isomorphic to the direct product of cyclic groups:

$$\mathbb{Z}_{n_1} \times \mathbb{Z}_{n_2} \times \cdots \times \mathbb{Z}_{n_k}, \qquad (3.21)$$

where $n_i > 1$ for all i and $n_i | n_{i+1}$ for $i = 1, 2, \ldots, k-1$. Moreover, n_k is the exponent of G.

Proof.

We will use the induction on $|G|$. Let n_1 be the exponent of G. Then, by the previous theorem $G \cong G_1 \times N$, where G_1 is a cyclic group of order n_1, N is a subgroup of order m and $n_1 | m$. It is clear that the exponent of N is a divisor of m. Since $|N| < |G|$, we can apply the indiction hypothesis for the subgroup N, and so the theorem will follow by induction. $\square$

From this theorem and Corollary 3.67, the fundamental theorem of finite Abelian groups follows immediately.

❋ Theorem 3.80.   (Fundamental Theorem of Finite Abelian Groups)

Each non-trivial finite Abelian group G is isomorphic to the direct product of cyclic groups:

$$G \cong \mathbb{Z}_{p_1^{\alpha_1}} \times \mathbb{Z}_{p_2^{\alpha_2}} \times \cdots \times \mathbb{Z}_{p_n^{\alpha_n}} \qquad (3.22)$$

where the p_i's are primes (not necessarily distinct).

Remark 3.81.

1. It can be shown that theorems 3.79 and 3.80 are equivalent, i.e., from theorem 3.80, theorem 3.79 also follows.

2. There is an easier and more elegant proof of theorem 3.80 which follows immediately from the structure theorem of finitely-generated modules over principal ideal domains, because each Abelian group can be regarded as a module over the ring of integers $\mathbb{Z}$.

3.8. Exercises

1. Find the number of inversions of permutations:

$$\sigma = \begin{pmatrix} 1 & 2 & 3 & 4 & 5 & 6 & 7 \\ 6 & 3 & 5 & 4 & 2 & 7 & 1 \end{pmatrix}, \quad \tau = \begin{pmatrix} 1 & 2 & 3 & 4 & 5 & 6 & 7 \\ 3 & 2 & 7 & 4 & 1 & 5 & 6 \end{pmatrix}$$

   Which of these permutations is even?

2. Write the permutation:

$$\pi = \begin{pmatrix} 1 & 2 & 3 & 4 & 5 & 6 & 7 & 8 \\ 3 & 4 & 5 & 2 & 1 & 8 & 7 & 6 \end{pmatrix} \in S_8$$

   in the form of disjoint cycles.

3. Verify the parity of the permutation:

$$\pi = \begin{pmatrix} 1 & 2 & 3 & 4 & 5 & 6 & 7 & 8 \\ 2 & 4 & 6 & 5 & 1 & 7 & 3 & 8 \end{pmatrix} \in S_8$$

   Find the order of the permutation π. Find π^{200}.

4. Write the permutation:

$$\pi = \begin{pmatrix} 1 & 2 & 3 & 4 & 5 & 6 & 7 & 8 \\ 4 & 1 & 8 & 2 & 7 & 3 & 6 & 5 \end{pmatrix} \in S_8$$

   in the form of disjoint cycles and find its order and parity. Find π^{152}.

5. Write all elements of the alternating groups A_3 and A_4.

6. Find all subgroups of the group of symmetries of a square.

7. Find the groups of symmetries of the figures presented in Figure 3.1 and Figure 3.3.

References

[1] Ash, R.B. 2007. Basic Abstract Algebra, Dover Publications.

[2] Dummit, D.S. and R.M. Foote. 2004. Abstract Algebra (3rd Ed.), John Wiley & Sons.

[3] Hall, M. (Jr.) 1959. The Theory of Groups, New York, Macmillan.

[4] Humphreys, J.F. 1996. A Course in Group Theory, Oxford University Press.

[5] Jacobson, N. 1985. Basic Algebra I, Freeman, San Francisco, H. Freeman & Co.

[6] Judson, Th.W. 1994. Abstract Algebra. Theory and Applications, PWS Publishing Company, Michigan.

[7] Lang, S. 1984. Algebra, Addison-Wesley.

[8] Oggier, F. 2011. Algebraic Methods.
http://www1.spms.ntu.edu.sg/~frederique/AA11.pdf.

[9] Oggier, F. and A.M. Bruckstein. 2013. Groups and Symmetries.
http://www1.spms.ntu.edu.sg/~frederique/groupsymmetryws.pdf.

[10] Rotman, J.J. 1994. An Introduction to the Theory of Groups, Springer-Verlag.

Chapter 4

Elements of Ring Theory

"Science is built up of facts, as a house is with stones. But a collection of facts is no more a science than a heap of stones is a house."
Henri Poincare

"The scientist does not study nature because it is useful; he studies it because he delights in it, and he delights in it because it is beautiful."
Henri Poincare

Ring theory is one of the basic branches of modern algebra. Associative rings and algebras (not necessarily commutative) are very interesting algebraic structures. A lot of natural examples of commutative rings are given in the theory of algebraic numbers, algebraic geometry and number theory. Non-commutative rings are completely different and more unusual, but now they are widely used in physics and mathematics along with the development of non-commutative geometry and quantum groups. The most important examples of non-commutative rings are square matrices and quaternions.

Historically, the theory of rings originates in the works of W.R. Hamilton, H. Grassmann, R. Dedekind, G. Frobenius, E. Artin and J.M. Wedderburn. In fact, the term "ring" was introduced by R. Dedekind and D. Hilbert in the end of the 19th century and only in the concrete setting of rings of algebraic integers, which are commutative rings. The first abstract (axiomatic) definition of a ring was given in 1914 by A. Fraenkel. However, for a long time, the rings of polynomials, algebraic numbers and hypercomplex numbers were central to the ring theory. Mainly thanks to significant works of outstanding mathematicians E. Artin (1898-1962) and Emmy Noether (1882-1935) in the 20s years of the XXth century ring theory has become a powerful modern theory. Emmy Noether is also well known due to the theorem often called

Emmy Noether
(1882-1935)

"The Noether Theorem" which plays a fundamental role in physics and shows the relationship between symmetries of physical laws and basic behavior of some physical constants.

In this chapter, we introduce a number of basic concepts of ring theory and prove some main properties of rings. We also give a number of different examples of rings.

4.1. Rings and Subrings

The important examples of algebraic structures with two inner binary operations are rings.

> ✳ **Definition 4.1.**
>   An algebraic structure $(R, +, \bullet)$ consisting of a non-empty set P and two inner binary operations $+$ and $\bullet$ is called a **ring** if the following conditions hold:
>
> - $(R, +)$ is an Abelian group;
>
> - the operation $\bullet$ is distributive with respect to the operation $+$, i.e.,
>
> $$a \bullet (b + c) = a \bullet b + a \bullet c \ \text{ and } \ (a + b) \bullet c = a \bullet c + b \bullet c$$
>
>   for all $a, b, c \in R$

Usually, the operation $+$ in a ring is called **addition**, and the second operation $\bullet$ is called **multiplication**, although, in general, they are not necessarily ordinary arithmetical operations. The symbol of multiplication $\bullet$ in a ring will often be omitted. The neutral element of a ring with respect to addition will be denoted by 0 and is called the **zero** of a ring. The inverse element to an element a with respect to addition will be written by $-a$. If $a, b \in R$, then we write $a - b$ instead of $a + (-b)$.

The neutral element with respect to multiplication is called the **identity** of a ring and is denoted by 1. From theorem A.11, it follows that a ring contains at most one identity element. A ring R is called a **ring with identity** if $R \neq \{0\}$ and there is the identity element $1 \in R$ such that

$$a \bullet 1 = 1 \bullet a = a, \ \ \forall (a \in R).$$

A ring R is called **associative** if the operation $\bullet$ is associative:

$$a \bullet (b \bullet c) = (a \bullet b) \bullet c, \ \ \forall (a, b, c \in R).$$

A ring R is called **commutative** if the operation $\bullet$ is commutative:

$$a \bullet b = b \bullet a, \ \ \forall (a, b \in R).$$

Otherwise, a ring is called **non-commutative**.

❋ **Remark 4.2.**

Further on in this book, we suppose that all rings are associative with identity unless otherwise stated.

An element a of a ring R is called **invertible** if it has the inverse with respect to multiplication $\bullet$, i.e., there exists an element $b \in R$ such that $a \bullet b = b \bullet a = 1$. If R is an associative ring with identity, then, from theorem A.13, it follows that there is at most one inverse element $b \in R$ to an element $a \in R$, which will be denoted by a^{-1}. The set of all invertible elements in R will be denoted by $U(R)$.

❋ **Lemma 4.3.**

The set $U(R)$ of all invertible elements of a ring R is a group under multiplication.

Proof.

1. The multiplication of elements in $U(R)$ is associative, since P is an associative ring.

2. The identity 1 is an invertible element in R and so $1 \in U(R)$.

3. If $x \in U(R)$, then there is an element $x^{-1} \in R$, which is also invertible in R because $(x^{-1})^{-1} = x$, which means that $x^{-1} \in U(R)$. □

❋ **Examples 4.4.**

1. The set of all integers $\{\mathbb{Z}, +, \cdot\}$ under the ordinary arithmetical operations of addition and multiplication is an associative commutative ring with identity 1. The group of invertible elements $U(\mathbb{Z}) = \{1, -1\}$ is often denoted by $\mathbb{Z}^*$.

2. The set of all even integers under the ordinary arithmetical operations of addition and multiplication is an associative commutative ring without identity.

3. The set of all functions defined on some line segment $[a, b]$ under the ordinary operations of addition and superposition of functions is an associative commutative ring with identity which is the identity function $1(x) = 1$ for all $x \in [a, b]$.

4. The set $\mathbb{Z}_m = \{0, 1, 2, \ldots, m - 1\}$, where $(m \in \mathbb{Z}) \wedge (m > 1)$, under operations $\oplus$ and $\bullet$ defined as follows:

$$a \oplus b = (a + b) \bmod m \tag{4.1}$$

$$a \bullet b = (ab) \bmod m \tag{4.2}$$

is an associative commutative ring with the identity 1. The group of invertible elements of the ring $\mathbb{Z}_m$ is a multiplicative group

$$(\mathbb{Z}_m)^* = \{k \ : \ k \in \mathbb{Z}, \ 1 \le k < m, \ \gcd(k, m) = 1\}.$$

5. The set $\mathbb{Z}/m\mathbb{Z} = \{[0], [1], [2], \ldots, [m-1]\}$, where $(m \in \mathbb{Z}) \wedge (m > 1)$, under operations $+$ and $\cdot$ defined as follows:

$$[a] + [b] = [a+b] \tag{4.3}$$

$$[a] \cdot [b] = [ab] \tag{4.4}$$

is an associative commutative ring with the identity $[1]$. The group of invertible elements of the ring $\mathbb{Z}/m\mathbb{Z}$ is a multiplicative group of residue classes modulo m:

$$(\mathbb{Z}/m\mathbb{Z})^* = \{[k] \ : \ k \in \mathbb{Z}, \ 1 \le k < m, \ \gcd(k,m) = 1\}.$$

6. The set of Gauss integers: $\mathbb{Z}[i] = \{a + ib \ : \ a, b \in \mathbb{Z}, \ i^2 = -1\}$ under operations of addition and multiplication is an associative commutative ring with the identity 1. The group of invertible elements of $\mathbb{Z}[i]$ is $(\mathbb{Z}[i])^* = \{1, -1, i, -i\}$.

7. The ring of square matrices $M_n(\mathbb{R}) = \{\mathbf{A} = (a_{ij}) \ : \ a_{ij} \in \mathbb{R}\}$ under operations of addition and multiplication of matrices:

$$(a_{ij}) + (b_{ij}) = (a_{ij} + b_{ij})$$

$$(a_{ij})(b_{ij}) = (c_{ij}), \quad \text{where} \quad c_{ij} = a_{i1}b_{1j} + a_{i2}b_{2j} + \cdots + a_{in}b_{nj}$$

is an associative non-commutative ring with identity which is the identity matrix. The group of invertible elements of the ring $M_n(\mathbb{R})$ is the group $GL(n, \mathbb{R})$.

8. The sets of polynomials in one variable $(\mathbb{Q}[x], +, \cdot)$, $(\mathbb{R}[x], +, \cdot)$, $(\mathbb{C}[x], +, \cdot)$ under ordinary operations of addition and multiplication of polynomials are associative commutative rings with the identity 1.

9. The set $(\wp(X), \oplus, \cap)$ of all subsets of a set X under operations of symmetrical difference of sets $\oplus$ defined as $A \oplus B = (A \backslash B) \cup (B \backslash A)$ and intersection of sets $\cap$ is an associative commutative ring with the identity 1 which is the whole set X.

✻ **Definition 4.5.**
    A nonempty subset B of a ring $(R, +, \bullet)$ is called a **subring** of R if B is a ring under operations $+$ and $\bullet$ defined in R, another words B is a subring of R if the following conditions hold:

- if $a, b \in B$, then $a + b \in B$;

- if $a, b \in B$, then $a \bullet b \in B$;

- if $a \in B$, then $-a \in B$;

    If B is a subring of a ring P with identity and $1 \in B$, then B is called a **subring with identity**.

✳ Examples 4.6.

1. The set of all integers $\{\mathbb{Z}, +, \cdot\}$ is a subring of the ring $\{\mathbb{Q}, +, \cdot\}$ of all rational numbers under addition and multiplication.

2. The set $\mathbb{Z}[\sqrt{2}] = \{a + b\sqrt{2} : a, b \in \mathbb{Z}\}$ is a subring of the ring $\mathbb{R}$ of all real numbers under addition and multiplication.

3. The set of all even integers $\{2\mathbb{Z}, +, \cdot\}$ is a subring of the ring $\{\mathbb{Z}, +, \cdot\}$.

4. The set of Gauss integers $\{\mathbb{Z}[i], +, \cdot\}$ is a subring of the ring of all complex numbers $\{\mathbb{C}, +, \cdot\}$.

✳ Definition 4.7.

A ring $(R, +, \bullet)$ with identity is called a **division ring** if $1 \neq 0$ and each of its non-zero elements is invertible.

✳ Examples 4.8.

1. The diagonal matrices of degree n over $\mathbb{R}$ or $\mathbb{C}$ form a division ring.

2. Let $\mathbb{H} = \{a_1 e + a_2 i + a_3 j + a_4 k : a_r \in \mathbb{R}, r = 1, 2, 3, 4\}$. The operation of addition is defined as follows:

$$x + y = (a_1 e + a_2 i + a_3 j + a_4 k) + (b_1 e + b_2 i + b_3 j + b_4 k) =$$

$$= (a_1 + b_1)e + (a_2 + b_2)i + (a_3 + b_3)j + (a_4 + b_4)k$$

for all elements $x = a_1 e + a_2 i + a_3 j + a_4 k \in \mathbb{H}$ and $y = b_1 e + b_2 i + b_3 j + b_4 k \in \mathbb{H}$.

The operation of multiplication in $\mathbb{H}$ we define first for elements e, i, j, k by the following table:

$\cdot$	e	i	j	k
e	e	i	j	k
i	i	$-e$	k	$-j$
j	j	$-k$	$-e$	i
k	k	j	$-i$	$-e$

Furthermore, using associative and distributive laws, we define multiplication in $\mathbb{H}$ as follows:

$$x \cdot y = (a_1 e + a_2 i + a_3 j + a_4 k) \cdot (b_1 e + b_2 i + b_3 j + b_4 k) =$$

$$= (a_1 b_1 - a_2 b_2 - a_3 b_3 - a_4 b_4)e+$$

$$+(a_1 b_2 + a_2 b_1 + a_3 b_4 - a_4 b_3)i+$$

$$+(a_1 b_3 - a_2 b_4 + a_3 b_1 + a_4 b_1)j+$$

$$+(a_1 b_4 + a_2 b_3 - a_3 b_2 a_4 b_1)k$$

for arbitrary elements $x = a_1 e + a_2 i + a_3 j + a_4 k \in \mathbb{H}$ and $y = b_1 e + b_2 i + b_3 j + b_4 k \in \mathbb{H}$. It is easy to verify that the set of elements $\mathbb{H}$ forms an associative non-commutative ring with identity $1 \cdot e + 0 \cdot i + 0 \cdot j + 0 \cdot k$ and zero element $0 \cdot e + 0 \cdot i + 0 \cdot j + 0 \cdot k$. If $x = a_1 e + a_2 i + a_3 j + a_4 k \in \mathbb{H}$, then we define $\bar{x} = a_1 e - a_2 i - a_3 j - a_4 k$. It is easy to verify that $x \cdot \bar{x} = a_1^2 + a_2^2 + a_3^2 + a_4^2 \in \mathbb{R}$. If $x \neq 0$, then also $x \cdot \bar{x} \neq 0$. In this case, x has the inverse element

$$x^{-1} = (a_1^2 + a_2^2 + a_3^2 + a_4^2)^{-1} \cdot \bar{x} \in \mathbb{H}.$$

Hence, $\mathbb{H}$ is a division ring which is called the ring of **quaternions**.

✳ Theorem 4.9.

A ring R with identity is a division ring if and only if each of equations:

$$ax = b \quad \text{and} \quad ya = b, \quad \forall[(a, b \in R) \wedge (a \neq 0)]$$

has a unique solution.

Proof.

$\Rightarrow$. If R is a division ring and $(a \in R) \wedge (a \neq 0)$, then a solution of an equation $ax = b$, where $a \neq 0$ is $x = ba^{-1}$. If x_1 is another solution of this equation, i.e., $ax_1 = b$, then $ax = ax_1$, or $a(x - x_1) = 0$. Since R is a division ring, this implies that $x - x_1 = 0$, or $x = x_1$. In a similar way, we can show that the second equation $ya = b$ also has only a unique solution.

$\Leftarrow$. Suppose that R is a ring with identity and each of equations $ax = b$ and $ya = b$, where $a, b \in R$, $a \neq 0$, has exactly one solution. Then, each of equations $ax = 1$ and $ya = 1$, where $a \in R$, $a \neq 0$, also has exactly one solution, which is the inverse element to the element a. This shows that R is a division ring. ☐

4.2. Integral Domains and Fields

In this section, we consider the important class of commutative rings which can be considered as a generalization of integers and real numbers.

✳ Definition 4.10.

A non-zero element a of a ring R is called a **left zero divisor** if there is a non-zero element $x \in R$ such that $ax = 0$. A non-zero element a of a ring R is called a **right zero divisor** if there is a non-zero element $y \in R$ such that $ya = 0$. If a non-zero element of a ring R is a right and left zero divisor, then a is called a **zero divisor**. If R is a commutative ring, then both notions of right and left zero divisors coincide.

✳ Examples 4.11.

Let $\mathbf{A} = \begin{pmatrix} 6 & 12 \\ 5 & 10 \end{pmatrix} \in M_2(\mathbb{R})$. Then, $\mathbf{A}$ is a left and right zero divisor

in $M_2(\mathbb{R})$ because $\mathbf{AX} = \mathbf{YA} = \begin{pmatrix} 0 & 0 \\ 0 & 0 \end{pmatrix}$, where $\mathbf{X} = \begin{pmatrix} -2 & -4 \\ 1 & 2 \end{pmatrix}$ and $\mathbf{Y} = \begin{pmatrix} 5 & -6 \\ -5 & 6 \end{pmatrix}$.

✳ Definition 4.12.

A ring without zero divisors is called a **domain**. A commutative ring with identity and without zero divisors is called an **integral domain**.

✳ Examples 4.13.

1. The ring of integers $(\mathbb{Z}, +, \cdot)$ is an integral domain.

2. The ring $\mathbb{Z}_5$ of least residue classes modulo 5 with operations defined in Example 4.4(4) is an integral domain.

3. The ring $\mathbb{Z}_6$ of least residue classes modulo 6 is not an integral domain since $2 \cdot 3 \equiv 0 \,(\mathrm{mod}\,6)$.

4. The ring of Gauss integers $\mathbb{Z}[i]$ is an integral domain since it is a subset of complex numbers.

5. The ring $(\wp(X), \oplus, \cap)$ of all subsets of a set X under the operations, as defined in Example 4.4(9), is not an integral domain.

6. The ring of square matrices $(M_n(\mathbb{R}), +, \cdot)$ is not an integral domain.

Recall that we consider now only associative rings with identity. Then, we can introduce the notion of an invertible element.

✳ Definition 4.14.

Let R be a ring with identity 1. An element $a \in R$ is called **left invertible** if there is an element $b \in R$ such that $ab = 1$, in this case, the element b is called a **left inverse** to a. An element $a \in R$ is called **right invertible** if there is an element $c \in R$ such that $ca = 1$, in this case, the element b is called a **right inverse** to a.

If a non-zero element $a \in R$ is a right and left invertible, then $c = c(ab) = (ca)b = b$. In this case, a is called an **invertible element** or a **unit** and the element $b = c$ is called the **inverse** to a and it is usually denoted by a^{-1}. If R is a commutative ring, then both notions of right and left invertible elements coincide.

✳ **Proposition 4.15.**

For any ring R, the following conditions hold:
1) A left zero divisor cannot be right invertible.
2) A right zero divisor cannot be left invertible.
3) A left invertible element cannot be a right zero divisor.
4) A right invertible element cannot be a left zero divisor.

Proof.

Let a non-zero element $a \in R$ be a left zero divisor. Then, there is a non-zero element $x \in R$ such that $ax = 0$. If a is right invertible, then there is an element $y \in R$ such that $ya = 1$. From these two equalities and using the associative law for multiplication, we obtain that $0 = y \cdot 0 = y(ax) = (ya)x = 1 \cdot x = x$, which is a contradiction.

Other parts of the proposition can be proved analogously. □

✳ **Definition 4.16.**

Let R be an integral domain and $b = aq$, where $a, b, q \in R$ and $a \neq 0$. Then, we say that a **divides** b (or a is a **divisor** of b) and we write $a|b$.

The properties of integral domains are represented in the form of the following theorem, whose proof is exactly the same as the proof of Theorem 1.3.

✳ **Theorem 4.17.**

Let R be an integral domain, then the following conditions hold:
1) $(a|b) \wedge (a|c) \Rightarrow (a|(b+c))$
2) $(a|b) \Rightarrow (a|(bc))$
3) $(a|b) \wedge (b|c) \Rightarrow (a|c)$
for all $a, b, c \in R$ and $a \neq 0$.

✳ **Lemma 4.18.**

Let a, b be non-zero elements of an integral domain R. If $a|b$ and $b|a$, then there exists an element $u \in U(R)$ such that $a = ub$.

Proof.

If $a|b$ and $b|a$, then $b = au$ and $a = bv$ for some $u, v \in R$. Hence, we obtain that $a = bv = auv$, i.e., $a(1 - uv) = 0$. Since R is an integral domain and $a \neq 0$, we obtain that $1 - uv = 0$. So $uv = 1$, which means that elements u, v are invertible, i.e., $u, v \in U(R)$. □

✳ **Definition 4.19.**

An associative commutative ring $(K, +, \bullet)$ with identity $1 \neq 0$ in which each non-zero element of K is invertible is called a **field**. In other words, a field is a commutative division ring.

A field which contains only a finite number of elements is called **finite**, otherwise it is called **infinite**. Finite fields have interesting properties and wide applications, which will be considered more carefully in Chapter 9.

✴ Examples 4.20.

1. The sets of rational numbers $\mathbb{Q}$, real numbers $\mathbb{R}$ and complex numbers $\mathbb{C}$ under ordinary arithmetical operations of addition and multiplications are fields.

2. The set of numbers of the form $a + b\sqrt{2}$, where $a, b \in \mathbb{Q}$, under ordinary operations of addition and multiplication is a field.

3. The set $F_2 = \{0, 1\}$ with operations $\oplus$ and $\bullet$ defined by the following tables:

$\oplus$	0	1
0	0	1
1	1	0

and

$\bullet$	0	1
0	0	0
1	0	1

   is a field.

4. Let $\mathbb{Z}_p = \{0, 1, 2 \ldots, p - 1\}$, where $p > 1$ is a prime, with operations $\oplus$ and $\bullet$ defined as follows:

$$a \oplus b = a + b \,(\mathrm{mod}\, p),$$

$$a \bullet b = ab \,(\mathrm{mod}\, p),$$

   where $n \,(\mathrm{mod}\, p)$ denotes the remainder of n on division by p. Then, $\mathbb{Z}_p$ is a commutative ring with identity. Since p is a prime, $\gcd(a, p) = 1$ for each $0 \neq a \in \mathbb{Z}_p$. Therefore, there exist integers $c, d \in \mathbb{Z}$ such that $ac + pd = 1$. Then, $b \equiv c \,(\mathrm{mod}\, p)$ is an inverse to a with respect to multiplication $\bullet$. Therefore, $\mathbb{Z}_p$ is a finite field. In particular, if $p = 2$, we obtain the field F_2 from the previous example.

5. The set of all integers $\mathbb{Z}$ with ordinary operations of addition and multiplication is not a field, because not each element in $\mathbb{Z}$ is invertible with respect to multiplication.

✴ Theorem 4.21.
Let $(K, +, \bullet)$ be a field. Then, there hold the conditions:
1) If $a \bullet b = 0$, then $a = 0$ or $b = 0$.
2) If $a \neq 0$ and $b \neq 0$, then $(a \bullet b)^{-1} = a^{-1} \bullet b^{-1}$
3) If $a \neq 0$, then there exists exactly one element $x \in K$ such that $a \bullet x = b$ for all $a, b \in K$.

Proof.

1) If $a \bullet b = 0$ and $a \neq 0$, then there is an element $a^{-1} \in K$. Then, $b = 1 \bullet b = (a^{-1} \bullet a) \bullet b = a^{-1} \bullet (a \bullet b) = a^{-1} \bullet 0 = 0$.

2) This follows from theorem 2.14.

3) This follows from theorem 4.9. □

✳ Corollary 4.22.

The following conditions are equivalent:

1) $\mathbb{Z}_n$ is a field.

2) $\mathbb{Z}_n$ is an integral domain.

3) an integer n is a prime.

Proof.

$1 \Rightarrow 2$. This follows from theorem 4.21(1).

$2 \Rightarrow 3$. If $n = uv$ then $uv \equiv 0 \,(\mathrm{mod}\, n)$. Hence, $\mathbb{Z}_n$ is not an integral domain.

$3 \Rightarrow 1$. This follows from Example 4.20(4). □

Theorem 4.21 states that each field is an integral domain. Unfortunately, the inverse statement is not true in general. There are a lot of integral domains which are not fields, for example the ring of integers $\mathbb{Z}$. Nevertheless, if an integral domain is finite, then it is a field.

✳ Theorem 4.23.

Each finite integral domain is a field.

Proof.

Let $(R, +, \cdot)$ be a finite integral domain and $0 \neq a \in R$. Consider a mapping $\varphi : R \to aR$, defined as follows: $\varphi(x) = ax$ for each $x \in R$. Then, the mapping φ is an injection because, from the equality $ax = ay$ in the integral domain R, it follows that $x = y$. Since each injection of finite sets is also a bijection and aR is a subset of R, we obtain that $aR = R$ for $a \neq 0$. This means that there is an element $b \in R$ such that $ab = 1$, i.e., the element $a \neq 0$ is invertible in R. This proves that R is a field. □

In a natural way, we can define the notion of a subfield.

✳ Definition 4.24.

A subring F of a field K which itself is a field is called a **subfield** of K.

For example, the field of rational numbers is a subfield of the field of real numbers.

4.3. Ideals and Ring Homomorphisms

One of the most important notions in ring theory is the notion of an ideal.

❋ **Definition 4.25.**
    A non-empty subset I of a ring R is called a **right** (resp. **left**) **ideal** of a ring R if the following conditions hold:

- if $a, b \in I$, then $a + b \in I$,

- if $x \in I$ and $a \in R$, then $xa \in I$ (resp. $ax \in I$).

    A non-empty subset I of a ring R is called a **two-sided ideal**, or simply an **ideal** of a ring R, if I is both a right and left ideal of R. If a ring is commutative, then all left and right ideals are two-sided ideals.

✻ **Examples 4.26.**

1. The set of even integers $2\mathbb{Z}$ is an ideal of the ring $\mathbb{Z}$.

2. The set of polynomials of the form $a_n x^n + a_{n-1} x^{n-1} + \cdots + a_1 x^1$, where $a_i \in \mathbb{R}$ is an ideal of the ring of polynomials $\mathbb{R}[x]$.

❋ **Definition 4.27.**
    Ideals $\{0\}$ and R of a ring R is called **trivial**. All other ideals are called **non-trivial**. An ideal I of a ring R is called **proper** if $I \neq R$. All other ideals are called **non-proper**.

    We write $I \leq R$ if I is an ideal of a ring R. If we want to emphasize that I is a proper ideal of R, we will write $I < R$.

✻ **Lemma 4.28.**
An ideal I of a ring R with the identity 1 is proper if and only if $1 \notin I$.

    Proof.
    Let $I \neq \{0\}$ be an ideal of a ring R with the identity 1. If $1 \in I$, then $a = a \cdot 1 \in I$ for any $a \in R$, by Definition 4.25. Hence, $I \subseteq R \subseteq I$. In this way, $I = R$. □

✻ **Theorem 4.29.**
A commutative ring R with $1 \neq 0$ is a field if and only is it does not contain non-trivial ideals.

    Proof.
    $\Rightarrow$. Let R be a field and $I \leq R$. If $I \neq \{0\}$ then there is a non-zero element $x \in R$. Since R is a field, there is the inverse $x^{-1} \in R$. Then, from

the definition of an ideal, it follows that $xx^{-1} = 1 \in I$, which means that $I = R$, by Lemma 4.28.

$\Leftarrow$. Let R be a commutative ring with $1 \neq 0$ which does not contain non-trivial ideals. For any non-zero element $x \in R$, consider a principal ideal xR generated by x. Since xR is a non-zero ideal, it is equal to R, i.e., $xR = R$. Hence, there is an element $y \in R$ such that $xy = 1$, which means that x is an invertible element. So, R is a field. □

✳ Lemma 4.30.

If I, J are ideals of a ring R, then:

1) $I + J = \{a + b : a \in I, b \in J\}$ is an ideal of R.

2) $IJ = \{\sum_{i,j} a_i b_j : a_i \in I, b_j \in J\}$ is an ideal of R.

3) $I \cap J = \{a : (a \in I) \wedge (a \in J)\}$ is an ideal of R.

4) $IJ \subseteq I \cap J$.

5) $(I \cap J)(I + J) \subseteq IJ + JI$.

6) If R is a commutative ring then $(I \cap J)(I + J) \subseteq IJ$.

Proof.

1. Let $a_1 + b_1, a_2 + b_2 \in I + J$. Since I, J are ideals, $a_1 + a_2 \in I$ and $b_1 + b_2 \in J$. Therefore, $(a_1 + b_1) + (a_2 + b_2) = (a_1 + a_2) + (b_1 + b_2) \in I + J$.

   Let $a \in I$, $b \in J$. Since I, J are ideals, $ax, xa \in I$ and $bx, xb \in J$ for any element $x \in R$. Therefore, $ax + bx = (a + b)x \in I + J$ and $xa + xb = x(a + b) \in I + J$. Hence, $I + J$ is an ideal of R.

2. Let $\sum a_i b_i, \sum c_i d_i \in IJ$ where $a_i, c_i \in I$, $b_i, d_i \in J$. It is obvious that $\sum a_i b_i + \sum c_i d_i \in IJ$. Since I, J are ideals, the elements $(\sum a_i b_i)x = \sum a_i(b_i x) \in IJ$ and $x(\sum a_i b_i) = \sum (x a_i) b_i \in IJ$ for any $x \in R$. Hence, IJ is an ideal of R.

3. Let $a, b \in I \cap J$. Then, $(a + b \in I) \wedge (a + b \in J)$. Hence, $a + b \in I \cap J$. Since $ax \in I$ and $ax \in J$ for each $x \in R$, the element $ax \in I \cap J$. Analogously, $xa \in I \cap J$. Therefore, $I \cap J$ is an ideal of R.

4. Let $\sum a_i b_i \in IJ$, where $a_i \in I$, $b_i \in J$. Since I, J are ideals, $a_i b_i \in I$ and $a_i b_i \in J$. Hence, $a_i b_i \in I \cap J$ for each i. Since, from part 3 of this lemma, $I \cap J$ is an ideal, $\sum a_i b_i \in I \cap J$. Therefore, $IJ \subseteq I \cap J$.

5. Let $x \in (I \cap J)(I + J)$. Then, $x = \sum a_i b_i$, where $a_i \in I \cap J$, $b_i \in I + J$. Let $b_i = c_i + d_i$, where $c_i \in I$ and $d_i \in J$. Then, $x = \sum a_i b_i = \sum a_i(c_i + d_i) = \sum a_i c_i + \sum a_i d_i \in JI + IJ$, since $(a_i \in J) \wedge (c_i \in I)$ and $(a_i \in I) \wedge (d_i \in J)$. Therefore $(I \cap J)(I + J) \subseteq IJ + JI$.

6. If R is a commutative ring, then $IJ = JI$ and, from part 5 of this lemma, we obtain that $(I \cap J)(I + J) \subseteq IJ$. □

Using the mathematical induction from this lemma, we can obtain the following corollary.

> ✳ **Corollary 4.31.**
> If $I_1, I_2, \ldots, I_n$ are ideals of a ring R, then the following sets are also ideals of R:
> 1) $I_1 + I_2 + \cdots + I_n = \{a_1 + a_2 + \cdots + a_n \ : \ a_k \in I_k, \ \forall(k \in \{1, 2, \ldots, n\})\}$;
> 2) $I_1 I_2 \cdots I_n = \{\sum a_1 a_2 \cdots a_n \ : \ a_k \in I_k, \ \forall(k \in \{1, 2, \ldots, n\})\}$;
> 3) $I_1 \cap I_2 \cap \cdots \cap I_n = \{a \ : \ a \in I_k, \ \forall(k \in \{1, 2, \ldots, n\})\}$.

Analogously, we can also define the intersection of an infinite number of ideals of a ring R:

$$I = \bigcap_{k \in \mathbb{N}} I_k = \{a \ : \ a \in I_k, \ k \in \mathbb{N}\}$$

and we can show that I is an ideal of R.

However, we need to take into account that, in general, the theoretically-set sum of ideals of a ring R:

$$I = \bigcup_{k \in \mathbb{N}} I_k = \{a \ : \ \exists(k \in \mathbb{N})\, [a \in I_k]\}$$

is not an ideal of R. Nevertheless, I is an ideal in the following particular case:

> ✳ **Theorem 4.32.**
> If $I_1, I_2, \ldots, I_n$ are proper ideals of a ring R and $I_n \subseteq I_{n+1}$ for each $n \in \mathbb{N}$ then $I = \bigcup_{k \in \mathbb{N}} I_k$ is a proper ideal of R.

Proof.
Let $x \in I$. Then, from the definition it follows that there is an index $n \in \mathbb{N}$ such that $x \in I_n$. At the time for each $a \in R$, we have that $ax \in I_n$ and $xa \in I_n$. Hence, $ax \in I$ and $xa \in I$. If $y \in I$, then there is an index $m \in \mathbb{N}$ such that $y \in I_m$. If $k = \max\{n, m\}$, then $I_n \subseteq I_k$ and $I_m \subseteq I_k$. Hence, $x + y \in I_k$ and so $x + y \in I$. Therefore, I is an ideal of R. If $I = R$, then there is an index $n \in \mathbb{N}$ such that $1 \in I_n$. In this way, I_n is not a proper ideal, by Lemma 4.28. This contradiction proves that I is a proper ideal of R. □

Similarly as for groups, we can introduce the notions of a homomorphism and an isomorphism of rings.

✳ **Definition 4.33.**

A mapping φ of a ring A to a ring B is called a **ring homomorphism** (or simply **homomorphism**) if it satisfies the following condition:

- $\varphi(a+b) = \varphi(a) + \varphi(b)$

- $\varphi(ab) = \varphi(a)\varphi(b)$,

for all $a, b \in A$.

Similar to groups, we can show that

1. $\varphi(1_A) = 1_B$, where 1_A and 1_B are the identities of rings A and B, correspondingly;

2. $\varphi(a^{-1}) = [\varphi(a)]^{-1}$ for $a \in A$.

We denote:

$$\mathrm{Ker}(\varphi) = \{x \in A \ : \ \varphi(x) = 0\},$$

$$\mathrm{Im}(\varphi) = \{y \in B \ : \ \exists (x \in A)[\varphi(x) = y]\}$$

which are called, correspondingly, the **kernel** and the **image** of a homomorphism $\varphi : A \to B$.

✳ **Example 4.34.**

The mapping $\varphi : \mathbb{Z} \to \mathbb{Z}/m\mathbb{Z}$, defined as follows: $\varphi(a) = [a]$, is a homomorphism of rings, which follows immediately from the definition of addition and multiplication in these rings.

✳ **Theorem 4.35.**

If $\varphi : A \to B$ is a homomorphism of rings A and B, then $\mathrm{Ker}(\varphi)$ is an ideal of A and $\mathrm{Im}(\varphi)$ is a subring of B.

Proof.

Since φ is also a homomorphism of additive groups, by Theorem 2.65, $\mathrm{Ker}(\varphi)$ is a subgroup of an additive group $(A, +)$. If $x \in \mathrm{Ker}(\varphi)$ and $a \in A$, then $\varphi(xa) = \varphi(x)\varphi(a) = 0$, therefore, $xa \in \mathrm{Ker}(\varphi)$. Similarly, $ax \in \mathrm{Ker}(\varphi)$, which proves that $\mathrm{Ker}(\varphi)$ is an ideal of A.

Analogously, by Theorem 2.67, $\mathrm{Im}(\varphi)$ is a subgroup of an additive group $(B, +)$. If $b_1, b_2 \in \mathrm{Im}(\varphi)$, then there are elements $a_1, a_2 \in A$ such that $\varphi(a_1) = b_1$ and $\varphi(a_2) = b_2$. Since φ is a homomorphism of rings, $b_1 b_2 = \varphi(a_1)\varphi(a_2) = \varphi(a_1 a_2) \in \mathrm{Im}(\varphi)$, which proves that $\mathrm{Im}(\varphi)$ is a subring of B. $\square$

* **Definition 4.36.**
 A homomorphism φ of a ring A to a ring B is called a **monomorphism** if φ is an injection, i.e., $\varphi(x) = \varphi(y)$ implies that $x = y$. A homomorphism φ of a ring A to a ring B is called an **epimorphism** if φ is a surjection, i.e., for each $b \in B$ there exists $a \in A$ such that $\varphi(a) = b$.

* **Definition 4.37.**
 A ring homomorphism $\varphi : A \to B$ is called an **isomorphism** if φ is a bijection, i.e., φ is both a monomorphism and an epimorphism. If there is an isomorphism $\varphi : A \to B$ of rings A and B, then A and B are called **isomorphic rings**.

* **Theorem 4.38.**
A ring homomorphism $\varphi : A \to B$ is an isomorphism if and only if $\mathrm{Ker}(\varphi) = \{0\}$ and $\mathrm{Im}(\varphi) = B$.

 Proof.
 Since a ring homomorphism is also a group homomorphism, the proof follows from Theorem 2.71. □

4.4. Quotient Rings

Since each ideal I of a ring R is also a normal subgroup of an additive group $(R, +)$, by Theorem 2.53, the factor set R/I forms an additive group under operation $(x + I) \oplus (y + I) = (x + y) + I$. In addition, we can define another operation on the set R/I as follows: $(x + I) \otimes (y + I) = (xy) + I$, since R is a ring. Then, the set R/I forms a ring under these operations, which the following theorem shows.

* **Theorem 4.39.**
If I is an ideal of a ring R, then the set $R/I = \{x + I \ : \ x \in R\}$ under operations defined as follows:

$$(x + I) \oplus (y + I) = (x + y) + I \tag{4.5}$$

$$(x + I) \otimes (y + I) = (xy) + I \tag{4.6}$$

forms a ring.

 Proof.
 Since, by Theorem 2.53, a factor set R/I forms a commutative additive group under operation (4.5), it suffices to show that the operation defined by (4.6) is well defined and axioms of associativity for operation (4.6) and distributivity with respect to operation (4.5) hold.

1) First, we will show that the operation defined by (4.6) is well defined. If $x + I = x_1 + I$ and $y + I = y_1 + I$, then $x - x_1 \in I$ and $y - y_1 \in I$. Therefore, $xy - x_1y_1 = xy - xy_1 + xy_1 - x_1y_1 = x(y - y_1) + (x - x_1)y_1 \in xI + Iy_1 \subseteq I$. Hence, $xy + I = x_1y_1 + I$.

2) Since $(xy)z = x(yz)$ in R, $((x + I) \otimes (y + I)) \otimes (z + I) = (x + I) \otimes ((y + I)) \otimes (z + I))$.

3) Since $x(y + z) = xy + xz$ in R, $(x + I) \otimes ((y + I) \oplus (z + I)) = (xy + I) \oplus (xz + I))$. The zero element (the neutral element with respect to addition $\oplus$) of the ring R/I is the element $0 + I = I$, and the identity element of R/I (the neutral element with respect to multiplication $\otimes$) is the element $1 + I$. $\quad \square$

> ✳ **Definition 4.40.**
> A ring R/I, defined by Theorem 4.39, is called a **quotient ring** (or **factor ring**).

✳ **Example 4.41.**

Let $m > 0$ be a positive integer. Then $m\mathbb{Z}$ is an ideal of the ring $\mathbb{Z}$. The quotient ring $\mathbb{Z}/m\mathbb{Z}$ consists of elements of the form $k + m\mathbb{Z}$, i.e., residue classes modulo m: $[0], [1], \ldots, [m - 1]$ with operations (4.3) and (4.5) defined in Example 4.4(5).

✴ **Theorem 4.42.**

If I is an ideal of a ring R, then a mapping $\varphi : R \to R/I$ defined as follows:

$$\varphi(x) = x + I$$

is a ring homomorphism. Moreover, $\mathrm{Ker}(\varphi) = I$ and $\mathrm{Im}(\varphi) = R/I$, i.e., φ is an epimorphism.

Proof.

Let $x, y \in R$ and $\varphi(x) = x + I$, $\varphi(y) = y + I$. Then, from the definition of operations (4.5), (4.6) in the quotient ring R/I, we have:

$$\varphi(x) \oplus \varphi(y) = (x + I) \oplus (y + I) = (x + y) + I = \varphi(x + y)$$

$$\varphi(x) \otimes \varphi(y) = (x + I) \otimes (y + I) = (xy) + I = \varphi(xy)$$

which shows that φ is a ring homomorphism.

Since $\varphi(x) = x + I = I$ if and only if $x \in I$ and the zero element of R/I is the ideal I, we obtain that $\mathrm{Ker}(\varphi) = I$.

Since $\varphi(x) = x + I$ for any element $x + I \in R/I$, we immediately obtain that $\mathrm{Im}(\varphi) = R/I$. Therefore, φ is a surjection, implying that φ is an epimorphism. $\quad \square$

> **❋ Theorem 4.43.**
> If $\varphi : A \to B$ is a homomorphism of rings A and B, then
> $$A/\text{Ker}(\varphi) \cong \text{Im}(\varphi). \tag{4.7}$$

Proof.
Consider the mapping $f : A/\text{Ker}(\varphi) \to \text{Im}(\varphi)$, defined by $f(x+\text{Ker}(\varphi)) = \varphi(x)$. This shows that f is surjective. We denote $\text{Ker}(\varphi) = I$. Then

$$f((x_1 + I) + (x_2 + I)) = f(x_1 + x_2 + I) = \varphi(x_1 + x_2) =$$
$$= \varphi(x_1) + \varphi(x_2) = f(x_1 + I) + f(x_2 + I)$$
$$f((x_1+I)(x_2+I)) = f(x_1 x_2 + I) = \varphi(x_1 x_2) = \varphi(x_1)\varphi(x_2) = f(x_1+I)f(x_2+I)$$

Hence, f is a ring homomorphism, which is obviously surjective, i.e., f is an epimorphism.

Suppose $f(x+\text{Ker}(\varphi)) = 0$, then $\varphi(x) = 0$, i.e., $x \in \text{Ker}(\varphi)$, which means that $x + \text{Ker}(\varphi) = \text{Ker}(\varphi)$. Therefore, f is a monomorphism. In this way, f is an isomorphism of rings. □

❋ Examples 4.44.
We show that two rings $\mathbb{Z}/n\mathbb{Z} = \{[0], [1], \ldots, [n-1]\}$ and $\mathbb{Z}_n = \{0, 1, \ldots, n-1\}$ are isomorphic.

Consider the mapping $\varphi : \mathbb{Z} \to \mathbb{Z}/n\mathbb{Z}$ given as follows: $\varphi(a) = r_n(a)$ for all $a \in \mathbb{Z}$. Then

$$\varphi(a + b) = r_n(a + b) = r_n(a) + r_n(b) = \varphi(a) + \varphi(b)$$
$$\varphi(ab) = r_n(ab) = r_n(a)r_n(b) = \varphi(a)\varphi(b)$$

which shows that φ is a ring homomorphism.

It is obvious that $\text{Im}(\varphi) = \mathbb{Z}_n$. We will now find the kernel of φ. If $a \in \text{Ker}(\varphi)$, then $r_n(a) = 0$, i.e., $n|a$, which means that $a \in n\mathbb{Z}$. Hence, $\text{Ker}(\varphi) = n\mathbb{Z}$. Then, from Theorem 4.43, it follows that $\mathbb{Z}/n\mathbb{Z} \simeq \mathbb{Z}_n$.

4.5. Maximal Ideals. Prime Ideals

In this section, we study the properties of two special classes of ideals.

> **✳ Definition 4.45.**
> An ideal I of a commutative ring R is called **principal** if it is generated by one element. This means that there is an element $a \in I$ such that each element $b \in I$ can be written in the form $b = ax$, where $x \in R$. In this case, we write this ideal as $I = \langle a \rangle$.

⁎ **Definition 4.46.**

A proper ideal M of a commutative ring R with $1 \neq 0$ is called **maximal** if, for each ideal $I \subseteq R$, the ascending chain of ideals $M \subseteq I \subseteq R$ implies that either $I = M$ or $I = R$.

⁕ **Theorem 4.47.**

An ideal I of a commutative ring R with $1 \neq 0$ is maximal if and only if a quotient ring R/I is a field.

Proof.

$\Rightarrow$. Let I be a maximal ideal of a commutative ring R. Consider a quotient ring R/I. Let $I \neq x + I \in R/I$, i.e., $x \notin I$. If $J = \langle x \rangle + I$, then $I \subset J \subseteq R$. Hence, $J = R$, since I is a maximal ideal. Therefore, $1 \in J = R$, i.e., there are elements $y \in R$ and $m \in I$ such that $1 = xy + m$. Since $(x + I)(y + I) = xy + I = (1 - m) + I = 1 + I$, the element $x + I$ is invertible in R/I, i.e., R/I is a field.

$\Leftarrow$. Let R/I be a field. Then, I is a proper ideal, otherwise R/I has only one element $0 = 1$.

Let J be an ideal of R such that $I \subseteq J \subseteq R$ and suppose that $J \neq R$. We will show that, in this case, $I = J$. Assume that $J \neq I$. Then, there is an element $x \in J$ and $x \notin I$. Since R/I is a field, the element $x + I \neq I$ is invertible in R/I, which means that there is an element $y + I$ such that $(x + I)(y + I) = 1 + I$. Therefore, there is an element $m \in I$ such that $1 = xy + m$. Therefore $1 \in J$, since $xy \in J$ and $m \in J$. Then, from Lemma 4.28 it follows that $J = R$, which contradicts the assumption. □

As a corollary of Theorems 4.29 and 4.47, we obtain the following theorem:

⁕ **Theorem 4.48.**

Let R be a commutative ring with $1 \neq 0$. Then, the following conditions are equivalent:
a) R is a field,
b) the ideal $\{0\}$ is maximal,
c) R has only trivial ideals: $\{0\}$ and R.

⁎ **Definition 4.49.**

A proper ideal I of a commutative ring R is called **prime** if

$$ab \in I \;\Rightarrow\; (a \in R) \vee (b \in R), \quad \forall(a, b \in R).$$

⁕ **Theorem 4.50.**

An ideal I of a commutative ring R is prime if and only if a quotient ring R/I is an integral domain.

Proof.

⇒. Let I be a prime ideal of a commutative ring R and $(a + I)(b + I) = ab + I = I$. Then, $ab \in I$. Hence, by Definition 4.49, $(a \in I) \vee (b \in I)$. This means that $(a + I = I) \vee (b + I = I)$. Therefore, R/I has no zero divisors, so it is an integral domain.

⇐. Assume that a quotient ring R/I is an integral domain and $ab \in I$. Then, $I = ab + I = (a + I)(b + I)$. Hence, $(a + I = I) \vee (b + I = I)$, which means that $(a \in I) \vee (b \in I)$, i.e., the ideal I of R is prime. □

As a corollary from Theorem 4.50, we obtain the following theorem.

❋ Theorem 4.51.

Each maximal ideal I of a commutative ring R is prime.

Note that the inverse statement is not true in general. There are prime ideals of commutative rings which are not maximal. Below, we show some such examples.

❋ Examples 4.52.

1. Consider the trivial ideal $I = \{0\}$ of the ring of integers $\mathbb{Z}$. Since $\mathbb{Z}/I \cong \mathbb{Z}$ is an integral domain but not a field, I is a prime ideal but not a maximal ideal.

2. Consider the ring of polynomials $\mathbb{Z}[x]$ over the ring of integers $\mathbb{Z}$ and the ideal $I = \langle x \rangle$. Since $\mathbb{Z}[x]/I \cong \mathbb{Z}$ is an integral domain but not a field, I is a prime ideal but not a maximal ideal.

❋ Lemma 4.53.

If R is an integral domain, then $a|b$ if and only if $\langle b \rangle \subseteq \langle a \rangle$.

Proof.

⇒. If $a|b$, then $b = ax$ for some $x \in R$. Hence, $by = axy \in \langle a \rangle$ for all $y \in R$. This means that $\langle b \rangle \subseteq \langle a \rangle$.

⇐. Suppose that $\langle b \rangle \subseteq \langle a \rangle$. Then, there exists an element $x \in R$ such that $b = ax$, which means that $a|b$. □

We say that a non-zero element p of a ring R is called irreducible if $p \notin U(R)$ and p cannot be written as a finite product of non-invertible elements of R.

❋ Lemma 4.54.

An element p is irreducible in an integral domain R if and only if the ideal $\langle p \rangle$ is maximal among all proper principal ideals of R.

Proof.

⇒. Suppose that R is an irreducible element in an integral domain R and $I = \langle p \rangle$ is not a maximal ideal among all proper principal ideals of R. Then, there is an element $a \in R$ such that $I \subseteq \langle a \rangle \subset R$. Then, from Lemma 4.53, it follows that $a|p$, i.e., there exists an element $b \in R$ such that $p = ab$. Since p is an irreducible element in R, there is one of two possibilities:

    1) $a \in U(R)$, then $\langle a \rangle = R$, that is not the case;

    2) $b \in U(R)$, then $a \in I$ and so $I = \langle a \rangle$.

    Therefore, $I = \langle p \rangle$ is a maximal ideal among all proper principal ideals of R.

    $\Leftarrow$. Let $I = \langle p \rangle$ be a maximal ideal among all proper principal ideals of R. Suppose that p is decomposable, then there are some elements $a, b \notin U(R)$ such that $p = ab$. By Lemma 4.53, $I \subset \langle a \rangle \subset R$, i.e., I is not maximal among all proper principal ideals of R. This contradiction shows that p is an irreducible element in R □

✻ **Definition 4.55.**

    A non-zero element p of a ring R is called **prime** if p is not invertible in R and $p|ab$ implies that either $p|a$ or $p|b$.

✻ **Lemma 4.56.**

1. Each prime element of an integral domain R is irreducible.

2. A non-zero element p of an integral domain R is prime if and only if a principal ideal $\langle p \rangle$ is prime.

Proof.

    1. Let a be a prime element in an integral domain R. If $a = bc$, then $a|b$ or $a|c$. Assume that $a|b$, then $b = ad = bcd$ for some $d \in R$. Hence, $0 = b - bcd = b(1 - cd)$. Since R is an integral domain, $1 - cd = 0$, which means that $c \in U(R)$.

    2. Let $p \in R$ be a prime element in an integral domain R, then, by the previous part of the lemma, the element p is irreducible. Suppose $ab \in \langle p \rangle$ for some $a, b \in R$, then, there is an element $c \in R$ such that $ab = pc$. Therefore $p|ab$ and so $(p|a) \vee (p|b)$, since p is a prime element. By Lemma 4.53, this means that $(\langle a \rangle \subseteq \langle p \rangle) \vee (\langle b \rangle \subseteq \langle p \rangle)$. Therefore, $(a \in \langle p \rangle) \vee (b \in \langle p \rangle)$. In this way, $\langle p \rangle$ is a prime ideal, by Definition 4.49.

    Inversely, let $I = \langle p \rangle$ be a prime ideal and $p \neq 0$. Suppose that $p = cd$ and $c, d \notin U(R)$. Therefore, $c + \langle p \rangle \neq \langle p \rangle$ and $d + \langle p \rangle \neq \langle p \rangle$, but $cd + \langle p \rangle = \langle p \rangle$, which contradicts Theorem 4.50. So, p is an irreducible element in R.

    If $p|ab$ for $a, b \in R$, then, from Lemma 4.53, it follows that $\langle ab \rangle \subseteq \langle p \rangle$, i.e., $ab \in \langle p \rangle$. Then, by Definition 4.54, $(a \in \langle p \rangle) \vee (b \in \langle p \rangle)$, which means that $(a = px) \vee (b = py)$ for some $x, y \in R$, i.e., $(p|a) \vee (p|b)$. Therefore, p is a prime element.   □

4.6. Principal Ideal Rings

In this section, we consider one of the most interesting classes of rings: Principal ideal rings.

✳ **Definition 4.57.**

A commutative ring R is called a **principal ideal ring** if each of its ideal I is principal, i.e., $I = \langle a \rangle$ is an ideal generated by one element $a \in R$. An integral domain R which is a principal ideal ring is called a **principal ideal domain**.

❇ **Theorem 4.58.**
The ring of integers $\mathbb{Z}$ is a principal ideal domain whose each ideal has the form $I = n\mathbb{Z}$.

Proof.

Let I be an ideal of the ring of integers $\mathbb{Z}$. Then, by the Well-Ordered Principal I as a subset of $\mathbb{Z}$ has a minimal positive element $n \neq 0$. Let m be an arbitrary element of I. If $m = 0$, then $m = n \cdot 0 \in I$. If $m \neq 0$, then there are numbers $q, r \in \mathbb{Z}$ such that $m = nq + r$, where $0 \leq r < n$. Hence, $r = n - mq \in I$. Since n is the least positive number of I, we obtain that $r = 0$, i.e., $m = nq$. So $I = n\mathbb{Z}$. □

❇ **Theorem 4.59.**
If R is a principal ideal domain, then an element $p \in R$ is irreducible if and only if $\langle p \rangle$ is a maximal ideal.

Proof.

Let R be an irreducible element of a principal ideal domain R. Then, $\langle p \rangle$ is a maximal ideal, by Lemma 4.54, since all ideal of R are principal.

Inversely, if $\langle p \rangle$ is a maximal ideal of R, then $\langle p \rangle$ is a prime ideal, by Theorem 4.51. Therefore, by Lemma 4.56(2), the element p is prime and, therefore, irreducible. □

❇ **Lemma 4.60.**
An element of a principal ideal domain is irreducible if and only if it is prime.

Proof.

⇒. Let p be an irreducible element of a principal ideal domain R. Then, $I = \langle p \rangle$ is a maximal ideal, by Theorem 4.59. Moreover, by Theorem 4.51, I is a prime ideal. So, p is a prime element, by Lemma 4.56(2).

⇐. This follows from Lemma 4.56(1). □

❇ **Theorem 4.61.**
Let R be a principal ideal ring. Then, the family $\{I_n \ : \ n \in \mathbb{N}\}$ of ideals of R with property $I_n \subseteq I_{n+1}$ for any $n \in \mathbb{N}$ has only a finite number of ideals, i.e., there is $k \in \mathbb{N}$ such that $I_n = I_k$ for each $n \geq k$.

Proof.

By Theorem 4.32, $I = \bigcup\limits_{k \in \mathbb{N}} I_k$ is an ideal of a ring R. Since R is a principal ideal ring, the ideal I is principal, i.e., there is an element $a \in I$ such that $I = \langle a \rangle$. Then, from the definition of I, it follows that there is a number $k \in \mathbb{N}$ such that $a \in I_k$. We will show that $I_n = I_k$ for all $n \geq k$. Suppose the opposite. Then, there is a number $n > k$ such that $I_k \subset I_n$ and $I_k \neq I_n$. Therefore there is an element $x \in I_n$ such that $x \notin I_k$. Since $x \in I = \langle a \rangle$, there is an element $b \in R$ such that $x = ab$. As $a \in I_k$ and I_k is an ideal, $ab = x \in I_k$. The obtained contradiction shows that $I_n = I_k$ for all $n \geq k$. $\square$

✳ Definition 4.62.

Let R be a commutative ring and $a, b \in R$ which not both are equal to zero. An element $d \in R$ is called the **greatest common divisor** of elements a, b and denoted by $\gcd(a, b)$ if it satisfies the following conditions:

- $d|a$ and $d|b$,

- if $h \in R$ and $(h|a) \wedge (h|b)$, then $h|d$.

By assumption we assume that $\gcd(0, 0) = 0$.

From this definition, it follows, in particular, that $\gcd(a, 0) = a$.

Note that $gdc(a, b)$ (if it exists) is defined uniquely up to factors which are invertible elements in a ring R.

The notion of the greatest common divisors can be generalized for the case of n elements of a ring.

✳ Definition 4.63.

An element $d \in R$ is called the **greatest common divisor** of elements $a_1, a_2, \ldots, a_n \in R$, which are not all equal to zero, and denoted by $\gcd(a_1, a_2, \ldots, a_n)$ if

- $d|a_i$ for all $i = 1, 2, \ldots, n$,

- if $h \in R$ and $h|a_i$ for all $i = 1, 2, \ldots, n$ then $h|d$.

✳ Theorem 4.64.

If R is a principal ideal domain, then, for any $a_1, a_2, \ldots, a_n \in R$, there exists their greatest common divisor $d = \gcd(a_1, a_2, \ldots, a_n)$. Moreover, there exist elements $x_1, x_2, \ldots, x_n \in R$ such that

$$d = x_1 a_1 + x_2 a_2 + \cdots + x_n a_n \qquad (4.8)$$

Proof.

Let $I = \langle a_1, a_2, \ldots, a_n \rangle$ be an ideal generated by elements $a_1, a_2, \ldots, a_n$. Since R is a principal ideal domain, there exists an element $d \in R$ such that $I = \langle d \rangle$. Therefore, there exist elements $x_1, x_2, \ldots, x_n \in R$ such that $d = x_1 a_1 + x_2 a_2 + \cdots + x_n a_n$. Since $a_i \in I = \langle d \rangle$, there exist exist elements $y_i \in R$ such that $a_i = d y_i$, i.e., $d | a_i$ for all $i = 1, 2, \ldots, n$. Suppose that $h \in R$ and $h | a_i$ for all $i = 1, 2, \ldots, n$. From equality $d = x_1 a_1 + x_2 a_2 + \cdots + x_n a_n$, it follows that $h | d$. Thus, d is the greatest common divisor of elements $a_1, a_2, \ldots, a_n$. $\square$

✳ **Definition 4.65.**

Elements $a_1, a_2, \ldots, a_n \in R$ are called **relatively prime** if $\gcd(a_1, a_2, \ldots, a_n) = 1$.

As a corollary of Theorem 4.64, we immediately obtain the following theorem:

✳ **Theorem 4.66.**

Let R be a principal ideal domain. Then, elements $a_1, a_2, \ldots, a_n \in R$ are relatively prime if and only if there exist elements $x_1, x_2, \ldots, x_n \in R$ such that

$$1 = x_1 a_1 + x_2 a_2 + \cdots + x_n a_n \tag{4.9}$$

4.7. Euclidean Domains. Euclidean Algorithm

In this section, we consider a special class of commutative rings possessing the Euclidean Algorithm.

✳ **Definition 4.67.**

An integral domain R is called a **Euclidean domain** if there is a map $\delta : R \backslash \{0\} \to \mathbb{Z}^+ \cup \{0\}$ satisfying the following conditions:

- $\delta(a) \leq \delta(ab)$ for all non-zero $a, b \in R$;

- for any $a, b \in R$, $b \neq 0$, there exist elements $q, r \in R$ such that $a = qb + r$ and either $r = 0$ or $\delta(r) \leq \delta(b)$.

✳ **Examples 4.68.**

1. The ring of integers $\mathbb{Z}$ is a Euclidean domain if we take $\delta(b) = |b|$ for all $b \in \mathbb{Z}$.

2. Each field F is a Euclidean domain if we take $\delta(b) = 1$ for all non-zero $b \in F$.

3. The ring $\mathbb{R}[x]$ of polynomials in one variable over the field of real numbers is a Euclidean domain if we take $\delta(f) = \deg(f)$ for all non-zero
   $f \in \mathbb{R}[x]$.

✳ Theorem 4.69.

Each Euclidean domain R is a principal ideal domain.

Proof.

Let I be an ideal of a Euclidean domain R. We choose an element $b \in I$
such that $\delta(b) = \min\{\delta(x) : x \in I\backslash\{0\}\}$. Then, for any non-zero element
$a \in I$ we have: $a = qb + r$, where either $r = 0$ or $\delta(r) < \delta(b)$. Hence,
$r = a - qb \in I$. From the choice of the element b, it follows that $r = 0$.
Therefore, $a = qb$, and so $I = \langle b \rangle$.   □

✳ Example 4.70.

The ring of polynomials $\mathbb{R}[x]$ over the field of real numbers is a Euclidean
domain with function $\delta(f) = \deg(f)$ for each $f \in \mathbb{R}[x]$. Therefore, by Theorem
4.64, $\mathbb{R}[x]$ is a principal ideal domain.

Taking into account Theorems 4.64, 4.66, and 4.69, we immediately obtain
the following results for Euclidean domains.

✳ Theorem 4.71.

If R is a Euclidean domain, then, for any $a_1, a_2, \ldots, a_n \in R$, there exists
their greatest common divisor $d = \gcd(a_1, a_2, \ldots, a_n)$. Moreover, there
exist elements $x_1, x_2, \ldots, x_n \in R$ such that

$$d = x_1 a_1 + x_2 a_2 + \cdots + x_n a_n \qquad (4.10)$$

✳ Theorem 4.72.

Let R be a Euclidean domain. Then, elements $a_1, a_2, \ldots, a_n \in R$ are
relatively prime if and only if there exist elements $x_1, x_2, \ldots, x_n \in R$ such
that

$$1 = x_1 a_1 + x_2 a_2 + \cdots + x_n a_n \qquad (4.11)$$

In order to find the greatest common divisor of elements of a Euclidean
domain, we can use the following Euclidean algorithm, which can be considered as a generalization of the Euclidean algorithm for integers.

Euclidean Algorithm

Let R be a Euclidean domain and $a, b \in R$, $b \neq 0$. Suppose that $\delta(b) \leq$
$\delta(a)$. If $b|a$, then $\gcd(a, b) = b$. Otherwise, we denote $r_0 = a$, $r_1 = b$. By
Definition 4.67, there exist elements $q_1, r_2 \in R$ such that $r_0 = q_1 r_1 + r_2$ and
either $r_2 = 0$ or $\delta(r_2) < \delta(r_1)$. Since b does not divide a, $r_2 \neq 0$, and we can
repeat this process:

$$
\begin{aligned}
r_0 &= q_1 r_1 + r_2, & r_2 &\neq 0, \quad \delta(r_2) < \delta(r_1) \\
r_1 &= q_2 r_2 + r_3, & r_3 &\neq 0, \quad \delta(r_3) < \delta(r_2) \\
\vdots \;\; \vdots \;\; \vdots & & \vdots & \\
r_{k-2} &= q_{k-1} r_{k-1} + r_k, & r_k &\neq 0, \quad \delta(r_k) < \delta(r_{k-1}) \\
r_{k-1} &= q_k r_k + 0, & r_{k+1} &= 0
\end{aligned} \tag{4.12}
$$

Then, analogously as in the case of Euclidean algorithm for integers, we can show that $r_k = \gcd(a, b)$. This element always exists because we have a decreasing bounded sequence of natural numbers:

$$
0 < \delta(r_k) < \delta(r_{k-1}) < \cdots < \delta(r_2) < \delta(r_1) = \delta(b) < \delta(a)
$$

which contains only a finite number of elements.

In order to find elements $s, t \in R$ such that $\gcd(a, b) = sa + tb$, we rewrite equalities (4.12) in the following form

$$
\begin{aligned}
r_2 &= r_0 - r_1 q_1 \\
r_3 &= r_1 - r_2 q_2 \\
r_4 &= r_2 - r_3 q_3 \\
\vdots \;\; \vdots \;\; \vdots & \\
r_{i+1} &= r_{i-1} - r_i q_i \\
\vdots \;\; \vdots \;\; \vdots & \\
r_{k-1} &= r_{k-3} - r_{k-2} q_{k-2} \\
r_k &= r_{k-2} - r_{k-1} q_{k-1}
\end{aligned} \tag{4.13}
$$

Using these equalities, we begin from the last equality $r_k = r_{k-2} - q_{k-1} r_{k-1}$ and, moving up step by step, successively substituting the equality above in it, we obtain the required elements $s, t \in R$ such that $\gcd(a, b) = sa + tb$.

❋ Definition 4.73.

Let R be an integral domain. An element $a \in R$ is called **decomposable** in R if a can be written as a product $a = xy$, where elements $x, y \in R$ are not invertible in R.

A non-zero element $a \in R$ is called **irreducible** (or **indecomposable**) if a is not invertible in R and a cannot be written as a product of non-invertible factors.

✳ Example 4.74.

The only irreducible numbers in the ring $\mathbb{Z}$ are up to the sign the prime numbers $2, 3, 5, 7, 11, \ldots$.

> ❋ **Theorem 4.75.**
> Let R be a Euclidean domain and $a \in R$. Then, a quotient ring R/I, where $I = \langle a \rangle$, is a field if and only if a is an irreducible element in R.

Proof.

$\Rightarrow$. Let a be an irreducible element in a Euclidean domain R, $I = \langle a \rangle$ and $b + I \in R/I$, where $b \notin \langle a \rangle$. Since a is an irreducible element, $\gcd(a, b) = 1$. Then, from Theorem 4.66, it follows that there are elements $x, y \in R$ such that $ax + by = 1$. Hence, $by = 1 - ax$, i.e., $b + I$ is invertible in the ring R/I.

$\Leftarrow$. Suppose that a is a decomposable element in R, i.e., there are non-invertible elements $s, t \in R$ such that $a = st$. In this case, $\delta(s) < \delta(st) = \delta(a)$ and $\delta(t) < \delta(st) = \delta(a)$. Therefore, elements s, t do not belong to $I = \langle a \rangle$. Therefore, $s + I$ and $t + I$ are non-zero elements in R/I. However, $(s + I)(t + I) = st + I = I$ is the zero element in R/I, i.e., R/I is not a field. $\square$

4.8. Unique Factorization Domains

✳ **Definition 4.76.**

Elements x, y of an integral domain R are called **associated elements**, or simply **associates**, if there exists an invertible element $\varepsilon \in U(R)$ such that $a = \varepsilon b$. Associated elements x, y will be written as $x \sim y$.

✳ **Definition 4.77.**

We say that a non-zero element $a \in R$ has a **unique factorization** in a ring R if from two factorizations:

$$a = u p_1 p_2 \cdots p_n = v q_1 q_2 \cdots q_m$$

where p_i and q_j are irreducible elements and $u, v \in U(R)$, it follows that $n = m$ and there is a permutation σ such that $p_i \sim q_{\sigma(i)}$ for all i.

✳ **Definition 4.78.**

An integral domain R is called a **unique factorization domain** if the following conditions hold:

1) each non-zero element $a \in R$ can be written in the form of a finite product of irreducible elements p_i with unit $u \in U(R)$:

$$a = u p_1 p_2 \cdots p_n$$

2) the above factorization is unique in R.

❋ Example 4.79.

The ring $\mathbb{Z}[\sqrt{-3}]$ is not a unique factorization domain because

$$4 = 2 \cdot 2 = (1 + \sqrt{-3}) \cdot (1 - \sqrt{-3})$$

and elements 2 and $(1 + \sqrt{-3})$ are not associates.

❋ Theorem 4.80.
If R is a principal ideal domain, then each element $0 \neq a \in R$ can be written as a finite product of irreducible elements.

Proof.

Suppose that there is an element $0 \neq a \in R$ which cannot be written as a finite product of irreducible elements. Then, there are non-invertible elements $a_1, b_1 \in R$ such that $a = a_1 b_1$. Moreover, at least one of these elements cannot be also written as a finite product of irreducible elements. Without loss of generality, we can assume that such an element is a_1. Since $a = a_1 b_1$ and b_1 is not invertible, from Lemma 4.53, it follows that $\langle a \rangle \subset \langle a_1 \rangle$. Since a_1 is not irreducible, there are non-invertible elements $a_2, b_2 \in R$ such that $a = a_2 b_2$. Again, at least one of these elements also cannot be written as a finite product of irreducible elements. Without loss of generality, we can assume that such an element is a_2. Since b_2 is not-invertible, we again obtain that $\langle a_1 \rangle \subset \langle a_2 \rangle$. Continuing this process, we obtain the sequence of non-invertible elements $a_0 = a, a_1, a_2, \ldots$ such that $a_{n+1} | a_n$ for $n = 0, 1, 2 \ldots$. Hence, we obtain the infinite family of principal ideals $\{\langle a_i \rangle \ : \ i \in \mathbb{N}\}$ which satisfy the condition $\langle a_n \rangle \subset \langle a_{n+1} \rangle$ for all $n \in \mathbb{N}$, which contradicts Theorem 4.61. $\qquad\square$

❋ Theorem 4.81.
Let R be an integral domain. Suppose that each element $0 \neq a \in R$ can be written as a finite product of irreducible elements. Then, the following conditions are equivalent:
1. Each irreducible element in R is prime.
2. R is a unique factorization domain.

Proof.

$1 \Rightarrow 2$. Suppose that each irreducible element in an integral domain R is prime and each non-zero element of R can be written as a finite product of irreducible elements. We use the mathematical induction on the number n of factors in factorization of elements.

For $n = 1$, the statement is trivial. Let $n > 1$. We assume that, for $k < n$, the statement is true, i.e., each element which can be written as a product of k irreducible factors for $k < n$ has unique factorization. Let $a = u p_1 p_2 \cdots p_n = v q_1 q_2 \cdots q_m \in R$, where $p_i, q_j \in R$ are irreducible elements and $u, v \in U(R)$. Since any irreducible element in R is prime, $p_1 | q_1 q_2 \cdots q_m$. Hence, there is an index j such that $p_1 | q_j$, i.e., $p_1 \sim q_j$. Without loss of generality, we can assume that $j = 1$. Then, $p_1 p_2 \cdots p_n = \varepsilon p_1 q_2 \cdots q_m$, which implies that $p_2 \cdots p_n = \varepsilon q_2 \cdots q_m$ because R is an integral domain. From induction assumption, we obtain that $n = m$ and $p_i \sim q_i$ for all i.

$2 \Rightarrow 1$. Let p be an irreducible element in a unique factorization domain R and $p|bc$, where $b, c \in R$. Then, $px = bc$ for some element $x \in R$. Since $p \neq 0$ and R is an integral domain, all elements x, b, c are non-zero. If x is an invertible element, then, from uniqueness of factorization, it follows immediately that $(p|b) \vee (p|c)$, i.e., p is a prime element. If b is an invertible element, then we obtain that $p|c$, and if c is invertible, then $p|b$. Therefore, we can assume that all elements x, b, c are not invertible. Then, each of these elements can be uniquely written as a finite product of irreducible elements: $x = x_1 x_2 \cdots x_n$, $b = b_1 b_2 \cdots b_m$, $c_1 c_2 \cdots c_k$. So, we have two factorizations:

$$px_1 x_2 \cdots x_n = b_1 b_2 \cdots b_m c_1 c_2 \cdots c_k.$$

From uniqueness of factorization, it follows that there is an element $b_i \sim p$ or an element $c_j \sim p$. In the first case, we have $p|b$, and in the second case, $p|c$. This means that p is a prime element.   □

✳ **Theorem 4.82.**
Each principal ideal domain is a unique factorization domain.

Proof.
    The proof of this theorem follows immediately from Theorem 4.81 and Theorem 4.80, taking into account that, by Lemma 4.60, each irreducible element of a principal ideal domain is a prime element.   □

✳ **Examples 4.83.**
    The ring of integers $\mathbb{Z}$ is a unique factorization domain, since it is a principal ideal domain.

✳ **Corollary 4.84.**
Each Euclidean ring R is a unique factorization domain.

4.9. Chinese Remainder Theorem

It is likely that the name "Chinese Remainder Theorem" originated from the question "How many soldiers were in the army of Han Xing if lining up them in three rows two soldiers were left, if lining up them in five rows three soldiers were left, and if lining up them in seven rows two soldiers were left. The number of "unpaired" soldiers in each arrangement (i.e., remainders on division of a whole number of soldiers by 3, 5 and 7) gives the possibility to find a required number of all soldiers, which is the solution of the following system of congruences:

$$\begin{cases} x \equiv 2 \,(\mathrm{mod}\,3) \\ x \equiv 3 \,(\mathrm{mod}\,5) \\ x \equiv 2 \,(\mathrm{mod}\,7) \end{cases}$$

This problem was solved by Chinese mathematician Sun Zi (the 3rd century AD) and was published later in 1247 by Qin Jiushao.

In this chapter, we consider the algebraic version of this theorem involving rings and ideals.

First of all, we consider the important construction of rings, which is a generalization of the direct product of groups.

✳ **Definition 4.85.**

Let $(R_1, +, \bullet)$ and $(R_2, +, \circ)$ be rings. The Cartesian product of sets $R_1 \times R_2$ with operations of addition $+$ and multiplication $\cdot$ defined componentwise as follows:

$$(g_1, h_1) + (g_2, h_2) = (g_1 + g_2, h_1 + h_2)$$

$$(g_1, h_1) \cdot (g_2, h_2) = (g_1 \bullet g_2, h_1 \circ h_2)$$

for all $g_1, g_2 \in R_1$ and $h_1, h_2 \in R_2$, is called a **external direct product** of rings R_1, R_2 and is denoted by $R_1 \times R_2$.

Analogously as for groups, we can show that the external direct product $R_1 \times R_2$ of rings R_1 and R_2 is also a ring. It is associative when the rings R_1, R_2 are associative, and it is commutative when these rings are commutative, the zero element of $R_1 \times R_2$ is $(0, 0)$ and the identity element is (e_1, e_2), where e_i is the identity element of R_i for $i = 1, 2$.

The notion of the external direct product of rings can be generalized to the case of arbitrary number of rings.

✳ **Definition 4.86.**

Let $R_1, R_2, \ldots, R_n$ be arbitrary rings. The Cartesian product of sets $R = R_1 \times R_2 \times \ldots \times R_n$ with componentwise addition and multiplication

$$(x_1, x_2, \ldots, x_n) + (y_1, y_2, \ldots, y_n) = (x_1 + y_1, x_2 + y_2, \ldots, x_n + y_n)$$

$$(x_1, x_2, \ldots, x_n)(y_1, y_2, \ldots, y_n) = (x_1 y_1, x_2 y_2, \ldots, x_n y_n)$$

is called the **external direct product** of rings $R_1, R_2, \ldots, R_n$ and we write $R = \prod_i^n R_i = R_1 \times R_2 \times \ldots \times R_n$. If $R = R_1 = R_2 = \ldots = R_n$, then we often write R^n instead of $R \times R \times \ldots \times R$.

Note, that the external direct product of rings is often called, for short, the direct product.

Now, we consider a generalization of the relatively prime integers to the case of ideals.

✳ Definition 4.87.

Ideals I, J of a commutative ring R are called **relatively prime** (or **comaximal**) if $I + J = R$.

✳ Example 4.88.

Let $R = \mathbb{Z}$, $I = \langle m \rangle$, $J = \langle n \rangle$, where m, n are positive integers. If $\gcd(m, n) = 1$, then there are integers x, y such that $1 = mx + ny$. Hence, $1 \in I + J$. Then, from Lemma 4.28, it follows that $I + J = R$, which means that I and J are relatively prime ideals.

✳ Lemma 4.89.

If ideals I, J of a commutative ring R are maximal and $I \neq J$, then ideals I, J are relatively prime.

Proof.

Consider an ideal $I + J$. Since $I \neq J$, $I + J \neq J$. Then, from the chain of ideals $J \subset I + J \subseteq R$, it follows that $I + J = R$, since J is a maximal ideal. □

✳ Lemma 4.90.

If ideals I, J are relatively prime ideals of a commutative ring R, then $IJ = I \cap J$.

Proof.

By Lemma 4.30, $IJ \subseteq I \cap J$ and $(I \cap J)(I + J) \subseteq IJ$. Since ideals I, J of a commutative ring R are relatively prime, $I + J = R$. Hence, $I \cap J \subseteq IJ$. In this way, $IJ = I \cap J$. □

Furthermore, we extend the relation of "modulo" to the case of ideals.

✳ Definition 4.91.

Two elements a, b of a ring R are called **congruent modulo ideal** I if and only if $a - b \in I$. We write $a \equiv b \,(\mathrm{mod}\, I)$.

✳ Theorem 4.92.

If I and J are relatively prime ideals of a commutative ring R, then:
1) For all $a, b \in R$, there is an element $x \in R$ such that

$$\begin{cases} x \equiv a \,(\mathrm{mod}\, I) \\ x \equiv b \,(\mathrm{mod}\, J) \end{cases} \tag{4.14}$$

2) If $y \in R$ is another solution of (4.14), then $y \equiv x \,(\mathrm{mod}\, IJ)$
3) There is an isomorphism of quotient rings:

$$R/IJ \cong R/I \times R/J \tag{4.15}$$

Proof.

Let I and J be relatively prime ideals of a commutative ring R.

1) Since $I + J = R$, there exist elements $u \in I$ and $v \in J$ such that $u + v = 1$. Then, the element $x = bu + av$ is the required element. Really,

$$x = bu + av = bu + a(1 - u) = a + (b - a)u \equiv a \,(\mathrm{mod}\, I)$$

Analogously,

$$x = bu + av = b(1 - v) + av = b + (a - b)v \equiv b \,(\mathrm{mod}\, J).$$

2) By assumption, x, y are solutions of (4.14), which implies that $(x \equiv y \,(\mathrm{mod}\, I)) \wedge (x \equiv y \,(\mathrm{mod}\, J))$. Therefore, $x - y \in I \cap J$. Since $I \cap J = IJ$, by Lemma 4.89, we get that $x \equiv y \,(\mathrm{mod}\, IJ)$.

3) Consider the mapping $f : R \to R/I \times R/J$, defined as follows: $f(x) = (x + I, x + J)$. We have:

a) $f(x+y) = (x+y+I, x+y+J) = ((x+I)+(y+I), (x+J)+(y+J)) = (x + I, x + J) + (y + I, y + J) = f(x) + f(y)$,

b) $f(xy) = (xy + I, xy + J) = ((x + I)(y + I), (x + J)(y + J)) = (x + I, x + J)(y + I, y + J) = f(x)f(y)$,

which shows that f is a homomorphism of rings. From the definition of f, it follows that f is an epimorphism.

Let $\mathrm{Ker}(f) = \{x \in R : f(x) = 0\}$. Then, $f(x) = (x+I, x+J) = (I, J)$ if and only if $(x \in I) \wedge (x \in J)$, i.e., $x \in I \cap J$. Since, by Lemma 4.89, $I \cap J = IJ$, we obtain that $\mathrm{Ker}(f) = IJ$. So, f is an epimorphism of rings with kernel IJ. Hence, by Corollary 2.77, we obtain that

$$R/IJ \cong R/I \times R/J.$$

$\square$

✳ **Definition 4.93.**

Ideals $I_1, I_2, \ldots, I_n$ of a commutative ring R are called **pairwise relatively prime** (or **pairwise comaximal**) if $I_i + I_j = R$ for all i, j with $i \neq j$.

✳ **Lemma 4.94.**
If $I_1, I_2, \ldots, I_n$ are pairwise relatively prime ideals of a commutative ring R, then $I_1 I_2 \ldots I_n = I_1 \cap I_2 \cap \ldots \cap I_n$.

Proof. The proof of this lemma follows from Lemma 4.90 by mathematical induction on the number of ideals. $\square$

Theorem 4.92 can be generalized by the following form:

✳ Theorem 4.95.

If $I_1, I_2, \ldots, I_n$ are pairwise relatively prime ideals of a commutative ring R, then:

1) For all $a_i \in R$, there is an element $x \in R$ such that

$$x \equiv a_i \,(\mathrm{mod}\, I_i), \quad \text{for all } \ i = 1, 2, \ldots, n \qquad (4.16)$$

2) If $y \in R$ is another solution of (4.16), then $y \equiv x \,(\mathrm{mod}\, I_1 I_2 \cdots I_n)$.

3) There is an isomorphism of quotient rings:

$$R/I_1 I_2 \cdots I_n \cong R/I_1 \times R/I_2 \times \cdots \times R/I_n \qquad (4.17)$$

Proof.

1) From Theorem 4.92(1), it follows that for each i there is an element c_i such that $c_i \equiv 1 \,(\mathrm{mod}\, I_i)$ and $c_i \equiv 0 \,(\mathrm{mod}\, I_j)$ for all i, j with $i \neq j$. Then, the element $x = a_1 c_1 + \cdots + a_n c_n$ is the required element. Really:

$$x - a_i = a_1 c_1 + \cdots + a_n c_n - a_i \equiv a_i c_i - a_i \equiv a_i - a_i \equiv 0 \,(\mathrm{mod}\, I_i)$$

for all i.

2) The proof is the same as for Theorem 4.92(2), taking into account Lemma 4.94.

3) Consider the mapping:

$$f : R \to R/I_1 \times R/I_2 \times \cdots \times R/I_n$$

defined as follows:

$$f(x) = (x + I_1, \ldots, x + I_n).$$

Then, similarly to the proof of Theorem 4.92(3), we can show that f is an epimorphism of rings with $\mathrm{Ker}(f) = I_1 I_2 \cdots I_n$. Therefore, by Corollary 2.77 we obtain the required isomorphism (4.17). □

Taking into account that for a principal ideal domain R:

$$a_1 R \cdot a_2 R \cdots a_n R = a_1 a_2 \cdots a_n R = \langle a_1 a_2 \ldots a_n \rangle,$$

where xR is a principal ideal $\langle x \rangle$ generated by an element $x \in R$, Theorem 4.95 for a principal ideal domain has the following form:

✳ Theorem 4.96.

If R is a principal ideal domain and $\gcd(a_i, a_j) = 1$ for all i, j with $i \neq j$, then

$$R/\langle a_1 a_2 \cdots a_n \rangle \cong R/a_1 R \times R/a_2 R \times \cdots \times R/a_n R \qquad (4.18)$$

In a particular case, when $R = \mathbb{Z}$, we obtain the well-known general form of the Chinese Remainder Theorem.

✳ **Theorem 4.97 (Generalized Chinese Remainder Theorem).**
Let $m = m_1 m_2 \cdots m_k$, where $m_1, \ldots, m_k$ are positive integers and $\gcd(m_i, m_j) = 1$ for all i, j with $i \neq j$. Then
1) System of linear congruences:

$$x \equiv a_i \, (\mathrm{mod} \, m_i), \text{ for all } i = 1, 2, \ldots, n$$

has a unique integer solution x modulo m for any $a_i \in \mathbb{Z}$.
2) There is an isomorphism of rings:

$$\mathbb{Z}/m\mathbb{Z} \cong \mathbb{Z}/m_1\mathbb{Z} \times \mathbb{Z}/m_2\mathbb{Z} \times \cdots \times \mathbb{Z}/m_k\mathbb{Z}. \tag{4.19}$$

For $k = 2$, Theorem 4.97 has historically original form for the Chinese Remainder Theorem:

✳ **Theorem 4.98 (Chinese Remainder Theorem) (Sum Zi, III AD)**
Let m, n be positive integers and $\gcd(m, n) = 1$. Then,
1) A system of linear congruences:

$$\begin{cases} x \equiv a_1 \, (\mathrm{mod} \, m) \\ x \equiv a_2 \, (\mathrm{mod} \, n) \end{cases}$$

has a unique integer solution x modulo mn for any $a_1, a_2 \in \mathbb{Z}$.
2) There is an isomorphism of rings:

$$\mathbb{Z}/mn\mathbb{Z} \cong \mathbb{Z}/m\mathbb{Z} \times \mathbb{Z}/n\mathbb{Z}. \tag{4.20}$$

✳ **Example 4.99.**
Solve the system of congruences:

$$\begin{cases} x \equiv 36 \, (\mathrm{mod} \, 41) \\ x \equiv 5 \, (\mathrm{mod} \, 17) \end{cases}$$

Here, we have $m_1 = 41$, $m_2 = 17$, $m = 41 \cdot 17 = 697$.

Since $\gcd(41, 17) = 1$, the system has an integer solution. From the first congruence $x \equiv 36 \, (\mathrm{mod} \, 41)$, we obtain that $x = 36 + 41t$, where $t \in \mathbb{Z}$. Substituting to the second congruence, we obtain $36 + 41t \equiv 5 \, (\mathrm{mod} \, 17)$. Hence, $41t \equiv -31 \, (\mathrm{mod} \, 17)$, implying that $7t \equiv 3 \, (\mathrm{mod} \, 17)$. This congruence has a solution $t = 15 + 17u$, where $u \in \mathbb{Z}$. Therefore, $x = 36 + 41 \cdot (15 + 17u) = 651 + 697u$. In this way, $x \equiv 651 \, (\mathrm{mod} \, 697)$ is a solution of the given system of congruences.

✳ Example 4.100.

Solve the system of congruences:

$$\begin{cases} x \equiv 4 \,(\mathrm{mod}\,5) \\ x \equiv 3 \,(\mathrm{mod}\,7) \\ x \equiv 1 \,(\mathrm{mod}\,9) \end{cases}$$

Here, we have $m_1 = 5$, $m_2 = 7$, $m_3 = 9$, $m = 5 \cdot 7 \cdot 9 = 315$.

Since $\gcd(m_i, m_j) = 1$ for all i, j with $i \neq j$, the system has a unique integer solution modulo $m = 315$. From the first congruence, $x \equiv 4 \,(\mathrm{mod}\,5)$, we obtain that $x = 4 + 5t$, where $t \in \mathbb{Z}$. Substituting it to the second congruence, we have: $4 + 5t \equiv 3 \,(\mathrm{mod}\,7)$, or $5t \equiv 6 \,(\mathrm{mod}\,7)$, which has a solution $t = 4 + 7u$, where $u \in \mathbb{Z}$. Hence, $x = 4 + 5(4 + 7u) = 24 + 35u$. Substituting x to the third congruence, we obtain: $24 + 35u \equiv 1 \,(\mathrm{mod}\,9)$ or $2u \equiv 1 \,(\mathrm{mod}\,9)$, having a solution $u = 5 + 9v$, where $v \in \mathbb{Z}$.

Hence, $x = 24 + 35 \cdot (5 + 9v) = 199 + 315v$, and $x \equiv 199 \,(\mathrm{mod}\,315)$ is a solution of the given system of congruences.

4.10. Exercises

1. Find zero divisors of the rings $\mathbb{Z}_4$, $\mathbb{Z}_8$, $\mathbb{Z}_{15}$.

2. Show that the set of Gauss integers $\mathbb{Z}[i] = \{a + bi \ : \ a, b \in \mathbb{Z}, i^2 = -1\}$ under operations of addition and multiplication is an integral domain.

3. Show that the set of numbers $\mathbb{Q}[\sqrt{2}] = \{a + b\sqrt{2} \ : \ a, b \in \mathbb{Q}\}$ under operations of addition and multiplication is a commutative ring.

4. Show that the set of numbers $\mathbb{Q}[i] = \{a + bi \ : \ a, b \in \mathbb{Q}, i^2 = -1\}$ under operations of addition and multiplication is a field.

5. Show that the set of numbers $\mathbb{Q}[\sqrt{2}] = (\{a + b\sqrt{2} \ : \ a, b \in \mathbb{Q}\}$ under operations of addition and multiplication is a subring of the ring $\mathbb{R}$.

6. Show that the ring $\mathbb{Q}[\sqrt{2}]$ is a field.

7. Show that the set of elements $\{a + b\sqrt{2} + c\sqrt{3} \ : \ a, b, c \in \mathbb{Q}\}$ under operations of addition and multiplication is not a ring.

8. Show that the set of all invertible elements of a ring with identity form a multiplicative group.

9. Find the set of all invertible elements of the ring $\mathbb{Z}_8$.

10. Show that if I, J are ideals and $I \subset J$, then $I + J = J$.

11. Solve the system of congruences:

$$\begin{cases} x \equiv 5 \,(\text{mod}\,7) \\ x \equiv 4 \,(\text{mod}\,6) \end{cases}$$

12. Solve the system of congruences:

$$\begin{cases} x \equiv 41 \,(\text{mod}\,65) \\ x \equiv 35 \,(\text{mod}\,72) \end{cases}$$

13. Solve the system of congruences:

$$\begin{cases} x \equiv 0 \,(\text{mod}\,2) \\ x \equiv 1 \,(\text{mod}\,3) \\ x \equiv 2 \,(\text{mod}\,5) \end{cases}$$

14. Solve the system of congruences:

$$\begin{cases} x \equiv 9 \,(\text{mod}\,12) \\ x \equiv 3 \,(\text{mod}\,13) \\ x \equiv 6 \,(\text{mod}\,25) \end{cases}$$

15. Solve the system of congruences:

$$\begin{cases} x \equiv 3 \,(\text{mod}\,4) \\ x \equiv 4 \,(\text{mod}\,5) \\ x \equiv 1 \,(\text{mod}\,9) \\ x \equiv 5 \,(\text{mod}\,7) \end{cases}$$

16. Show that the determinant of the matrix

$$\mathbf{A} = \begin{pmatrix} 320 & 461 & 5264 & 72 \\ 702 & 1008 & -967 & -44 \\ -91 & 2333 & 46 & 127 \\ 164 & -216 & 1862 & 469 \end{pmatrix}$$

is not equal to 0.

17. Find the determinant of the matrix

$$\mathbf{B} = \begin{pmatrix} 676 & 117 & 522 \\ 375 & 65 & 290 \\ 825 & 143 & 639 \end{pmatrix}$$

knowing that it is a positive number which is less than 100.

References

[1] Ash, R.B. 2007. Basic Abstract Algebra, Dover Publications.

[2] Ding, C., D. Pei and A. Salomaa. 1996. Chinese Remainder Theorem: Applications in Computing, Coding and Cryptography, World Scientific Publishing.

[3] Dummit, D.S. and R.M. Foote. 2004. Abstract Algebra (3rd Ed.), John Wiley & Sons.

[4] Hazewinkel, M., N. Gubareni and V.V. Kirichenko. 2004. Algebras, Rings and Modules. Volume 1. Kluwer Acad. Publ.

[5] Jacobson, N. 1985. Basic Algebra I, Freeman, San Francisco, H. Freeman & Co.

[6] Judson, Th.W. 1994. Abstract Algebra. Theory and Applications, PWS Publishing Company, Michigan.

[7] Lang, S. 1984. Algebra, Addison-Wesley.

[8] Lidl, R. and G. Pilz. 1997. Applied Abstract Algebra, (2nd Ed.), Springer-Verlag, New York.

[9] Oggier, F. 2011. Algebraic Methods.
http://www1.spms.ntu.edu.sg/~frederique/AA11.pdf.

[10] Zarisky, O. and P. Samuel. 1958. Commutative Algebra, Vol. I.

Chapter 5

Polynomial Rings in One Variable

"'Obvious' is the most dangerous word in mathematics."
E.T. Bell

"The most painful thing about mathematics is how far away you are from being able to use it after you have learned it."
James Newman

The concept of a polynomial is one of the central notions in mathematics. Though a polynomial ring is a particular example of a ring, it has its own features. Also, it should be noted that properties of polynomials depend on a ring which coefficients of polynomials belong to. For example, a polynomial may be decomposable over one ring and irreducible over another.

In this chapter, we assume that coefficients of polynomials belong to a ring K which is an integral domain. We consider division with remainder for polynomials, the greatest common divisors of polynomials and the Euclidean algorithm for finding them.

We also consider the concept of irreducible polynomials. In general, the problem of verifying irreducibility of a given polynomial is not trivial. For some classes of polynomials over rational numbers, there is an interesting method for verifying irreducibility. This is Eisenstein's criterion, which gives a sufficient condition of irreducibility for some class of polynomials over rational numbers. This criterion was first proved by the German mathematician G. Eisenstein (1823-1852), a student of C.F. Gauss. In section 5.6, we consider this criterion. At the end of

Gottholb Eisenstein
(1823-1852)

this chapter, we describe the very important construction of factor rings of polynomials and show when this construction can lead to fields.

Note that, everywhere in this chapter, we assume that a ring K, to which all coefficients of polynomials belong, is an integral domain, unless otherwise stated. In particular, we may consider that K is a field or the ring of integers $\mathbb{Z}$.

5.1. Basic Definitions and Properties

We start this section from the traditional definition of polynomials.

> ✳ **Definition 5.1.**
>   Any expression of the form:
>
> $$P(x) = a_n x^n + a_{n-1} x^{n-1} + \cdots + a_1 x + a_0, \qquad (5.1)$$
>
> where n is an arbitrary positive integer, $a_i \in K$ for $i = 0, 1, \ldots, n$, and $a_n \neq 0$, is called a **polynomial** in one variable (or indeterminate) x over a ring K. Elements $a_i \in K$, for $i = 0, 1, \ldots, n$, are called **coefficients** of the polynomial $P(x)$, an element a_0 is called a **constant term** and a_n is called a **leading coefficient**. The number n is called the **degree** of $P(x)$ and is denoted by $\deg(P)$. We assume that the degree of a polynomial which is a non-zero element of K is equal to zero, and $\deg(0) = \infty$. Polynomials of degree 0 and ∞ are called **constant**.
>   We also assume that $x^0 = 1$ and $1 \cdot x^i = x^i \cdot 1 = x^i$ $i = 0, 1, \ldots, n$ and write usually x instead of x^1.

Consider two polynomials

$$P(x) = a_n x^n + a_{n-1} x^{n-1} + \cdots + a_1 x + a_0$$

and

$$Q(x) = b_m x^m + b_{m-1} x^{m-1} + \cdots + b_1 x + b_0$$

with $\deg(P) = n$ and $\deg(Q) = m$. They are said to be equal if and only if $n = m$ and $a_i = b_i$ for all $i = 0, 1, 2, \ldots, n$.

We denote by $K[x]$ the set of all polynomials in the indeterminate x over a ring K. In a natural way, we can define operations of addition, substraction and multiplication of polynomials:

$$(P \pm Q)(x) = P(x) \pm Q(x) = \sum_{i=0}^{\max(m,n)} (a_i \pm b_i) x^i,$$

$$(P \cdot Q)(x) = P(x) \cdot Q(x) = \sum_{i=0}^{m+n} c_i x^i, \quad \text{where} \quad c_i = \sum_{j=0}^{i} a_j b_{i-j}.$$

✳ Examples 5.2.

- $P(x) = 2x^4 - 1/3x + 12$ is the polynomial of degree 4 over the field of real numbers.

- $P(x) = 2ix^7 + 3/4x^3 - (1 + 4i)x^2 + (2 - 3i)$ is the polynomial of degree 7 over the field of complex numbers.

✳ Example 5.3.
Let $P(x) = 3x^2 + x - 5$ and $Q(x) = 2x + 3$ be polynomials over the field of real numbers. Then,
$$(P + Q)(x) = 3x^2 + 3x - 2,$$
$$(P - Q)(x) = 3x^2 - x - 8,$$
$$(PQ)(x) = 6x^3 + 11x^2 - 7x - 15.$$

✳ Theorem 5.4.
The set $K[x]$ of all polynomials in the indeterminate x over a ring K under addition and multiplication is a commutative ring with identity, which is called the **ring of polynomials**, or the **polynomial ring**.

Proof.
The operation of addition is obviously commutative and associative. The inverse element to a polynomial $P(x) = a_n x^n + a_{n-1} x^{n-1} + \cdots + a_1 x + a_0$ with respect to addition is a polynomial $Q(x) = -P(x) = -a_n x^n - a_{n-1} x^{n-1} - \cdots - a_1 x - a_0$, since $P(x) + Q(x) = 0$. The neutral element with respect to addition is the element 0, and the neutral element with respect to multiplication is 1, i.e., for any polynomial $P(x)$, the equalities $P(x) + 0 = 0 + P(x) = P(x)$ and $P(x) \cdot 1 = 1 \cdot P(x) = P(x)$ hold. Since K is a commutative ring, from the definition of multiplication in $K[x]$, it is easy to verify that multiplication is also commutative. We show the associativity of multiplication. Let $P(x) = a_n x^n + a_{n-1} x^{n-1} + \cdots + a_1 x + a_0$, $Q(x) = b_m x^m + b_{m-1} x^{m-1} + \cdots + b_1 x + b_0$, $R(x) = c_k x^k + c_{k-1} x^{k-1} + \cdots + c_1 x + c_0$. Suppose that
$$(P(x)Q(x))R(x) = d_s x^s + d_{s-1} x^{s-1} + \cdots + d_1 x + d_0$$
and
$$P(x)(Q(x)R(x)) = h_s x^s + h_{s-1} x^{s-1} + \cdots + h_1 x + h_0.$$
Then, from the definition of multiplication, we have:
$$d_i = \sum_{j+r=i} \left(\sum_{p+t=j} a_p b_t \right) c_r = \sum_{p+t+r=i} (a_p b_t) c_r,$$
and on the other hand:
$$h_i = \sum_{p+j=i} a_p \left(\sum_{r+t=j} b_t c_r \right) = \sum_{p+t+r=i} a_p (b_t c_r).$$

Hence, we obtain that $d_i = h_i$, which means that $(P(x)Q(x))R(x) = P(x)(Q(x)R(x))$. The distributive law of multiplication with respect to addition can be proved analogously. So, the set $K[x]$ with operations of addition and multiplication satisfies all axioms of commutative and associative ring with identity.   □

Note that elements of a ring K can be considered as polynomials ax^0 of a ring $K[x]$. It is easy to verify that the mapping $\varphi : K \to K[x]$ given as follows $\varphi(a) = ax^0 = a$ is a monomorphism of rings which allows the ring K to be considered as a subring of the ring of polynomials $K[x]$.

✻ Remark 5.5.

In the formal way, a polynomial over a ring K can be introduced as an infinity sequence of the form $(a_0, a_1, \ldots, a_n \ldots)$ such that $a_i \in K$ and only a finite number of elements a_i is not equal to zero. The greatest number n, for which $a_n \neq 0$, is called the **degree** of the polynomial. The operations of addition and multiplication of polynomials are defined by the following way:

$$(a_0, a_1, \ldots, a_n, \ldots) + (b_0, b_1, \ldots, b_n, \ldots) =$$

$$= (a_0 + b_0, a_1 + b_1, \ldots, a_n + b_n, \ldots) \tag{5.2}$$

$$(a_0, a_1, \ldots, a_n, \ldots) \cdot (b_0, b_1, \ldots, b_n, \ldots) = (c_0, c_1, \ldots, c_n, \ldots) \tag{5.3}$$

where $c_i = \sum\limits_{j=0}^{i} a_j b_{i-j}$.

The inverse polynomial to a polynomial $f = (a_0, a_1, \ldots, a_n \ldots)$ with respect to addition is a polynomial $-f = (-a_0, -a_1, \ldots, -a_n \ldots)$.

Two polynomials, $f = (a_0, a_1, \ldots, a_n \ldots)$ and $g = (b_0, b_1, \ldots, b_n \ldots)$, are said to be equal if and only if $a_i = b_i$ for all $i = 0, 1, 2, \ldots$.

We denote by K' the set of such sequences with operations of addition and multiplication defined by (5.2) and (5.3). In this set, the neutral element with respect to addition is the element $(0, 0, \ldots, 0, \ldots)$, and the neutral element with respect to multiplication is the element $(1, 0, \ldots, 0 \ldots)$. Similarly to Theorem 5.4, we can prove that the set K' is a commutative associative ring with identity.

We now compare these two definitions of polynomials. Denote:

$$x = (0, 1, 0, \ldots, 0, \ldots).$$

Then, by definition, we have:

$$x^2 = (0, 0, 1, 0, \ldots, 0, \ldots),$$

$$x^3 = (0, 0, 0, 1, 0, \ldots, 0, \ldots),$$

$$\cdots\cdots\cdots\cdots\cdots\cdots\cdots\cdots\cdots$$

$$x^n = (\underbrace{0, 0, \ldots, 0}_{n}, 1, 0, \ldots, 0, \ldots) \quad n \geq 1.$$

It is easy to verify that the mapping $\varphi : K \to K'$, defined as follows: $\varphi(a) = (a, 0, 0, \ldots)$, is a monomorphism of rings and, therefore, we can identify elements of K with elements of K' of the form $(a, 0, 0, \ldots)$. Then, elements of K' can be represented in the following common form:

$$(a_0, a_1, a_2, \ldots, a_n, 0, 0 \ldots) = (a_0, 0, 0, \ldots)(1, 0, \ldots) + (a_1, 0, \ldots)(0, 1, 0, \ldots) + \cdots +$$

$$+(a_n, 0, \ldots)\underbrace{(0, 0, \ldots, 0, 1, 0, \ldots)}_{n} = a_0 + a_1 x + a_2 x^2 + \cdots + a_n x^n.$$

Therefore, it is easy to show that the mapping $\delta : K[x] \to K'$ given by $\delta(a_0 + a_1 x + a_2 x^2 + \cdots + a_n x^n) = (a_0, a_1, \ldots, a_n, 0, 0, \ldots)$ is an isomorphism of the rings $K[x]$ and K', which means that both definitions of polynomials are equivalent. Hence, it follows that the element x from Definition 5.1 only plays an auxiliary role.

Furthermore, we will write polynomials in the common form as elements of the ring $K[x]$.

✻ Theorem 5.6.
Let $P(x), Q(x)$ be non-zero polynomials over an integral domain K. Then,
1) $\deg(P + Q) \leq \max(\deg P, \deg Q)$.
2) $\deg(PQ) = \deg P + \deg Q$.

Proof.
Let $P(x) = a_n x^n + a_{n-1} x^{n-1} + \cdots + a_1 x + a_0$ and $Q(x) = b_m x^m + b_{m-1} x^{m-1} + \cdots + b_1 x + b_0$, where $a_n \neq 0$ and $b_m \neq 0$. Then, part 1 of the theorem follows directly from the definition of addition of polynomials. The second part follows from the fact that $P(x)Q(x)$ contains a non-zero monomial $a_n b_m x^{n+m}$. Since its coefficient $a_n b_m \neq 0$ as a product of non-zero elements a_n and b_m of an integral domain K, $a_n b_m x^{n+m} \neq 0$ as well. $\square$

✻ Theorem 5.7.
A polynomial ring $K[x]$ over an integral domain K is an integral domain.

Proof.
Let $P(x), Q(x)$ be non-zero polynomials in $K[x]$. Then, by Theorem 5.6, $\deg(PQ) = \deg P + \deg Q > 0$. Hence, $P(x)Q(x) \neq 0$. $\square$

Note that the ring $K[x]$ is not a field, since no polynomial $P(x) \in K[x]$ has the inverse.

✻ Theorem 5.8.
A polynomial $P(x)$ over an integral domain K is invertible in $K[x]$ if and only if $\deg(P(x)) = 0$, i.e., $P(x) = c = \text{const} \in K$ and c is an invertible element in K. In particular, if K is a field then a polynomial $P(x) \in K[x]$ is invertible if and only if $\deg(P(x)) = 0$.

Proof.

Let $P(x)$ be an invertible element in $K[x]$. Then, there is an element $Q(x) \in K[x]$ such that $P(x)Q(x) = 1$. By Theorem 5.6, $\deg P + \deg Q = 0$, implying that $\deg P = \deg Q = 0$, i.e., $P(x) = c \in K$ and $P(x)$ is invertible in $K[x]$ if and only if it is invertible in K. □

5.2. Division with Remainder

In the ring $K[x]$ for two arbitrary polynomials $P(x), Q(x) \in K[x]$, where $Q(x) \neq 0$, there does not always exist a polynomial $S(x)$ such that $P(x) = Q(x)S(x)$, i.e., not every polynomial $P(x)$ can be divided by an arbitrary non-zero polynomial $Q(x)$.

In this section, we show that if K is a field, then in the ring $K[x]$ we can always perform division with remainder, i.e., for arbitrary non-zero polynomials $P(x), Q(x) \in K[x]$, $Q(x) \neq 0$, there are polynomials $S(x), R(x) \in K[x]$ such that:

$$P(x) = Q(x)S(x) + R(x) \tag{5.4}$$

where either $R(x) = 0$ or $0 \leq \deg R(x) < \deg Q(x)$. Polynomials $S(x)$ and $R(x)$ are called the **quotient** and the **remainder**, respectively.

✳ Theorem 5.9 (Remainder Theorem).

Let K be a field and $P(x), Q(x) \in K[x]$, $Q(x) \neq 0$ and $\deg P(x) \geq \deg Q(x)$. Then, there are uniquely defined polynomials $S(x), R(x) \in K[x]$ such that

$$P(x) = Q(x)S(x) + R(x)$$

and either $R(x) = 0$ or $0 \leq \deg R(x) < \deg Q(x)$.

Proof.

If $\deg P(x) < \deg Q(x)$ or $P(x) = 0$, then we can consider that $S(x) = 0$ and $R(x) = P(x)$, since $P(x) = 0 \cdot Q(x) + P(x)$ and so the theorem is proved.

Let $P(x) = a_n x^n + a_{n-1}x^{n-1} + \cdots + a_1 x + a_0 \neq 0$ and $Q(x) = b_m x^m + b_{m-1}x^{m-1} + \cdots + b_1 x + b_0 \neq 0$ with $a_n \neq 0$, $b_m \neq 0$. Suppose that $n \geq m$. We will perform the proof using the mathematical induction on the number $n = \deg P(x)$.

If $n = 0$, then $P(x) = a_0 \neq 0$. Since $m \leq n$, $m = 0$ and so $Q(x) = b_0 \neq 0$. Therefore, there exists $b_0^{-1} \in K$ and we can write $P(x) = (a_0 b_0^{-1})b_0 + 0$. Hence, $S(x) = a_0 b_0^{-1}$ and $R(x) = 0$.

Assume that the theorem is proved for all polynomials of degree $< n$. Consider the polynomial $H(x) = P(x) - (a_n b_m^{-1})x^{n-m}Q(x)$. Since this polynomial is a difference of two polynomials of the same degree n with the same leading coefficients a_n, we obtain that $\deg H(x) < n$. Then, by inductive assumption, there are polynomials $T(x), R(x)$ such that

$$H(x) = Q(x)T(x) + R(x),$$

where $T(x), R(x) \in K[x]$ and either $R(x) = 0$ or $0 \leq \deg R(x) < m$. Therefore, $P(x) = H(x) + (a_n b_m^{-1})x^{n-m}Q(x) = Q(x)T(x) + R(x) + (a_n b_m^{-1})x^{n-m}Q(x) = Q(x)[T(x) + (a_n b_m^{-1})x^{n-m}] + R(x) = Q(x)S(x) + R(x)$, where $S(x) = T(x) + (a_n b_m^{-1})x^{n-m}$.

We will show that polynomials $S(x)$ and $R(x)$ are defined uniquely. Suppose that

$$P(x) = Q(x)S(x) + R(x) = Q(x)S_1(x) + R_1(x).$$

Then, $Q(x)[S(x) - S_1(x)] = R_1(x) - R(x)$. If $S(x) - S_1(x) \neq 0$, then $\deg Q(x)[S(x) - S_1(x)] \geq m$, meanwhile $\deg[R_1(x) - R(x)] < m$. This contradiction shows that $S(x) - S_1(x) = 0$, hence, $S(x) = S_1(x)$ and $R(x) = R_1(x)$. $\square$

❋ Example 5.10.

Divide the polynomial $P(x) = x^3 + 2x^2 + x + 2$ by $Q(x) = x^2 + 2$ in the ring $\mathbb{Z}_3[x]$.

We have $x^3 + 2x^2 + x + 2 = (x^2 + 2)(x + 2) + (2x + 1)$, where $S(x) = x + 2$ is the quotient and $R(x) = 2x + 1$ is the remainder.

❋ Definition 5.11.

A polynomial $P(x)$ is said to be **divisible** by a polynomial $Q(x) \in K[x]$, or $Q(x)$ is a **divisor** of $P(x)$, that is denoted $Q(x)|P(x)$, if there exists a polynomial $S(x) \in K[x]$ such that $P(x) = Q(x)S(x)$.

Note that if a polynomial $P(x)$ is divisible by a polynomial $Q(x)$ then we can divide with remainder and, in this case, we can choose polynomials $S(x)$, $R(x)$ in such a way that $R(x) = 0$.

❋ Definition 5.12.

A non-zero polynomial $P(x) \in K[x]$ is called **monic** if its leading coefficient is equal to 1, i.e.,

$$P(x) = x^n + a_{n-1}x^{n-1} + \cdots + a_1x + a_0$$

❋ Remark 5.13.

From the proof of Theorem 5.9, it follows that division with remainder also takes place if K is an integral domain and the leading coefficient of a polynomial $Q(x)$ is an invertible element in K, for example, if $Q(x)$ is a monic polynomial. Taking this fact into account, we immediately obtain the following theorem:

❋ **Theorem 5.14.**

Let K be an integral domain, $a \in K$, $Q(x) = x - a$ and $P(x) \in K[x]$ with $\deg P(x) \geq 1$. Then, there is a unique polynomial $S(x) \in K[x]$ and an element $r \in K$ such that

$$P(x) = (x - a)S(x) + r \qquad (5.5)$$

❋ **Example 5.15.**

If $P(x) = 2x^4 - x^3 + 5x + 3 \in \mathbb{Z}[x]$ and $Q(x) = 2 + x \in \mathbb{Z}[x]$, then

$$2x^4 - x^3 + 5x + 3 = (x + 2)(2x^3 - 5x^2 + 10x - 15) + 33$$

where $S(x) = 2x^3 - 5x^2 + 10x - 15$ and $r = 33$.

From Theorems 5.6, 5.7 and 5.9, it follows that the ring of polynomials over a field K is a Euclidean ring considering the degree of polynomials as a function $\delta : K[x]\backslash\{0\} \to \mathbb{Z}^+ \cap \{0\}$, i.e., $\delta(f(x)) = \deg f(x)$. Therefore, taking Theorem 4.72 into account, we have the following statement.

❋ **Theorem 5.16.**

If K is a field, then a ring $K[x]$ is:
1) a Euclidean ring,
2) a principal ideal domain.

5.3. Greatest Common Divisor of Polynomials

As in Theorem 5.16, a ring $K[x]$ over a field is a Euclidean ring, and so a principal ideal domain, we can use all notions which were introduced in Sections 4.6 and 4.7, and we can use all statements proved in these sections. Therefore, in this section, we only introduce some notions of these sections for polynomials and rewrite the corresponding results.

✳ **Definition 5.17.**

The **greatest common divisor** of polynomials $P(x), Q(x) \in K[x]$ is a polynomial $D(x) \in K[x]$ that satisfies the following conditions:

- $D(x)|P(x)$ and $D(x)|Q(x)$,

- if $H(x) \in K[x]$ and $(H(x)|P(x)) \wedge (H(x)|Q(x))$, then $H(x)|D(x)$.

From this definition, it follows that if $D(x)$ is the greatest common divisor of polynomials $P(x)$ and $Q(x)$, then each polynomial $\varepsilon D(x)$, where $\varepsilon \in U(P)$ is an invertible element of K, is also the greatest common divisor of these polynomials. So, the greatest common divisor of polynomials is defined uniquely up to factors which are invertible elements in K. If K is a field, then in order to uniquely define the g.c.d. of polynomials, we choose in the set of all g.c.d of polynomials $P(x), Q(x)$, a monic polynomial. This uniquely determined g.c.d of polynomials $P(x), Q(x)$ will be denoted by $\gcd(P, Q)$.

In order to find the greatest common divisor of two polynomials, we can use the Euclidean algorithm in the same form as for integers.

✳ Example 5.18.
Find $\gcd(P(x), Q(x))$, where $P(x) = x^5 - x^4 + x^3 - 2x^2 + 2x - 2$, $Q(x) = x^5 - 1 \in \mathbb{R}[x]$.

Since
$$x^5 - x^4 + x^3 - 2x^2 + 2x - 2 = (x^5 - 1) \cdot 1 + (-x^4 + x^3 - x^2 + 2x - 1),$$
$$x^5 - 1 = (-x^4 + x^3 - x^2 + 2x - 1)(-x - 1) + (x^2 + x - 2),$$
$$-x^4 + x^3 - x^2 + 2x - 1 = (x^2 + x - 2)(x^2 + 2x - 5) + 11(x - 1),$$
$$x^2 + x - 2 = (x - 1)(x + 2),$$
we obtain that $\gcd(P(x), Q(x)) = x - 1$.

The notion of the greatest common divisors for two polynomials can be generalized for n polynomials.

✳ Definition 5.19.
The **greatest** **common** **divisor** of polynomials $P_1(x), P_2(x), \ldots, P_n(x) \in K[x]$ is a polynomial $D(x) \in K[x]$ that satisfies the following conditions:

- $D(x) | P_i(x)$ for all $i = 1, \ldots, n$.

- if $H(x) \in K[x]$ and $H(x) | P_i(x)$ for all $i = 1, \ldots, n$, then $H(x) | D(x)$.

From Theorem 4.64, we obtain as corollaries the following theorem:

✳ Theorem 5.20.
Let K be a field and let $P_i(x) \in K[x]$ be non-zero polynomials for $i = 1, 2, \ldots, n$. Then, there exists the greatest common divisor of these polynomials $\gcd(P_1, P_2, \ldots, P_n)$ and there exist uniquely defined polynomials $S_i(x) \in K[x]$ for $i = 1, 2, \ldots, n$ such that

$$\gcd(P_1, P_2, \ldots, P_n) = S_1(x)P_1(x) + S_2(x)P_2(x) + \cdots + S_n(x)P_n(x) \quad (5.6)$$

✳ Example 5.21.
Find $\gcd(P(x), Q(x)) = D(x)$, where $P(x) = x^4 + 5x^3 + 2x + 6$, $Q(x) = x^3 + 4x^2 + 4x + 5 \in \mathbb{Z}_7[x]$, and polynomials $S(x), T(x)$ such that $D(x) = S(x)P(x) + T(x)Q(x)$.

Since
$$x^4 + 5x^3 + 2x + 6 = (x^3 + 4x^2 + 4x + 5)(x + 1) + (6x^2 + 1),$$
$$x^3 + 4x^2 + 4x + 5 = (6x^2 + 1)(6x + 3) + (5x + 2),$$
$$6x^2 + 1 = (5x + 2)(4x + 4)$$
we obtain that $D(x) = \gcd(P(x), Q(x)) = 5x + 2$.

In order to find $S(x), T(x)$, we use the Extended Euclidean Algorithm:

$$6x^2 + 1 = (x^4 + 5x^3 + 2x + 6) - (x^3 + 4x^2 + 4x + 5)(x+1) = P(x) - Q(x)(x+1)$$

$$5x + 2 = (x^3 + 4x^2 + 4x + 5) - (6x^2 + 1)(6x + 3) = Q(x) - (6x^2 + 1)(6x + 3)$$

$$= Q(x) - [P(x) - Q(x)(x+1)](6x + 3) = P(x)(x + 4) + Q(x)(6x^2 + 2x + 4)$$

Therefore, $S(x) = x + 4$ and $T(x) = 6x^2 + 2x + 4$.

> ✳ **Definition 5.22.**
> Polynomials $P(x), Q(x) \in K[x]$ are called **relatively prime** if $\gcd(P, Q) = 1$.

This notion can also be generalized for n polynomials.

> ✳ **Definition 5.23.**
> Polynomials $P_1, P_2, \ldots, P_n \in K[x]$ are called **relatively prime** if $\gcd(P_1, P_2, \ldots, P_n) = 1$.

As an immediate corollary of Theorem 5.20, we obtain the following theorem:

✳ **Theorem 5.24.**

Let K be a field. Then, non-zero polynomials $P_i(x) \in K[x]$ for $i = 1, 2, \ldots, n$ are relatively prime if and only if there exist uniquely defined polynomials $S_i(x) \in K[x]$ such that

$$S_1(x)P_1(x) + S_2(x)P_2(x) + \cdots + S_n(x)P_n(x) = 1 \qquad (5.7)$$

✳ **Example 5.25.**
Show that polynomials $P(x) = x^5 + 2$, $Q(x) = 2x^4 + 2 \in \mathbb{Z}_5[x]$ are relatively prime.

Since
$x^5 + 2 = (2x)(2x^4 + 2) + (2x + 2)$,
$2x^4 + 2 = (x^3 + 2x^2 + x + 2)(2x + 2) + 1$,
$2x + 2 = (2x + 2) \cdot 1 + 0$,
we obtain that $\gcd(P(x), Q(x)) = 1$ and so these polynomials are relatively prime.

5.4. Factorization of Polynomials. Irreducible Polynomials

One of the most important theorems in the theory of polynomials is the theorem of uniqueness of the factorization of a polynomial into a finite product of irreducible polynomials.

Recall that the only invertible elements in the ring $K[x]$ are invertible elements of the ring K. If K is a field, then invertible elements of $K[x]$ are all elements of the field K and only these elements.

✳ **Definition 5.26.**

A polynomial $P(x) \in K[x]$ is called **decomposable** (or **reducible**) over a ring K if $P(x)$ can be written in the form of a product of polynomials $P(x) = S(x)R(x)$, where $S(x), R(x)$ are not invertible elements of the ring $K[x]$. A polynomial $P(x)$ is called **irreducible** (or **undecomposable**) over K if for every factorization $P(x) = S(x)R(x)$ at least one of the polynomials $S(x)$ or $R(x)$ is invertible in the ring $K[x]$.

Note that reducibility or irreducibility of a polynomial $P(x)$ depends on a ring of coefficients of this polynomial.

✳ **Examples 5.27.**

1. The polynomial $P(x) = x^2 + 1$ is irreducible over the field of real numbers and is decomposable over the field of complex numbers:

$$P(x) = (x - i)(x + i).$$

2. The polynomial $Q(x) = x^3 - 2$ is irreducible over the field of rational numbers and is decomposable over the field of real numbers:

$$Q(x) = (x - \sqrt[3]{2})(x^2 + \sqrt[3]{2}x + \sqrt[3]{4}).$$

Since the ring of polynomials $K[x]$ over a field K is a principal ideal domain, from Lemma 4.60, it follows the following theorem.

✻ **Theorem 5.28.**
Let K be a field. If $P(x), Q(x), H(x) \in K[x]$, $P(x)$ is an irreducible polynomial over K and $P(x)|Q(x)H(x)$, then $P(x)|Q(x)$ or $P(x)|H(x)$.

Using mathematical induction from the last theorem, we obtain the following corollary:

✻ **Theorem 5.29.**
Let K be a field. If $P(x), H_i(x) \in K[x]$, $i = 1, 2, \ldots, n$, $P(x)$ is an irreducible polynomial over K and $P(x)|(H_1(x)H_2(x)\cdots H_n(x))$, then there is an index $i = 1, 2, \ldots, n$ such that $P(x)|H_i(x)$.

✳ **Definition 5.30.**

If $P(x)|Q(x)$ and $Q(x)|P(x)$, then polynomials $P(x)$ and $Q(x)$ are called **associated**.

✳ **Theorem 5.31.**
If $P(x), Q(x) \in K[x]$ are associated polynomials, then $P(x) = \alpha Q(x)$ where $\alpha \in U(K)$.

Proof.

If $P(x), Q(x) \in K[x]$ are associated polynomials, then $P(x) = Q(x)S(x)$ and $Q(x) = P(x)R(x)$. Hence, $P(x) = P(x)R(x)S(x)$, or $R(x)S(x) = 1$, because $K[x]$ is an integral domain. Since, in the ring $K[x]$, the only invertible elements are invertible elements of the ring K, $R(x), S(x) \in U(K)$. □

✳ **Theorem 5.32.**
Let K be a field. If $P(x), Q(x), H(x) \in K[x]$, $P(x)|Q(x)H(x)$ and $\gcd(P, Q) = 1$, then $P(x)|H(x)$.

Proof.

If $\gcd(P(x), Q(x)) = 1$, then, by Theorem 5.24, there exist polynomials $R(x), S(x) \in K[x]$ such that $P(x)R(x) + Q(x)S(x) = 1$. Multiplying both sides of this equality by $H(x)$, we obtain that $P(x)R(x)H(x) + Q(x)S(x)H(x) = H(x)$. Therefore, $P(x)|H(x)$, because $P(x)|Q(x)H(x)$ by assumption of the theorem. □

The next theorem is known as the theorem of the factorization uniqueness of polynomials. It follows immediately from Theorem 4.82, since a polynomial ring $K[x]$ over a field K is a principal ideal domain. In fact, this theorem only has theoretical value. It is typical theorem of existence, because it states only the existence of this factorization, but it does not show in what way we can obtain this factorization.

✳ **Theorem 5.33.**
Let K be a field. If $P(x)$ is a non-zero polynomial of the ring $K[x]$, then $P(x)$ can be written in the form of a product of polynomials:

$$P(x) = H_1(x)H_2(x)\cdots H_s(x) \qquad (5.8)$$

where each polynomial H_i, for $i = 1, 2, \ldots, s$, is irreducible in $K[x]$.
If $P(x)$ has two factorizations:

$$P(x) = H_1(x)H_2(x)\cdots H_s(x) = Q_1(x)Q_2(x)\cdots Q_t(x) \qquad (5.9)$$

where each polynomial $H_i(x)$, for $i = 1, 2, \ldots, s$, and each polynomial $Q_j(x)$, for $j = 1, 2, \ldots, t$, is irreducible in $K[x]$, then $s = t$ and there is a permutation σ such that $H_i = \alpha_{\sigma(i)}Q_{\sigma(i)}$, where $\alpha_{\sigma(i)} \in U(K)$ for each $i = 1, 2, \ldots, s$.

Proof. Suppose that in factorization (5.8) there exist a_i polynomials equal to $H_i(x)$. Then, we can group the equal polynomials and obtain the canonical form of the factorization of $P(x)$ into a product of irreducible polynomials:

$$P(x) = cH_1^{a_1}(x)H_2^{a_2}(x)\cdots H_r^{a_r}(x), \qquad (5.10)$$

where $r \le s$ and $0 \neq c \in K$.   □

An exponent a_i is called the **multiplicity** of an irreducible polynomial $H_i(x)$.

✻ Example 5.34.
The polynomial $P(x) = 7(x^3 - 1)^2 \in \mathbb{R}[x]$ has the following canonical form of the factorization into the product of irreducible polynomials over $\mathbb{R}$:

$$P(x) = 7(x^2 + x + 1)^2(x - 1)^2.$$

❋ Theorem 5.35.
Let K be a field and $P(x), Q(x)$ non-zero polynomials in the ring $K[x]$. Let

$$P(x) = H_1^{a_1}(x)H_2^{a_2}(x)\cdots H_s^{a_s}(x) \text{ and } Q(x) = H_1^{b_1}(x)H_2^{b_2}(x)\cdots H_s^{b_s}(x),$$

where each polynomial H_i is irreducible in $K[x]$ and $a_i, b_i \ge 0$ for $i = 1, 2, \ldots, s$. Then, the greatest common divisor $D(x) = \gcd(P, Q)$ of polynomials $P(x)$ and $Q(x)$ has the following form:

$$D(x) = \alpha H_1^{c_1}(x)H_2^{c_2}(x)\cdots H_s^{c_s}(x), \qquad (5.11)$$

where $c_i = \min(a_i, b_i)$ for $i = 1, 2, \ldots, s$, $\alpha \in U(K)$.

Proof.
It is obvious that $D(x)|P(x)$ and $D(x)|Q(x)$. Suppose there is another common divisor $H(x)$ of polynomials $P(x)$ and $Q(x)$, i.e., $H(x)|P(x)$ and $H(x)|Q(x)$. Then, from Theorem 5.33, it follows that

$$H(x) = \varepsilon H_1^{d_1}(x)H_2^{d_2}(x)\cdots H_s^{d_s}(x),$$

where $\varepsilon \in U(K)$ and $(d_i \le a_i) \wedge (d_i \le b_i)$ for $i = 1, 2, \ldots, n$. Therefore, $H(x)|D(x)$ and $D(x)$ is the greatest common divisor of polynomials $P(x)$ and $Q(x)$.   □

❋ Corollary 5.36.
If polynomials $P(x)$ and $Q(x)$ are decomposed into products of irreducible factors $P(x) = P_1^{k_1}P_2^{k_2}\cdots P_n^{k_n}$ and $Q(x) = Q_1^{s_1}Q_2^{s_2}\cdots Q_m^{s_m}$, then they are relatively prime if and only if no factor P_i is associated with any factor Q_j.

5.5. Roots of Polynomials

By the **value** of a polynomial $P(x) = a_n x^n + a_{n-1} x^{n-1} + \cdots + a_1 x + a_0 \in K[x]$ at an element $c \in K$, we mean the result of substituting c for x in $P(x)$, i.e., an element $a_n c^n + a_{n-1} c^{n-1} + \cdots + a_1 c + a_0 \in K$ and we denote it by $P(c)$.

For all polynomials $P(x) = a_n x^n + a_{n-1} x^{n-1} + \cdots + a_1 x + a_0$ and $Q(x) = b_m x^m + b_{m-1} x^{m-1} + \cdots + b_1 x + b_0$, the following equalities hold:

$$(P + Q)(c) = P(c) + Q(c)$$

$$(P \cdot Q)(c) = P(c) \cdot Q(c)$$

for all $c \in K$. This means that the mapping $\varphi : K[x] \to K$, given as follows: $\varphi(P(x)) = P(c)$, is a homomorphism of $K[x]$ to K. The image of this homomorphism is denoted by $K[c]$.

✳ **Theorem 5.37.**
The remainder on division of a polynomial $P(x)$ by $(x - c)$ is equal to $P(c)$.

Proof.
By Theorem 5.14, there exist a polynomial $Q(x)$ and an element $r \in K$ such that

$$P(x) = (x - c)Q(x) + r \tag{5.12}$$

Substituting c instead of x in (5.12), we obtain that $r = P(c)$. $\square$

✳ **Definition 5.38.**
An element $x_0 \in K$ is called a **root** of a polynomial $P(x) \in K[x]$ if $P(x_0) = 0$.

✳ **Example 5.39.**
Let $P(x) = x^4 - 1$ be the polynomial over the field of real numbers. Then $x_1 = 1$ and $x_2 = -1$ are roots of $P(x)$.

If we consider $P(x) = x^4 - 1$ as a polynomial over the field of complex numbers, then numbers $x_1 = 1$, $x_2 = -1$, $x_3 = i$ and $x_4 = -i$ are roots of $P(x)$.

✳ **Theorem 5.40.**
An element $x_0 \in K$ is a root of a polynomial $P(x) \in K[x]$ if and only if there is a polynomial $Q(x) \in K[x]$ such that

$$P(x) = (x - x_0)Q(x) \tag{5.13}$$

Proof.
Let $P(x_0) = 0$. Dividing the polynomial $P(x)$ by $x - x_0$ we obtain:

$$P(x) = (x - x_0)Q(x) + r,$$

where $r \in K$. Hence, $r = 0$ because $P(x_0) = 0$. Therefore, $P(x) = (x - x_0)Q(x)$.

On the other hand, if $P(x) = (x - x_0)Q(x)$ then $P(x_0) = 0$, which means that x_0 is a root of $P(x)$. □

To verify whether an element $c \in K$ is a root of a polynomial $P(x) \in K[x]$, there is an effective algorithm called the **Horner scheme**.

Let $P(x) = a_n x^n + a_{n-1} x^{n-1} + \cdots + a_1 x + a_0$ be a non-zero polynomial in $K[x]$ with $a_n \neq 0$ and $Q(x) = x - c \in K[x]$. By the division algorithm, there is a polynomial $S(x) \in K[x]$ and an element $r \in K$ such that

$$P(x) = (x - c)S(x) + r, \quad \text{where} \quad S(x) = b_{n-1} x^{n-1} + \cdots + b_1 x + b_0.$$

From this equality we obtain that:

$a_n = b_{n-1}$
$a_k = b_{k-1} - cb_k$ for $k = 1, 2, \ldots, n - 1$
$a_0 = r - cb_0$
Hence,
$b_{n-1} = a_n$
$b_{k-1} = cb_k + a_k$ for $k = 1, 2, \ldots, n - 1$.
$r = cb_0 + a_0$

These results can be written in the form of the following table, which is called the **Horner scheme**:

	a_n	a_{n-1}	$\cdots$	a_1	a_0
c	$b_{n-1} = a_n$	$b_{n-2} = cb_{n-1} + a_{n-1}$	$\cdots$	$b_0 = cb_1 + a_1$	$r = cb_0 + a_0$

In the bottom line of this table, we have the coefficients b_i of the polynomial $S(x)$. The last element of this line is the remainder r. From Theorem 5.37, it follows that $r = P(c)$, so that if $r = 0$, then c is a root of $P(x)$.

✳ Examples 5.41.

1. Divide the polynomial $P(x) = 2x^5 - 9x^4 - x^2 + 27 \in \mathbb{R}[x]$ by the polynomial $Q(x) = x - 4$. We use the Horner scheme:

	2	-9	4	-1	0	27
4	2	-1	0	-1	-4	11

   From this scheme, we get that the quotient $S(x) = 2x^4 - x^3 - x - 4$ and the remainder $r = 11$. So, $P(x) = (x - 4)(2x^4 - x^3 - x - 4) + 11$.

2. Verify if the element $c = 2 \in \mathbb{R}$ is a root of the polynomial $P(x) = x^4 - 5x^2 + 6x - 8 \in \mathbb{R}[x]$. We use the Horner scheme:

	1	0	-5	6	-8
2	1	2	-1	4	0

This table shows that the quotient $S(x) = x^3 + 2x^2 - x + 4$ and the remainder $r = 0$. Therefore, $P(x) = (x - 2)(x^3 + 2x^2 - x + 4)$, where $r = P(2) = 0$. Thus, $c = 2$ is a root of the polynomial $P(x)$.

3. Verify if the element $c = 3 \in \mathbb{Z}_5$ is a root of the polynomial $P(x) = 2x^5 + x^4 + 4x^3 + x^2 + 3 \in \mathbb{Z}_5[x]$. Using the Horner scheme in the following form:

	2	1	4	1	0	3
3	2	2	0	1	3	2

we obtain that $Q(x) = 2x^4 + 2x^3 + x + 3$ and so $P(x) = (x-3)(2x^4 + 2x^3 + x + 3) + 2 = (x+2)(2x^4 + 2x^3 + x + 3) + 2$. In this way, $r = P(3) = 2 \neq 0$. Thus, $c = 3$ is not a root of the polynomial $P(x)$.

* **Definition 5.42.**

An element $x_0 \in K$ is called a **root of multiplicity** m of a polynomial $P(x) \in K[x]$ if and only if there is a polynomial $Q(x) \in K[x]$ such that

$$P(x) = (x - x_0)^m Q(x) \qquad (5.14)$$

and $Q(x_0) \neq 0$. If $m = 1$, then x_0 is called a **simple root**.

* **Theorem 5.43.**
Let $P(x) \in K[x]$ be a polynomial of degree n. Then, the polynomial $P(x)$ possesses at most n roots in K and the sum of their multiplicities is $\leq n$.

Proof.
We prove this theorem using mathematical induction on degree of polynomials. The statement is trivial, if $\deg P(x) = n = 0$. Suppose that the statement is proved for all polynomials with degree $< n$. Let $P(x) \in K[x]$ be a polynomial of degree n. If $P(x)$ has no roots, then the statement holds. Let $a_1, a_2, \ldots, a_s \in K$ be all roots of $P(x)$ of multiplicity correspondingly $k_1, k_2, \ldots, k_s$. Therefore, by Theorem 5.40, $P(x) = (x - a_1)^{k_1} Q(x)$ and $\deg Q(x) < n - k_1$. Then, by inductive assumption, $Q(x)$ possesses roots $a_2, \ldots, a_s$ of multiplicity correspondingly $k_2, \ldots, k_s$ and $k_2 + k_3 + \cdots + k_s \leq n - k_1$, hence, $k_1 + k_2 + \cdots + k_s \leq n$. □

* **Corollary 5.44.**
Let $P(x), Q(x) \in K[x]$ be polynomials of degree $\leq n$ which have the same values at least for $n + 1$ elements of K. Then, polynomials $P(x)$ and $Q(x)$ are equal one to other.

Proof.
A polynomial $H(x) = P(x) - Q(x)$ is of degree $\leq n$ and has $> n$ different roots. Then, from Theorem 5.43, it follows that $H(x) = 0$, i.e., $P(x) = Q(x)$. □

The multiplicity of a root can be efficiently verified using the Horner scheme.

✳ Example 5.45.
Find the multiplicity of the root $x = 2$ of the polynomial:

$$P(x) = x^5 - 5x^4 + 7x^3 - 2x^2 + 4x - 8 \in \mathbb{R}[x].$$

We use the Horner scheme a few times:

	1	−5	7	−2	4	−8
2	1	−3	1	0	4	0
2	1	−1	−1	−2	0	
2	1	1	1	0		
2	1	3	7			

Since the last elements are 0 in the first three lines (except the top line), and in the fourth line the last element $7 \neq 0$, $x = 2$ is a root of multiplicity 3.

✳ Example 5.46.
Find the multiplicity of the root $x = 1$ of the polynomial:

$$P(x) = x^4 + 4x^3 + 3x^2 + 3x + 4 \in \mathbb{Z}_5[x].$$

We use the Horner scheme:

	1	4	3	3	4
1	1	0	3	1	0
1	1	1	4	0	
1	1	2	1		

Since the last elements are 0 in the first two lines (except the top line), and in the third row the last element $1 \neq 0$, $x = 1$ is a root of multiplicity 2.

The notion of a root with multiplicity is closely related to the concept of the derivative of a polynomial.

❋ Definition 5.47.
The **derivative** of a polynomial $P(x) = a_n x^n + a_{n-1} x^{n-1} + \cdots + a_1 x + a_0 \in K[x]$ is a polynomial of the form:

$$P' = na_n x^{n-1} + (n-1)a_{n-1} x^{n-2} + \cdots + a_1 \qquad (5.15)$$

Higher order derivatives of a polynomial $P(x)$ are defined by the following recursive formula:

$$P^{(m)}(x) = (P^{(m-1)}(x))' \quad \text{for} \ \ m \in \mathbb{N}. \qquad (5.16)$$

✳ Examples 5.48.

- Let $P(x) = 2x^3 - 3x^2 - 7x + 15 \in \mathbb{R}[x]$, then $P'(x) = 6x^2 - 6x - 7$, $P''(x) = 12x - 6$, $P'''(x) = 12$, $P^{(4)} = 0$.

- Let $P(x) = 4x^5 - 2x + 34 \in \mathbb{R}[x]$, then $P'(x) = 20x^4 - 2$, $P''(x) = 80x^3$, $P'''(x) = 240x^2$, $P^{(4)} = 480x$, $P^{(5)} = 480$, $P^{(4)} = 0$.

✳ Theorem 5.49.
For any polynomials $P(x), Q(x) \in K[x]$, the following equalities hold:

$$(P + Q)' = P' + Q' \tag{5.17}$$

$$(PQ)' = P'Q + PQ' \tag{5.18}$$

Proof.

Let $P(x) = a_n x^n + a_{n-1} x^{n-1} + \cdots + a_1 x + a_0$ and $Q(x) = b_m x^m + b_{m-1} x^{n-1} + \cdots + b_1 x + b_0$. Suppose that $n \geq m$. Then,

$$P(x) + Q(x) = a_n x^n + a_{n-1} x^{n-1} + \cdots + a_{m+1} x^{m+1} + (a_m + b_m) x^m +$$

$$+ (a_{m-1} + b_{m-1}) x^{m-1} + \cdots + (a_1 + b_1) x + (a_0 + b_0),$$

hence,

$$(P(x) + Q(x))' = n a^n x^{n-1} + (n-1) a_{n-1} x^{n-2} + \cdots + (m+1) a_{m+1} x^m +$$

$$+ m(a_m + b_m) x^{m-1} + (m-1)(a_{m-1} + b_{m-1}) x^{m-2} + \cdots +$$

$$+ 2(a_2 + b_2) x + (a_1 + b_1) \tag{5.19}$$

On the other hand,

$$P'(x) = n a_n x^{n-1} + (n-1) a_{n-1} x^{n-2} + \cdots + a_1,$$

$$Q'(x) = m b_m x^{m-1} + (m-1) b_{m-1} x^{m-2} + \cdots + b_1,$$

hence,

$$P'(x) + Q'(x) = n a_n x^{n-1} + (n-1) a_{n-1} x^{n-2} + \cdots + + (m+1) a_{m+1} x^m +$$

$$+ m(a_m + b_m) x^{m-1} + (m-1)(a_{m-1} + b_{m-1}) x^{m-2} + \cdots +$$

$$+ 2(a_2 + b_2) x + (a_1 + b_1) \tag{5.20}$$

Comparing equalities (5.19) and (5.20), we obtain that $(P + Q)' = P' + Q'$. The second part of the theorem is proved analogously. □

✳ Theorem 5.50.
If an element $x_0 \in K$ is a root of multiplicity m of a polynomial $P(x) \in K[x]$, then, for $m > 1$, the element x_0 is a root of multiplicity $m - 1$ of the derivative P' of the polynomial $P(x)$. If $m = 1$, then the element x_0 is not a root of P'.

Proof.

Let an element $x_0 \in K$ be a root of multiplicity m of a polynomial $P(x)$, then $P(x) = (x - x_0)^m Q(x)$ and $Q(x_0) \neq 0$. We use Theorem 5.46 to find the derivative of $P(x)$:

$$P'(x) = (x - x_0)^m Q'(x) + m(x - x_0)^{m-1} Q(x) =$$

$$= (x - x_0)^{m-1} [(x - x_0) Q'(x) + m Q(x)] \qquad (5.21)$$

1. If $m = 1$, then $P'(x) = (x-x_0)Q'(x)+Q(x)$. Then, $P'(x_0) = Q(x_0) \neq 0$, which means, by Theorem 5.40, that x_0 is not a root of $P'(x)$.

2. Let $m > 1$. Since the first summand of the sum $(x-x_0)Q'(x)+mQ(x)$ is divisible by $(x-x_0)$ and the second summand is not divisible by $(x-x_0)$, the whole sum is not divisible by $(x - x_0)$. Therefore, the element x_0 is a root of multiplicity $m - 1$ of the polynomial $P(x)$.

$\square$

Using Theorem 5.50 a few times, we obtain the following corollary:

✳ **Corollary 5.51.**
An element $x_0 \in K$ is a root of multiplicity m of a polynomial $P(x) \in K[x]$ if and only if

$$P(x_0) = P'(x_0) = \cdots = P^{(m-1)}(x_0) = 0 \text{ and } P^{(m)}(x_0) \neq 0.$$

✳ **Example 5.52.**
Determine the multiplicity of the root $x = 1$ of the polynomial:

$$P(x) = x^4 + 4x^3 + 3x^2 + 3x + 4 \in \mathbb{Z}_5[x]$$

We have
$P(1) = 0$
$P'(x) = 4x^3 + 2x^2 + x + 3 \implies P'(1) = 4 + 2 + 1 + 3 = 0$
$P''(x) = 2x^2 + 4x + 1 \implies P''(1) = 2 \neq 0$
Therefore, $x = 1$ is the root of multiplicity 2.

5.6. Polynomials over Rational Numbers

Note that each polynomial $P(x) = a_n x^n + a_{n-1} x^{n-1} + \cdots + a_1 x + a_0 \in \mathbb{Q}[x]$ over the field of rational numbers can be multiplied by some integer $b \in \mathbb{Z}$ (e.g., the least common multiple of all denominators of numbers $a_n, a_{n-1}, \ldots, a_1, a_0$) such that $bP(x) \in \mathbb{Z}[x]$ and rational roots of $P(x)$ and $bP(x)$ will be the same. Therefore, everywhere in this section, we will assume that $P(x) \in \mathbb{Z}[x]$.

> **✳ Theorem 5.53.**
> Let $P(x) = a_n x^n + a_{n-1} x^{n-1} + \cdots + a_1 x + a_0 \in \mathbb{Z}$. If an integer $c \neq 0$ is a root of $P(x)$, then c is a divisor of the constant term a_0.

Proof.
Let $c \neq 0$ be a root of a polynomial $P(x) \in \mathbb{Z}[x]$. This means that

$$P(c) = a_n c^n + a_{n-1} c^{n-1} + \cdots + a_1 c + a_0 = 0.$$

Hence, $a_0 = -(a_n c^n + a_{n-1} c^{n-1} + \cdots + a_1 c) = -c(a_n c^{n-1} + a_{n-1} c^{n-2} + \cdots + a_1)$. In this way, the integer c is a divisor of a_0. □

✳ Remark 5.54.
In order to find integer roots using Theorem 5.53, we need to verify all divisors of the constant term a_0. The number of these divisors may be large enough and such calculations may be quite time consuming. To simplify these calculations, we can use the following observation. Since $P(x) \in \mathbb{Z}[x]$ and $c \in \mathbb{Z}$ is a root of $P(x)$, by Theorem 5.40, there exists a polynomial $Q(x) \in \mathbb{Z}[x]$ such that $P(x) = (x - c)Q(x)$. Hence, $P(1) = (1 - c)Q(1)$ and $P(-1) = (-1 - c)Q(-1)$. If ± 1 are not roots of $P(x)$, then also

$$\frac{P(1)}{1 - c}, \quad \frac{P(-1)}{1 + c} \in \mathbb{Z} \tag{5.22}$$

since $Q(1), Q(-1) \in \mathbb{Z}$.

Therefore, in order for the number $c \in \mathbb{Z}$ to be a root of $P(x) \in \mathbb{Z}[x]$, provided $P(\pm 1) \neq 0$, conditions (5.22) are necessary.

✳ Example 5.55.
Find all integer roots of the polynomial $P(x) = x^4 - 2x^3 - 8x^2 + 13x - 24 \in \mathbb{Z}[x]$.

The divisors of the constant term -24 are integers: $\pm 1, \pm 2, \pm 3, \pm 4, \pm 8, \pm 12, \pm 24$.

Since $P(1) = -20$, $P(-1) = -42$, integers ± 1 are not roots of $P(x)$.

We now verify the conditions (5.22) for integers $\pm 2, \pm 3, \pm 4, \pm 8, \pm 12, \pm 24$. Since

$P(1)/(1 - (-2)) = -20/3 \notin \mathbb{Z}$,
$P(-1)/(1 + 3) = -42/4 \notin \mathbb{Z}$,
$P(1)(1 - 4) = 20/3 \notin \mathbb{Z}$,
$P(-1)/(1 + (-6)) = 42/5 \notin \mathbb{Z}$,
$P(1)/(1 - 8) = 20/7 \notin \mathbb{Z}$,
$P(1)/(1 - 12) = 20/11 \notin \mathbb{Z}$,
$P(1)/(1 - (-12)) = -20/13 \notin \mathbb{Z}$,
$P(1)/(1 - 24) = 20/23 \notin \mathbb{Z}$,
$P(1)/(1 - (-24)) = -20/25 \notin \mathbb{Z}$,

the integers $-2, 3, 4, -6, \pm 8, \pm 12, \pm 24$ are not integer roots of the polynomial $P(x)$.

Therefore, it only the integers 2, -3, -4, 6 remain to be verified. We use the Horner scheme for these numbers.

	1	-2	-8	13	-24
2	1	0	-8	3	-18

From this table, it follows that $x = 2$ is not a root of $P(x)$.

	1	-2	-8	13	-24
-3	1	-5	7	-8	0

From this table, it follows that $x = -3$ is a root of $P(x)$.

	1	-2	-8	13	-24
-4	1	-6	16	-51	180

From this table, it follows that $x = -4$ is not a root of $P(x)$.

	1	-2	-8	13	-24
6	1	4	16	109	630

From this table, it follows that $x = 6$ is not a root of $P(x)$.

Thus, the only integer root of the polynomial $P(x)$ is $x = -3$.

✳ Theorem 5.56.

Let $P(x) = a_n x^n + a_{n-1}x^{n-1} + \cdots + a_1 x + a_0 \in \mathbb{Z}[x]$. If a rational number $p/q \neq 0$, where p, q are relatively primes, is a rational root of $P(x)$, then p is a divisor of the constant term a_0 and q is a divisor of the leading coefficient a_n of this polynomial.

Proof.

Let $\gcd(p, q) = 1$ and let $p/q \neq 0$ be a root of a polynomial $P(x) \in \mathbb{Z}[x]$, which means that

$$P(p/q) = a_n(p/q)^n + a_{n-1}(p/q)^{n-1} + \cdots + a_1(p/q) + a_0 = 0$$

or

$$a_n p^n + a_{n-1}p^{n-1}q + \cdots + a_1 pq^{n-1} + a_0 q^n = 0.$$

Hence, we obtain that

$$a_0 q^n = -(a_n p^n + a_{n-1}p^{n-1}q + \cdots + a_1 pq^{n-1}) =$$

$$= -p(a_n p^{n-1} + a_{n-1}p^{n-2}q + \cdots + a_1 q^{n-1}) \qquad (5.23)$$

and

$$a_n p^n = -(a_{n-1}p^{n-1}q + \cdots + a_1 pq^{n-1} + a_0 q^n) =$$

$$= -q(a_{n-1}p^{n-1} + \cdots + a_1 pq^{n-2} + a_0 q^{n-1}) \qquad (5.24)$$

From equality (5.23), it follows that p is a divisor of $a_0 q^n$, and so $p|a_0$ because $\gcd(p, q) = 1$. Analogously, from equality (5.24), it follows that q is a divisor of $a_n p^n$, and so $q|a_n$. $\qquad\square$

✳ Example 5.57.
Find all rational roots of the polynomial $P(x) = 2x^4 - 7x^3 + 4x^2 - 2x - 3 \in \mathbb{Z}[x]$. By Theorem 5.56, the rational roots of this polynomial can be only numbers: $\pm 1, \pm 3, \pm 1/2, \pm 3/2$.

We have: $P(1) = -7$, $P(-1) = 12$, $P(3) = 0$, so $x = 3$ is an integer root of the polynomial $P(x)$. Then, by Theorem 5.40, this polynomial is divisible by $x - 3$:

$$P(x) = 2x^4 - 7x^3 + 4x^2 - 2x - 3 = (x - 3)(2x^3 - x^2 + x + 1).$$

Hence, other rational roots of $P(x)$ are roots of the polynomial $Q(x) = 2x^3 - x^2 + x + 1$, which can be only the numbers $\pm 1/2$. Since $Q(1/2) = 3/2$ and $Q(-1/2) = 0$, the rational roots of $P(x)$ are only the numbers $x_1 = 3$ and $x_2 = -1/2$.

The finding of all rational roots of polynomials with integer coefficients by directly using Theorem 5.56 may be quite time consuming. To avoid this, we can use another method. First, we note that, from Theorem 5.56, we can immediately obtain the following theorem.

✳ Theorem 5.58.
Let $P(x) = x^n + a_{n-1}x^{n-1} + \cdots + a_1 x + a_0 \in \mathbb{Z}[x]$. If the polynomial $P(x)$ has a rational root, then this root is an integer.

In this way, the method of finding rational roots of a polynomial $P(x) = a_n x^n + a_{n-1}x^{n-1} + \cdots + a_1 x + a_0 \in \mathbb{Z}[x]$ consists of finding integer roots of another polynomial, namely a monic polynomial:

$$(a_n)^{n-1}P(x) = Q(y) = y^n + a_{n-1}y^{n-1} + \cdots + (a_n)^{n-2}a_1 y + (a_n)^{n-1}a_0$$

where $y = a_n x$.

Since any rational root of $P(x)$ is also a rational root of the monic polynomial $Q(y)$, which is an integer root, by Theorem 5.58, this method reduces to finding integer roots of monic polynomials.

✳ Example 5.59.
Find all rational roots of the polynomial $P(x) = 2x^3 + 3x^2 + 6x - 4 \in \mathbb{Q}[x]$. Multiplying $P(x)$ by 4 and making the substitution $y = 2x$, we obtain a new polynomial $Q(y) = y^3 + 3y^2 + 12y - 16$. We will find integer roots of polynomial $Q(y)$. These roots are among the divisors of the constant term 16: $\pm 1, \pm 2, \pm 4, \pm 8, \pm 16$. Since $Q(1) = 0$, $y = 1$ is a root of $Q(y)$. Then, $Q(y) = (y - 1)(y^2 + 4y + 16)$. Since the polynomial $y^2 + 4y + 16$ has no real roots, the only integer root of $Q(y)$ is $y = 1$, hence, $x = 1/2$ is the only rational root of the polynomial $P(x)$.

At the end of this section, we consider irreducible polynomials with integer coefficients.

> ✳ **Definition 5.60.**
> Let $P(x) = a_n x^n + a_{n-1} x^{n-1} + \cdots + a_1 x + a_0 \in \mathbb{Z}[x]$. Then, $\gcd(a_n, a_{n-1}, \ldots, a_1, a_0)$ is called a **content** of the polynomial $P(x)$ and denoted by $\mathrm{cont}(P)$.

It is obvious that $P(x) = \mathrm{cont}(P) \cdot Q(x)$, where $\mathrm{cont}(Q) = 1$.

> ✳ **Lemma 5.61 (Gauss' Lemma).**
> If $P(x), Q(x) \in \mathbb{Z}[x]$ then $\mathrm{cont}(PQ) = \mathrm{cont}(P)\mathrm{cont}(Q)$.

Proof.
In order to prove this lemma it suffices to consider the case $\mathrm{cont}(P) = \mathrm{cont}(Q) = 1$, otherwise we can divide $P(x)$ and $Q(x)$, respectively, by $\mathrm{cont}(P)$ and $\mathrm{cont}(Q)$.

Let $P(x) = a_n x^n + a_{n-1} x^{n-1} + \cdots + a_1 x + a_0$, $Q(x) = b_m x^m + b_{m-1} x^{m-1} + \cdots + b_1 x + b_0$ and $P(x)Q(x) = c_k x^k + c_{k-1} x^{k-1} + \cdots + c_1 x + c_0$, where $c_i = \sum_{j=0}^{i} a_j b_{i-j}$. Suppose that $\mathrm{cont}(PQ) = d > 1$ and a prime p is a divisor of d. Since $\mathrm{cont}(P) = \mathrm{cont}(Q) = 1$, not all coefficients of polynomials P and Q are divided by p. Let a_r be the first among coefficients of $P(x)$ which is not divided by p, and b_s the first among coefficients of $Q(x)$ which is not divided by p. Then,

$$c_{r+s} = a_r b_s + a_{r+1} b_{s-1} + a_{r+2} b_{s-2} + \cdots + a_{r-1} b_{s+1} + \cdots \equiv a_r b_s \,(\mathrm{mod}\, p),$$

since $b_{s-1} \equiv b_{s-2} \equiv \cdots \equiv b_1 \equiv b_0 \equiv 0 \,(\mathrm{mod}\, p)$ and $a_{r-1} \equiv a_{r-2} \equiv \cdots \equiv a_1 \equiv a_0 \equiv 0 \,(\mathrm{mod}\, p)$. Therefore, $c_{r+s} \equiv a_r b_s \not\equiv 0 \,(\mathrm{mod}\, p)$. This contradicts the assumption that $p | \mathrm{cont}(PQ)$. Thus, $\mathrm{cont}(PQ) = 1$. □

> ✳ **Theorem 5.62.**
> A polynomial $P(x) \in \mathbb{Z}[x]$ is irreducible over $\mathbb{Z}$ if and only if it is irreducible over $\mathbb{Q}$.

Proof.
⇒. Suppose that a polynomial $P(x) \in \mathbb{Z}[x]$ is irreducible over $\mathbb{Z}$ and is decomposable over $\mathbb{Q}$, i.e., $P(x) = G(x)H(x)$, where $G(x), H(x) \in \mathbb{Q}[x]$. Without loss of generality, we can assume that $\mathrm{cont}(P) = 1$. For polynomials $G(x), H(x) \in \mathbb{Q}[x]$, there are natural numbers m, n such that $mG(x), nH(x) \in \mathbb{Z}[x]$. Let $k = \mathrm{cont}(mG)$ and $h = \mathrm{cont}(nH)$. Then, $r = m/k$, $s = n/h \in \mathbb{Z}$ and $rG(x), sH(x) \in \mathbb{Z}[x]$, $\mathrm{cont}(rG) = \mathrm{cont}(sH) = 1$.

By Lemma 5.61, $\mathrm{cont}(rG)\mathrm{cont}(sH) = \mathrm{cont}(rsGH) = 1$, i.e., $\mathrm{cont}(rsP) = 1$. Hence, $rs = 1$. This means that $P(x) = (rG(x))(sH(x))$ is a decomposition of $P(x)$ in $\mathbb{Z}[x]$. The obtained contradiction shows that $P(x)$ is also irreducible in $\mathbb{Q}[x]$.

$\Leftarrow$. This is obvious. $\qquad\square$

One of the most known and effective criteria of irreducibility for some kinds of polynomials with integer coefficients is Eisenstein's criterion:

✳ **Theorem 5.63 (Eisenstein's criterion).**
Let $P(x) = a_n x^n + a_{n-1} x^{n-1} + \cdots + a_1 x + a_0 \in \mathbb{Z}[x]$. If p is a prime such that p divides all coefficients except the leading coefficient a_n, but p^2 does not divide the constant term a_0, then $P(x)$ is irreducible in $\mathbb{Q}[x]$.

Proof.
By Theorem 5.62, it suffices to show that a polynomial $P(x) = a_n x^n + a_{n-1} x^{n-1} + \cdots + a_1 x + a_0 \in \mathbb{Z}[x]$ is irreducible over $\mathbb{Z}$.

Suppose that $P(x) = G(x)H(x)$, where $G(x) = b_m x^m + b_{m-1} x^{m-1} + \cdots + b_1 x + b_0$ and $H(x) = c_k x^k + c_{k-1} x^{k-1} + \cdots + c_1 x + c_0$, where $a_i = \sum_{j=0}^{i} b_i c_{i-j}$.
Since, by assumption, p divides $a_0 = b_0 c_0$ not divided by p^2, but p^2 does not divide a_0, p does not divide either b_0 or c_0. Without loss of generality we assume that b_0 is not divided by p. If all numbers c_i are divided by p, then also a_n must be divided by p. Since this is not the case, there exists a coefficient c_i which is not divided by p. Let i be the least among indexes c_i which is not divided by p. Then, by assumption, the a_i is divided by p. On the other hand,

$$a_i = b_i c_0 + b_{i-1} c_1 + \cdots + b_1 c_{i-1} + b_0 c_i \equiv b_0 c_i \not\equiv 0 \,(\mathrm{mod}\,p),$$

since $c_{i-1} \equiv c_{i-2} \equiv \cdots \equiv c_1 \equiv c_0 \equiv 0 \,(\mathrm{mod}\,p)$ and $b_0 \not\equiv 0 \,(\mathrm{mod}\,p)$, $c_i \not\equiv 0 \,(\mathrm{mod}\,p)$. This contradiction shows that the polynomial $P(x)$ is irreducible over $\mathbb{Z}$, so it is irreducible over $\mathbb{Q}$, by Theorem 5.62. $\qquad\square$

✳ **Example 5.64.**
Verify whether or not the polynomial $P(x) = 7x^5 + 8x^4 - 4x^2 + 6x - 2 \in \mathbb{Z}[x]$ is irreducible over $\mathbb{Q}$. We pick the prime $p = 2$. Since all coefficients except the leading coefficient of the polynomial $P(x)$ are divided by $p = 2$ and the constant term -2 is not divided by $p^2 = 4$, by Eisenstein's criterion, the polynomial $P(x)$ is irreducible over $\mathbb{Q}$.

5.7. Quotient Rings of Polynomial Rings

Let I be an ideal of a polynomial ring $K[x]$ over a field K. Consider the quotient ring

$$K[x]/I = \{f + I \; : \; f \in K[x]\}$$

with operations of addition and multiplications given as follows:

$$(a + I) + (b + I) = (a + b) + I,$$

$$(a + I) \cdot (b + I) = (ab) + I$$

for $a, b \in K[x]$.

Since $K[x]$ is a principal ideal domain, each ideal I is principal, i.e., $I = \langle p(x) \rangle$ for some polynomial $p(x) \in K[x]$.

We will write $f(x) \equiv g(x) \,(\mathrm{mod}\, I) \Leftrightarrow f(x) - g(x) \in I$.

✳ Lemma 5.65.
Let $p(x) \in K[x]$ and $I = \langle p(x) \rangle$. Then, $f(x) \equiv g(x) \,(\mathrm{mod}\, I)$ if and only if $f(x)$ and $g(x)$ has the same remainders on division by $p(x)$.

Proof.
Let $f(x) = q(x)p(x) + r(x)$, where either $r(x) = 0$ or $0 \leq \deg r(x) < \deg p(x)$, and $g(x) = s(x)p(x) + t(x)$, where either $r(x) = 0$ or $0 \leq \deg t(x) < \deg p(x)$. Using these equalities, we obtain: $f(x) - g(x) = (q(x) - s(x))p(x) + (r(x) - t(x))$ or

$$t(x) - r(x) = (q(x) - s(x))p(x) - (f(x) - g(x)) \tag{5.25}$$

$\Rightarrow$. Let $f(x) \equiv g(x) \,(\mathrm{mod}\, I)$, where $I = \langle p(x) \rangle$. This means that $p(x)|(f(x) - g(x))$. Then, from (5.25), it follows that $p(x)|(r(x) - t(x))$. If $r(x) = t(x) = 0$, then we are done. Otherwise, $r(x) - t(x) = p(x)h(x)$ for some $h(x) \in K[x]$. Since $\deg(r(x) - t(x)) < \deg p(x)$, we obtain $r(x) - t(x) = 0$ and so $r(x) = t(x)$.

$\Leftarrow$. This is obvious. □

✳ Theorem 5.66.
Let K be a field, $p(x) = a_n x^n + a_{n-1} x^{n-1} + \cdots + a_1 x + a_0 \in K[x]$ and let $I = \langle p(x) \rangle$ be an ideal of $K[x]$. Then, each element of $K[x]/I$ is of the form:

$$b_0 + b_1 x + b_2 x^2 + \cdots + b_{n-1} x^{n-1} + I \tag{5.26}$$

where $b_i \in K$.

Proof.
The proof of the theorem follows immediately from the definition of a quotient ring and from Lemma 5.65. □

✳ Example 5.67.
Find all cosets of the ideal $I = \langle p(x) \rangle = \langle x^2 + x + 1 \rangle$ of the ring $\mathbb{Z}_2[x]$.
We construct tables of operations in the quotient ring $P = \mathbb{Z}_2[x]/I$.
The quotient ring has 4 cosets: $P = \{I, 1 + I, x + I, 1 + x + I\}$.

Since $x^2 \equiv 1 + x \,(\text{mod}\, I)$, $(x+1)^2 \equiv x \,(\text{mod}\, I)$, and $x(x+1) \equiv 1 + x \,(\text{mod}\, I)$, we have the following tables of addition and multiplication in P:

$+$	I	$1+I$	$x+I$	$1+x+I$
I	I	$1+I$	$x+I$	$1+x+I$
$1+I$	$1+I$	I	$1+x+I$	$x+I$
$x+I$	$x+I$	$1+x+I$	I	$1+I$
$1+x+I$	$1+x+I$	$x+I$	$1+I$	I

and

$\cdot$	I	$1+I$	$x+I$	$1+x+I$
I	I	I	I	I
$1+I$	I	$1+I$	$x+I$	$1+x+I$
$x+I$	I	$x+I$	$1+x+I$	$1+I$
$1+x+I$	I	$1+x+I$	$1+I$	$x+I$

✳ Example 5.68.

Let $P = \mathbb{Q}[x]/I$, where $I = \langle p(x) \rangle = \langle x^2 - 2 \rangle$. Then, $P = \{a + bx + I : a, b \in \mathbb{Q}\}$.

In this ring, we have, for example, $(3x + 4 + I)(5x - 6 + I) = 15x^2 + 20x - 18x - 24 + I = 15x^2 + 2x - 24 + I = 15(x^2 - 2) + 2x + 6 + I = 2x + 6 + I$.

✳ Example 5.69.

Show that there is an isomorphism: $\mathbb{Q}[x]/\langle x^2 - 2 \rangle \cong \mathbb{Q}(\sqrt{2})$.

Consider the mapping $\varphi : \mathbb{Q}[x] \to \mathbb{Q}(\sqrt{2})$ given as follows: $\varphi(f(x)) = f(\sqrt{2})$.

Then, φ is a ring homomorphism. It is an epimorphism, i.e., $\text{Im}(\varphi) = \mathbb{Q}(\sqrt{2})$, since $\varphi(a + bx) = a + b\sqrt{2}$ for all $a, b \in \mathbb{Q}$.

$\text{Ker}(\varphi) = \{f(x) \in \mathbb{Q}[x] : f(\sqrt{2}) = 0\}$.

Since $f(\sqrt{2}) = 0 \Leftrightarrow (x^2 - 2)|f(x)$, we obtain that $f(x) \in I = \langle x^2 - 2 \rangle$, i.e., $\text{Ker}(\varphi) = \langle x^2 - 2 \rangle$. Hence, by Theorem 4.43, it follows that $\mathbb{Q}[x]/\langle x^2 - 2 \rangle \cong \mathbb{Q}(\sqrt{2})$.

If K is a field, then $K[x]$ is a Euclidean domain. Therefore, from Theorem 4.75 we obtain the following theorem.

✳ Theorem 5.70.

Let K be a field. Then, a quotient ring $K[x]/\langle p(x) \rangle$ is a field if and only if $p(x)$ is an irreducible polynomial in $K[x]$.

✳ Example 5.71.

Let $I = \langle x^3 - 5 \rangle$. Since the polynomial $x^3 - 5$ is irreducible in $\mathbb{Q}[x]$, by Theorem 5.70, the quotient ring

$$\mathbb{Q}[x]/I = \{a_0 + a_1 x + a_2 x^2 + I \; : \; a_i \in \mathbb{Q}\}$$

is a field.

✳ Example 5.72.

Let $I = \langle x^3 + 2x + 1 \rangle$, then $\mathbb{Z}_3[x]/I = \{a_0 + a_1 + a_2 x^2 + I \; : \; a_i \in \mathbb{Z}_3\}$. For each element of $\mathbb{Z}_3 = \{0, 1, 2\}$, the value of the polynomial $x^3 + 2x + 1$ is equal to $1 \neq 0$. Therefore, the polynomial $x^3 + 2x + 1$ is irreducible over $\mathbb{Z}_3$. Then, by Theorem 5.70, the quotient ring $\mathbb{Z}_3[x]/I$ is a field. The field $\mathbb{Z}_3[x]/I$ has $3^3 = 27$ elements.

5.8. Exercises

1. Find the remainder on division of the polynomial $P(x) = x^4 - 1$ by the polynomial $Q(x) = x - 2$.

2. Find all values of the polynomial $x^5 + 2x^4 + x^3 + 4x^2 + 3x + 1 \in \mathbb{Z}_5[x]$.

3. Divide with remainder the polynomial $2x^4 + x^3 + x^2 + 3x + 3$ by the polynomial $3x^2 + x + 4$ in the ring $\mathbb{Z}_5[x]$.

4. Find $\gcd(f, g)$ where $f(x) = 3x^4 + x^3 + 2x^2 + 1$, $g(x) = x^2 + 4x + 2 \in \mathbb{R}[x]$.

5. Find the multiplicity of the root $x = 2$ of the polynomial $P(x) = x^5 - 5x^4 + 7x^3 - 2x^2 + 4x - 8 \in \mathbb{R}[x]$.

6. Divide with remainder the polynomial $x^4 + x + 2$ by the polynomial $x + 4$ in the ring $\mathbb{Z}_3[x]$ using the Horner scheme.

7. Find the multiplicity of the root $x = 3$ of the polynomial $P(x) = 2x^4 + 3x^3 + 3x + 3$ in the ring $\mathbb{Z}_5[x]$ using the Horner scheme.

8. Knowing that $x_1 = 1 - 2i$ is a root of the polynomial $P(x) = x^4 - x^3 + x^2 + 9x - 10$, find another roots of this polynomial in $\mathbb{C}$.

9. Find $\gcd(f(x), g(x))$, where $f(x) = 5(x + 1)^6 (x + 2)(x - 3)(x^2 + 7)^2$, $g(x) = 8(x - 1)(x + 2)^3 (x - 3)^2 (x^2 + 4) \in \mathbb{R}[x]$.

10. Verify whether or not the polynomial $f(x) = 7x^4 - 6x^3 + 8x^2 + 12x - 10$ is irreducible in the ring $\mathbb{Q}[x]$.

11. Find all higher order derivatives of the polynomial $f(x) = 3x^4 + 5x^3 + 7x^2 + 9x + 6$ in the ring $\mathbb{Z}_{11}[x]$.

12. Find $a, b \in \mathbb{R}$, for which the polynomial $2x^3 - 3x^2 - ax + b \in \mathbb{R}[x]$ has the remainder 7 on division by $x+1$ and has the remainder 5 on division by $x - 1$.

13. Find all integer roots of the polynomial $P(x) = x^4 - 7x^3 + 4x^2 + 3 \in \mathbb{Z}[x]$.

14. Find all rational roots of the polynomial $P(x) = 4x^3 - 18x^2 - 2x + 5 \in \mathbb{Z}[x]$.

15. Show that $K = \mathbb{Z}_3/\langle x^2 + x + 2 \rangle$ is a field.

16. Find all cosets of the ideal $I = (x^2+1)\mathbb{Z}_2[x]$ of the ring $\mathbb{Z}_2[x]$. Construct the table of operations in the quotient ring $\mathbb{Z}_2[x]/I$.

17. Find all cosets of the ideal $I = (x^2 + x + 1)\mathbb{Z}_2[x]$ of the ring $\mathbb{Z}_2[x]$. Construct the table of operations in the quotient ring $\mathbb{Z}_2[x]/I$.

18. Verify whether or not the quotient ring $\mathbb{Z}_5[x]/I$, where $I = (x^2+1)\mathbb{Z}_5[x]$, is an integral domain.

19. Show that there is an isomorphism of rings: $\mathbb{R}[x]/ < x^2 + 1 > \cong \mathbb{C}$.

References

[1] Ash, R.B. 2007. Basic Abstract Algebra, Dover Publications.

[2] Dummit, D.S. and R.M. Foote. 2004. Abstract Algebra (3rd Ed.), John Wiley & Sons.

[3] Joyner, D., R. Kreminski and J. Turisco. 2004. Applied Abstract Algebra, The Johns Hopkins University Press.

[4] Judson, Th.W. 1994. Abstract Algebra. Theory and Applications, PWS Publishing Company, Michigan.

[5] Zarisky, O. and P. Samuel. 1958. Commutative Algebra, Vol. I.

Chapter 6

Elements of Field Theory

"Pure mathematics is on the whole distinctly more useful than applied. For what is useful above all is technique, and mathematical technique is taught mainly through pure mathematics."

G.H. Hardy

Abstract field theory originates from the works of outstanding French mathematicians E. Galois and J.L. Lagrange, which were devoted to solvability of polynomial equations, and the works related to number theory of brilliant German mathematician C.F. Gauss. The concept of a field is one of the most natural notions in mathematics, as the most known sets of numbers, such as rational, real and complex numbers, are examples of fields. The notion of a field was formed at the turn of 18th and 19th centuries, in the works of German mathematicians L. Kronecker, E.E. Kummer and R. Dedekind. The abstract definition of a field was given in 1871 by J.W.R. Dedekind who essentially completed and expanded the theory of fields.

Some of the basic problems of field theory in algebra are the following: Finding extensions of a given field, i.e., fields that contain a given field as a subfield; construction of new fields and their classification with up to isomorphism; finding groups of automorphisms of given fields.

In this section, we show the construction of fields of fractions of integral domains, study the structure of extensions of fields, and consider algebraic extensions and splitting fields. At the end of this chapter, we give some important properties of polynomials over fields of real and complex numbers.

6.1. A Field of Fractions of an Integral Domain

Recall that a field K is an associative commutative ring with division in which $1 \neq 0$. This means that the set $K^* = K \backslash \{0\}$ of non-zero elements of K is an Abelian group under multiplication. The important examples of fields are fields of rational numbers, real numbers and complex numbers. Also, the example of a finite field is the field $\mathbb{Z}_p$, where p is a prime.

There are different methods for constructing new fields, the most important of which are:

- Quotient rings of polynomials by means of irreducible polynomials.

- The fields of fractions of integral domains.

The first method was considered in the previous chapter. The second method we consider in this section.

✳ Theorem 6.1.

For an integral domain R, a field $Q(R)$ can be constructed such that:
1) A field $Q(R)$ contains a subring P which is isomorphic to the ring R.
2) Each non-zero element of the field $Q(R)$ can be written in the form ab^{-1}, where $a, b \in P$, $b \neq 0$.

Proof.

Consider the set $S = R \times R^* = \{(a, b) \ : \ a \in R, b \in R^*\}$, where $R^* = R\backslash\{0\}$. Define a relation $\sim$ on the set S as follows:

$$(a, b) \sim (c, d) \iff ad = bc. \tag{6.1}$$

It is easy to show that $\sim$ is an equivalence relation on the set S, i.e., it is reflexive, symmetric and transitive. With respect to this relation, we obtain the factor set $S/\sim$ whose elements are equivalence classes. We denote this set by $Q(R)$. The equivalence class containing an element $(a, b) \in S$ will be denoted by $[a, b]$.

On the set $Q(R)$, we can define two operations, called addition and multiplication. Namely, we define addition and multiplication as follows:

$$[a, b] + [c, d] = [ad + bc, bd] \tag{6.2}$$

$$[a, b] \cdot [c, d] = [ac, bd] \tag{6.3}$$

Since $b, d \in R^*$ and R is an integral domain, $bd \in R^*$ as well. So, addition and multiplication defined by (6.2)-(6.3) are operations on S.

First, we will show that these two operations are well-defined, i.e., they do not depend on the choice of representatives of classes.

Let $[a, b] = [a_1, b_1]$ and $[c, d] = [c_1, d_1]$. We need to show that $[ad + bc, bd] = [a_1 d_1 + b_1 c_1, b_1 d_1]$, i.e., $(ad + bc)(b_1 d_1) = (a_1 d_1 + b_1 c_1)(bd)$ and $[ac, bd] = [a_1 c_1, b_1 d_1]$, i.e., $(ac)(b_1 d_1) = (a_1 c_1)(bd)$.

Since $[a, b] = [a_1, b_1]$ and $[c, d] = [c_1, d_1]$, we have that $ab_1 = a_1 b$ and $cd_1 = c_1 d$. Hence, we obtain:

$$(ad + bc)(b_1 d_1) = adb_1 d_1 + bcb_1 d_1 = a_1 bdd_1 + bb_1 c_1 d = (a_1 d_1 + b_1 c_1)(bd)$$

$$(ac)(b_1 d_1) = acb_1 d_1 = (ab_1)(cd_1) = (a_1 b)(c_1 d) = (a_1 c_1)(bd).$$

We will show that $Q(R)$, under operations (6.2) and (6.3), is a field.

By direct verification, it can be shown that addition is commutative and associative with neutral element $[0, 1]$. From the definition of multiplication, it immediately follows that multiplication is commutative and associative with neutral element $[1, 1]$.

We show that addition and multiplication are interconnected by distributive low:

$$([a, b] + [c, d])[f, g] = [a, b][f, g] + [c, d][f, g].$$

Really, the left part of this equality: $([a, b] + [c, d])[f, g] = [ad + bc, bd][f, g] = [(ad+bc)f, (bd)g] = [adf + bcf, bdg]$ and the right one: $[a, b][f, g] + [c, d][f, g] = [af, bg] + [cf, dg] = [afdg + bgcf, bgdg]$

Since $(adf + bcf)(bgdg) = (afdg + bgcg)(bdg)$, $(afdg + bgcf, bgdg) \sim (adf + bcf, bdg)$. Hence, $[adf + bcf, bdg] = [afdg + bgcf, bgdf]$ and so $([a, b] + [c, d])[f, g] = [a, b][f, g] + [c, d][f, g]$.

Therefore, the set $Q(R)$ under operations (6.2), (6.3) is a ring. The zero of this ring is the element $[0, 1]$ and the identity is the element $[1, 1]$. It is obvious that $[0, b] = [0, 1]$ for all $b \in R$, as $(0, b) \sim (0, 1)$. Moreover, all elements $[a, b]$ with $a \neq 0$ are non-zero. An inverse element to a non-zero element $[a, b]$ is an element $[b, a]$, because $[a, b][b, a] = [ab, ab]$ and $(ab, ab) \sim (1, 1)$.

Consider the subset $P = \{[x, 1] : x \in R\} \subset Q(R)$. Then, for all elements $a, b \in R$, we have:

$$[a, 1] + [b, 1] = [a + b, 1] \in P$$

$$[a, 1][b, 1] = [ab, 1] \in P$$

$$[0, 1], \ [1, 1] \in P$$

So, P is a subring of $Q(R)$ and the mapping $\varphi : R \to P$ given as follows: $\varphi(x) = [x, 1]$ for all $x \in R$, is an isomorphism of rings. Really,

$$\varphi(x + y) = [x + y, 1] = [x, 1] + [y, 1] = \varphi(x) + \varphi(y)$$

$$\varphi(xy) = [xy, 1] = [x, 1][y, 1] = \varphi(x)\varphi(y)$$

Therefore, φ is a ring homomorphism. Since

$$\varphi(x) = \varphi(y) \Rightarrow [x, 1] = [y, 1] \Rightarrow (x, 1) \sim (y, 1) \Rightarrow x = y$$

we obtain that φ is an injection. It is also obvious that $\text{Im}(\varphi) = P$. Therefore, φ is an isomorphism of rings.

Note that each non-zero element $[a, b] \in Q(R)$ can be written in the form:

$$[a, b] = [a, 1][1, b] = [a, 1][b, 1]^{-1} = \varphi(a)[\varphi(b)]^{-1},$$

since $[1, b][b, 1] = [1, 1]$. $\qquad\qquad\qquad\qquad\qquad\qquad\qquad\qquad \square$

✳ **Definition 6.2.**

The field $Q(R)$ constructed in theorem 6.1 is called the **field of fractions** of a ring R.

✳ Examples 6.3.

1. If $\mathbb{Z}$ is the ring of integers, then its field of fractions is $\mathbb{Q}$. Really:

$$Q(\mathbb{Z}) = \{[a, b] \ : \ a, b \in \mathbb{Z}, b \neq 0\} = \mathbb{Q}$$

is the field of rational numbers, and an element $[a, b] \in Q(\mathbb{Z})$ is denoted by $\dfrac{a}{b}$.

2. If K is an integral domain, then the ring of polynomials $K[x]$ is also an integral domain. The field of fractions of $K[x]$ is:

$$Q(K[x]) = \{[f(x), g(x)] \ : \ f(x), g(x) \in K[x], g(x) \neq 0\},$$

which is called the **field of rational functions** with coefficients in K and it is denoted by $K(x)$. An element $[f(x), g(x)] \in K(x)$ is denoted by $\dfrac{f(x)}{g(x)}$.

3. If $K = F_q$ in a finite field having q elements, then the ring of polynomials $F_q[x]$ is an integral domain. Then, its field of fractions is:

$$Q(F_q[x]) = \{[f(x), g(x)] \ : \ f(x), g(x) \in F_q[x], g(x) \neq 0\},$$

which is called the **field of rational functions** with coefficients in F_q and it is denoted by $F_q(x)$. An element $[f(x), g(x)] \in F_q(x)$ is denoted by $\dfrac{f(x)}{g(x)}$.

6.2. The Characteristic of a Field

The important property of a field is its characteristic. The characteristic can be introduced for a field or an arbitrary ring with identity.

✳ Definition 6.4.

The **characteristic** of a field (or a ring with identity) K is a minimal positive integer n such that the sum of n copies of the identity 1 is equal to 0:

$$n \cdot 1 = 1 + \cdots + 1 = 0 \tag{6.4}$$

In this case, we write char $K = n$. If such a number does not exist, we assume that the characteristic of K is equal to 0 and write char $K = 0$.

✳ Examples 6.5.

1. The fields of rational numbers, real numbers and complex numbers have the characteristic zero:

$$\text{char } \mathbb{Q} = \text{char } \mathbb{R} = \text{char } \mathbb{C} = 0.$$

2. The field $\mathbb{Z}_p$, where p is a prime, has the characteristic p: char $\mathbb{Z}_p = p$.

3. The field of rational functions $\mathbb{Z}_p(x)$ over the field $\mathbb{Z}_p$ has the characteristic p: char $\mathbb{Z}_p(x) = p$.

✳ Theorem 6.6.

The characteristic of a field K is equal to either zero or a prime p. If char $K = p$ and $p|n$, where $n \in \mathbb{Z}^+$, then $na = 0$ for all $a \in K$.

Proof.

Suppose that char $K \neq 0$. If char $K = p$ is not a prime, then $p = nm$, where $m, n \in \mathbb{Z}$ with $m, n \neq 1$. Then, $(nm) \cdot 1 = 0 = n(m \cdot 1)$. Since K is a field and $n \neq 0$, $m \cdot 1 = 0$, which contradicts the definition of p as a minimal positive integer with this property.

If $a \in K$, char $K = p$ and $n = mp$, then $na = mp(1 \cdot a) = m(p \cdot 1)a = 0$.
□

✳ Theorem 6.7.

If char $K = 0$, then K contains a subfield isomorphic to the field of rational numbers $\mathbb{Q}$. If the characteristic of a field K is a prime p, then K contains a subfield isomorphic to the field $\mathbb{Z}_p$.

Proof.

Consider the mapping $f : \mathbb{Z} \to K$ such that $f(n) = n \cdot 1 = 1 + \cdots + 1$ for all $n \in \mathbb{Z}^+$. We also define $f(-n) = (-n) \cdot = (-1) + \cdots + (-1)$ for all $n \in \mathbb{Z}^+$, and $f(0) = 0$. Then, f is a ring homomorphism and we have two possibilities:

1. char $K = 0$. Then, $\text{Ker}(f) = \{0\}$ and $f(\mathbb{Z}) \cong \mathbb{Z}$. In this case, K contains a field of fractions of the ring $f(\mathbb{Z})$. Really, let $S = \{xy^{-1} \in K : x, y \in f(\mathbb{Z})\}$. Then, S is a subfield of K and the mapping $g : \mathbb{Q} \to S$, where $f(a/b) = f(a)[f(b)]^{-1}$, is an isomorphism of fields. This shows that $S \cong \mathbb{Q}$.

2. char $K = p$. Then, $\text{Ker}(f) = p\mathbb{Z}$ and $f(\mathbb{Z}) \cong \mathbb{Z}/p\mathbb{Z} \cong \mathbb{Z}_p$ is a minimal subfield of the field K.

□

✳ Definition 6.8.

A field K is called **simple** if it does not contain any other fields except itself.

From Theorem 6.7, it follows that the only simple fields are the fields $\mathbb{Q}$ and $\mathbb{Z}_p$, where p is a prime.

6.3. Field Extensions

> **✳ Definition 6.9.**
> A field F is called an **extension** of a field K, provided that K is a subfield of F. In this case, we write $K \leq F$. If $F \neq K$, then F is called a **proper extension** of K, otherwise F is called a **non-proper extension** of K.

In particular, each field is an extension of a simple field which contains as a subfield. So, each field can be considered as an extension of either $\mathbb{Q}$ or $\mathbb{Z}$.

> **✳ Theorem 6.10.**
> Each homomorphism of fields $f : K \to F$ is a monomorphism.

Proof.
By Theorem 4.29, a field does not contain non-trivial ideals, so $\text{Ker}(f) = \{0\}$, i.e., f is a monomorphism. □

If F is an extension of a field K, then F is an Abelian group under addition and each element $x \in F$ can be multiplied by an element $\lambda \in K$. In this way, all axioms of a vector space hold. Therefore, we have the following theorem:

> **✳ Theorem 6.11.**
> If F is an extension of a field K, then F is a vector space over K.

> **✳ Definition 6.12.**
> The **degree of extension** F over a field K is a dimension of the field F as a vector space over K and it is denoted by $[F : K]$. An extension F of K is called **finite** if $[F : K] < \infty$.

> **✳ Theorem 6.13.**
> If F is an extension of a field K and E is an extension of a field F such that $[E : F] < \infty$, $[F : K] < \infty$, then E is a finite extension of K and
> $$[E : K] = [E : F] \cdot [F : K] \tag{6.5}$$

Proof.
Let $[E : F] = n$, $[F : K] = m$ and let $\{\alpha_1, \alpha_2, \ldots, \alpha_n\}$ be a basis of a field F over a field K, $\{\beta_1, \beta_2, \ldots, \beta_m\}$ a basis of a field E over a field F. Then, an element $x \in E$ can be written as:
$$x = b_1\beta_1 + b_2\beta_2 + \cdots + b_m\beta_m, \quad \text{where} \quad b_1, b_2, \ldots, b_m \in F,$$
and each element $b_s \in F$ can be written in the form:
$$b_s = c_{s1}\alpha_1 + c_{s2}\alpha_2 + \cdots + c_{sn}\alpha_n, \quad \text{where} \quad c_{s1}, c_{s2}, \ldots, c_{sn} \in K.$$

Therefore, an element $x \in E$ can be written in the following form:

$$x = \sum_{s,k} c_{sk} \alpha_k \beta_s.$$

So nm elements $\alpha_k \beta_s$ form a linear span of the field E as a vector space over a field K. We show that these elements are linear independent. Suppose that

$$\sum_{s,k} c_{sk} \alpha_k \beta_s = 0. \qquad (6.6)$$

Since elements $c_{sk} \alpha_k \in F$ for all s, k and elements $b_1, b_2, \ldots, b_m$ are linear independent over F, from (6.6), it follows that:

$$c_{s1} \alpha_1 + c_{s2} \alpha_2 + \cdots + c_{sn} \alpha_n = 0$$

for all s. Since elements $\alpha_1, \alpha_2, \ldots, \alpha_n$ are linear independent over K, from this equality, it follows that $c_{s1} = c_{s2} = \cdots = c_{sn} = 0$ for all s. So, the set $\{\alpha_i \beta_j : i = 1, \ldots, n; j = 1, \ldots, m\}$ forms a basis of the field E over the field K. $\qquad \square$

✴ **Theorem 6.14.**
Let K be a field and $I = \langle p(x) \rangle$, where

$$p(x) = a_0 + a_1 x + a_2 x^2 + \cdots + a_{n-1} x^{n-1} + a_n x^n \in K[x], \quad a_n \neq 0$$

is an irreducible polynomial of degree n over K. Then, $F = K[x]/I$ is an extension of a field K. An element $\theta = x + I \in F$ is a root of the polynomial $p(x)$ in a field F and elements $1, \theta, \theta^2, \ldots, \theta^{n-1}$ form a basis of F over K. Moreover, $[F : K] = n$ and

$$F = \{c_0 + c_1 \theta + c_2 \theta^2 + \cdots + c_{n-1} \theta^{n-1} : c_i \in K, \ p(\theta) = 0\}.$$

Proof.
Since $p(x) \in K[x]$ is an irreducible polynomial over K, by Theorem 5.70, the factor ring $K[x]/I$, where $I = < p(x) >$, is a field. We will show that the field $K[x]/I$ contains a subfield which is isomorphic to the field K.

Consider a homomorphism of fields $h : K \to K[x]/I$, given as follows: $h(b) = b + I$, where $b \in K$. By Theorem 6.10, h is a monomorphism. So, we can identify K with a subfield $\mathrm{Im}\,(h)$ in $K[x]/I$, i.e., we can consider that the field $F = K[x]/I$ is an extension of the field K.

If $\theta = x + I \in F$, then

$$p(\theta) = a_0 + a_1 \theta + a_2 \theta^2 + \cdots + a_{n-1} \theta^{n-1} + a_n \theta^n =$$

$$= a_0 + a_1(x + I) + a_2(x + I)^2 + \cdots + a_{n-1}(x + I)^{n-1} + a_n(x + I)^n =$$

$$= a_0 + a_1 x + a_2 x^2 + \cdots + a_{n-1} x^{n-1} + a_n x^n + I = p(x) + I \equiv I \,(\mathrm{mod}\,I),$$

which is zero in F. In this way, θ is a root of the polynomial $p(x)$ in the field F.

We will show that elements $1, \theta, \theta^2, \ldots, \theta^{n-1}$ form a basis of F over the field K. First, we will show that these elements are linear independent over the field K. Suppose that there are elements $b_0, b_1, b_2, \ldots, b_{n-1} \in K$ such that

$$b_0 + b_1\theta + b_2\theta^2 + \ldots + b_{n-1}\theta^{n-1} = 0.$$

Then:

$$b_0 + b_1x + b_2x^2 + \ldots + b_{n-1}x^{n-1} + I = I,$$

hence, $h(x) = b_0 + b_1x + b_2x^2 + \ldots + b_{n-1}x^{n-1} \in I$, which means that $h(x) = p(x)g(x)$ for some $g(x) \in K[x]$. Since $\deg h(x) = n-1 < n = \deg p(x)$, we get that $h(x) = 0$, i.e., $b_0 = b_1 = b_2 = \cdots = b_{n-1} = 0$. This shows that elements $1, \theta, \theta^2, \ldots, \theta^{n-1}$ are linear independent over K.

Furthermore, we will show that elements $1, \theta, \theta^2, \ldots, \theta^{n-1}$ form a linear span of the field F over K, i.e., $F = \mathrm{Lin}_K\{1, \theta, \theta^2, \ldots, \theta^{n-1}\}$. Consider an element $q(x) + I \in K[x]/I$. Since $K[x]$ is an Euclidean domain, by Theorem 5.9, there are polynomials $g(x), r(x) \in K[x]$ such that $q(x) = g(x)p(x) + r(x)$ and either $r(x) = 0$ or $0 \le \deg r(x) < \deg p(x) = n$. If $r(x) = 0$, then $q(x) \in I$, i.e., the element $q(x) + I = I$ in F, hence, $q(\theta) = 0$ in F. If $r(x) \ne 0$, then $r(x) = c_0 + c_1x + c_2x^2 + \ldots + c_{n-1}x^{n-1}$, where $c_i \in K$. Then, $q(x) + I = r(x) + I$ and so

$$q(\theta) = c_0 + c_1\theta + c_2\theta^2 + \ldots + c_{n-1}\theta^{n-1} \in \mathrm{Lin}_K\{1, \theta, \theta^2, \ldots, \theta^{n-1}\}.$$

Hence, $[F : K] = n$ and

$$F = \{c_0 + c_1\theta + c_2\theta^2 + \ldots + c_{n-1}\theta^{n-1} \; : \; c_i \in K, p(\theta) = 0\}.$$

The operations of addition and multiplication in F are given as follows:

$$r_1(\theta) + r_2(\theta) = (r_1 + r_2)(\theta),$$

$$r_1(\theta)r_2(\theta) = r(\theta),$$

where $r_1(x), r_2(x) \in K[x]$, $0 \le \deg r_1(x), \deg r_2(x) < n$, and $r(x) \in K[x]$ is the remainder on division of $r_1(x)r_2(x)$ by $p(x)$. $\qquad\square$

✳ Example 6.15.
Consider the quotient ring $F = \mathbb{R}[x]/\langle x^2+1 \rangle$. Since the polynomial x^2+1 is irreducible over $\mathbb{R}$, F is a field and $[F : \mathbb{R}] = 2$. By Theorem 6.14,

$$F = \{a + b\theta \; : \; a, b \in \mathbb{R}; \; \theta^2 + 1 = 0\}$$

The operations of addition and multiplication are defined as follows:

$$(a + b\theta) + (c + d\theta) = (a + c) + (b + d)\theta,$$

$$(a + b\theta)(c + d\theta) = ac + (ad + bc)\theta + bd\theta^2 = (ac - bd) + (ad + bc)\theta.$$

Consider the mapping $\varphi : F \to \mathbb{C}$ given as follows:

$$\varphi(a + b\theta) = a + bi.$$

Then, it is easy to verify that φ is an isomorphism of fields F and $\mathbb{C}$.

✳ Example 6.16.

Consider the quotient ring $F = \mathbb{Z}_3[x]/\langle p(x)\rangle$, where $p(x) = x^3 + 2x + 1$. Because $p(0) = 1 \neq 0$, $p(1) = 1 \neq 0$, $p(2) = 1 \neq 0$, then $p(x)$ has no roots in $\mathbb{Z}_3$, so $p(x)$ is an irreducible polynomial in $\mathbb{Z}_3[x]$. Therefore, F is a field and $[F : \mathbb{Z}_3] = 3$.

$$F = \{a + b\theta + c\theta^2 \; : \; a, b, c \in \mathbb{Z}_3, \theta^3 + 2\theta + 1 = 0\}.$$

Hence, we obtain that $\theta^3 = \theta + 2$ and $|F| = 3^3 = 27$. The operations of addition are defined as follows:

$$(a + b\theta + c\theta^2) + (d + u\theta + v\theta^2) = (a + d) + (b + u)\theta + (c + v)\theta^2.$$

The operation of multiplication is defined in a natural way, taking into account that $\theta^3 = \theta + 2$. The following example shows multiplication of elements in F:
$(2+\theta+\theta^2)(1+\theta^2) = 2+\theta+\theta^2+2\theta^2+\theta^3+\theta^4 = 2+\theta+\theta+2+\theta(\theta+2) = 1+\theta+\theta^2.$
Theorem 6.14 can be rewritten in the following form:

✳ Theorem 6.17.
For each non-constant polynomial $f(x) \in K[x]$, there is a finite extension F of a field K and an element $\alpha \in F$ such that $f(\alpha) = 0$.

Proof.

Since $K[x]$ is a unique factorization domain, each polynomial $f(x) \in K[x]$ can be uniquely up to factors in K^* decomposed in a product of irreducible polynomials. Furthermore, we can use Theorem 6.14. □

6.4. Algebraic Elements. Algebraic Extensions

✳ Definition 6.18.

Let a field F be an extension of a field K. An element $\alpha \in F$ is called **algebraic** over a field K if there is a non-zero polynomial $p(x) \in K[x]$ such that $p(\alpha) = 0$. Otherwise, α is called **transcendental** over K.

In a particular case, when $K = \mathbb{Q}$ is the field of rational numbers and $F = \mathbb{C}$ is the field of complex numbers, algebraic elements of $\mathbb{C}$ over $\mathbb{Q}$ are called **algebraic numbers** and transcendental elements are called **transcendental numbers**.

The existence of some transcendental numbers and their construction were discovered in 1844 by J. Liouville. G. Cantor in 1872 showed that the set of all algebraic numbers is enumerable, hence, the existence of transcendental numbers and their infiniteness immediately follow.

✳ Examples 6.19.

1. The numbers 5, $\sqrt[5]{7}$ and i, where $i^2 = -1$, are algebraic over $\mathbb{Q}$ since they are, respectively, the roots of polynomials:

$$x - 5, \quad x^5 - 7, \quad x^2 + 1.$$

2. A rational number p/q is algebraic over $\mathbb{Q}$ since it is a root of the polynomial $qx - p \in \mathbb{Q}[x]$.

3. The Gauss numbers $a + bi$, where $a, b \in \mathbb{Q}$, are algebraic over $\mathbb{Q}$.

4. The number π is not algebraic over $\mathbb{Q}$, it is transcendental over $\mathbb{Q}$, that was proved for the first time by F. Lindemann in 1882.

5. Ch. Hertmite proved for the first time in 1873 that Euler's number e is transcendental over $\mathbb{Q}$.

✳ Examples 6.20.
Show that the number $\sqrt[3]{5} + \sqrt{7}$ is algebraic over the field $\mathbb{Q}$ and find the polynomial whose root is this number.
Let $x = \sqrt[3]{5} + \sqrt{7}$. Then $x - \sqrt{7} = \sqrt[3]{5}$, hence $(x - \sqrt{7})^3 = 5$ or $x^3 - 3\sqrt{7}x^2 + 21x - 7\sqrt{7} = 125$. Therefore $x^3 + 21x - 125 = (7 - 3x^2)\sqrt{7}$ or $(x^3 + 21x - 125)^2 = 7(7 - 3x^2)^2$.
Thus, $\sqrt[3]{5} + \sqrt{7}$ is the root of the polynomial:

$$p(x) = x^6 - 21x^4 - 250x^3 + 735x^2 - 5250x + 15282$$

and so it is an algebraic number.

✶ Definition 6.21.
An extension F of a field K is called **algebraic** if each element of F is algebraic over K.

✻ Theorem 6.22.
If F is a finite extension of a field K, then F is an algebraic extension of K.

Proof.
Let $[F : K] = n$ and $\alpha \in F$. Then, elements $1, \alpha, \alpha^2, \ldots, \alpha^{n-1}, \alpha^n$ are linear dependent over a field K, i.e., there are elements $b_i \in K$ such that $b_0 + b_1\alpha + b_2\alpha^2 + \cdots + b_{n-1}\alpha^{n-1} + b_n\alpha^n = 0$, which means that α is a root of the polynomial $p(x) = b_0 + b_1x + b_2x^2 + \cdots + b_{n-1}x^{n-1} + b_nx^n \in K[x]$, i.e., α is an algebraic element over the field K. ☐

> **✹ Theorem 6.23.**
> If α is an algebraic element over a field K, then:
> 1) there is a unique monic irreducible polynomial $p(x) \in K[x]$ such that $p(\alpha) = 0$;
> 2) if $f(x) \in K[x]$, then $f(\alpha) = 0 \Longleftrightarrow p(x)|f(x)$.

Proof.
Let α be an algebraic element over a field K and let:

$$I = \{f(x) \in K[x] \ : \ f(\alpha) = 0\} \subset K[x].$$

If $f(x), g(x) \in I$, then it is obvious that $f(x) + g(x) \in I$ and $f(x)h(x) \in K[x]$ for all $h(x) \in K[x]$. Therefore, I is an ideal of the ring $K[x]$. Since $K[x]$ is a principal ideal ring, there is a polynomial of smallest degree $p(x) \in I$ such that $I = \langle p(x) \rangle$. Multiplying $p(x)$ by an element of the field K, we can assume that the polynomial $p(x)$ is monic.

Suppose that $p(x)$ is decomposable, i.e., $p(x) = h(x)g(x)$, where $\deg g(x) < \deg p(x)$ and $\deg g(x) < \deg p(x)$. Then, $h(\alpha) = 0$ or $g(\alpha) = 0$, hence, $h(x) \in I$ or $g(x) \in I$. Therefore, $p(x)|h(x)$ or $p(x)|g(x)$. However, this contradicts the choice of $p(x)$, since $\deg g(x) < \deg p(x)$ and $\deg h(x) < \deg p(x)$. Therefore, $p(x)$ is an irreducible polynomial.

If $f(x)$ is another polynomial and $f(\alpha) = 0$, then $f(x) \in I$ and so $p(x)|f(x)$. Conversely, if $p(x)|f(x)$, then $f(\alpha) = 0$. In particular, if $g(x)$ is another monic irreducible polynomial such that $g(\alpha) = 0$, then $p(x)|g(x)$. Hence, $g(x) = p(x)$. $\qquad\square$

> **✳ Definition 6.24.**
> Let α be an algebraic element over a field K. Then a non-zero monic irreducible polynomial of smallest degree $p(x) \in K[x]$, such that $p(\alpha) = 0$, is called the **minimal polynomial** of α. The degree of the minimal polynomial $p(x)$ is called the **degree** of the algebraic element α.

> **✹ Example 6.25.**
> The minimal polynomials of the algebraic numbers -3, $\sqrt[n]{2}$, i are, respectively, the following polynomials:
>
> $$x + 3, \ x^n - 2, \ x^2 + 1.$$

> **✳ Definition 6.26.**
> Let F be an extension of a field K and $\alpha \in F$. The intersection of all subfields of F containing K and α is called the **extension** of K by an element α or a field generated by α and K, and it is written as $K(\alpha)$.
>
> Alternatively, $K(\alpha)$ can be described as the smallest subfield of F containing both K and α.

Note, that $K(\alpha)$ can be described as the set of all rational functions of the form

$$\frac{g(\alpha)}{h(\alpha)} = \frac{b_0 + b_1\alpha + b_2\alpha^2 + \cdots + b_{n-1}\alpha^{n-1}}{c_0 + c_1\alpha + c_2\alpha^2 + \cdots + c_{m-1}\alpha^{m-1}},$$

where $g(x), h(x) \in K[x]$ and $h(\alpha) \neq 0$.

✱ **Examples 6.27.**

     1. $\mathbb{R}(i) = \mathbb{C}$.

     2. $\mathbb{Q}(\sqrt{2}) = \{a + b\sqrt{2} \;:\; a,b \in \mathbb{Q}\}$,   since $\dfrac{x + y\sqrt{2}}{x_1 + y_1\sqrt{2}} =$

$\dfrac{(x + y\sqrt{2})(x_1 - y_1\sqrt{2})}{x_1^2 - 2y_1^2} = x_2 + y_2\sqrt{2}$, where $x, y, x_1, y_1, x_2, y_2 \in \mathbb{Q}$.

     3. $\mathbb{Q}(\sqrt[3]{2}) = \left\{ \dfrac{a + b\sqrt[3]{2} + c\sqrt[3]{4}}{a_1 + b_1\sqrt[3]{2} + c_1\sqrt[3]{4}} \;:\; a, b, c, a_1, b_1, c_1 \in \mathbb{Q} \right\}$.

✱ **Theorem 6.28.**

Let α be an algebraic element over a field K with minimal polynomial $p(x) \in K[x]$. If $\deg p(x) = n$, then

$$K(\alpha) \cong K[x]/\langle p(x) \rangle$$

and $K(\alpha)$ is an n-dimensional vector space over K with a basis $\{1, \alpha, \alpha^2, \ldots, \alpha^{n-1}\}$, i.e., each element $\beta \in K(\alpha)$ can be uniquely written in the form:

$$\beta = b_0 + b_1\alpha + b_2\alpha^2 + \cdots + b_{n-1}\alpha^{n-1}, \quad \text{where} \;\; b_i \in K.$$

Proof.

An arbitrary element $\beta \in K(\alpha)$ has the form: $\beta = g(\alpha)/h(\alpha)$. Since $p(x)$ is an irreducible polynomial and $h(\alpha) \neq 0$, we have:

$$\gcd(p(x), h(x)) = 1.$$

Hence, by Theorem 5.24, there exist polynomials $u(x), v(x) \in K[x]$ such that $1 = p(x)u(x) + h(x)v(x)$. Therefore, $h(\alpha)v(\alpha) = 1$, and so $\beta = g(\alpha)v(\alpha)$. Let $g(x)v(x) = p(x)s(x) + r(x)$, where either $r(x) = 0$ or $0 \leq \deg r(x) < n$.

If $r(x) = 0$, then $\beta = 0$. Otherwise, $\beta = g(\alpha)v(\alpha) = r(\alpha) = a_0 + a_1\alpha + a_2\alpha^2 + \cdots + a_{n-1}\alpha^{n-1}$, where $a_i \in K$.

Thus, each element $\beta \in K(\alpha)$ has the form $\beta = r(\alpha) = a_0 + a_1\alpha + a_2\alpha^2 + \cdots + a_{n-1}\alpha^{n-1}$ and either $r(x) = 0$ or $0 \leq \deg r(x) < n$. By Theorem 5.70, the quotient ring $K[x]/I$, where $I = \langle p(x) \rangle$ is a field. Consider a homomorphism of fields $\varphi : K(\alpha) \to K[x]/I$, given as follows $\varphi(r(\alpha)) = r(x) + I$. Then, by Theorems 6.10 and 5.66, φ is an isomorphism. Since $\varphi(\alpha) = x + I$, from Theorem 6.14, it follows that elements $1, \alpha, \alpha^2, \ldots, \alpha^{n-1}$ form a basis of the vector space $K(\alpha)$ over the field K. In this way, $[K(\alpha) : K] = n$.   □

✳ **Corollary 6.29.**
If α is an algebraic element over a field K with minimal polynomial $p(x) \in K[x]$ then $[K(\alpha) : K] = \deg p(x)$. If α is a transcendental element over a field K, then $[K(\alpha) : K] = \infty$.

6.5. Splitting Fields

✳ **Definition 6.30.**
   Let F be an extension of a field K and $f(x) \in K[x]$. We say that $f(x)$ splits over K if $f(x)$ can be expressed as a product of linear factors:

$$f(x) = a(x - \alpha_1)(x - \alpha_2) \cdots (x - \alpha_n),$$

where $\alpha_1, \alpha_2, \ldots, \alpha_n \in F$ and $a \in K$.
   We say that F is a **splitting field** for the polynomial $f(x)$ over K if f splits over K but does not split over any proper subfield of F containing K.

✳ **Theorem 6.31.**
For any polynomial $f(x) \in K[x]$ of degree n, there is a finite extension F of the field K which is a splitting field of $f(x)$ and $[F : K] \leq n!$.

Proof.
   If $f(x) = \mathrm{const} \in K$, then we can take $F = K$. Suppose that $\deg(f(x)) = n \geq 1$. We will perform the proof by mathematical induction on the degree of a polynomial $f(x)$. By Theorem 6.17, for polynomial $f(x)$ there is an extension of K containing an element α_1 such that $f(\alpha_1) = 0$. Consider $F_1 = K(\alpha_1)$. Then, $[K(\alpha_1) : K] \leq n$. By Theorem 5.40,

$$f(x) = (x - \alpha_1)^{m_1} g(x),$$

where $m_1 \geq 1$, $g(\alpha_1) \neq 0$ and $\deg(g(x)) \leq n - 1$. If $g(x)$ is constant, we are done. In this case, $K(\alpha_1)$ is a splitting field of K and $[K(\alpha_1) : K] \leq n$.
   Otherwise, we can find an extension of $K(\alpha_1)$ containing an element α_2 such that $g(\alpha_2) = 0$. Let $F_2 = K(\alpha_1)(\alpha_2) = K(\alpha_1, \alpha_2)$, then $[K(\alpha_1, \alpha_2) : K(\alpha_1)] \leq n - 1$. Therefore,

$$[K(\alpha_1, \alpha_2) : K] = [K(\alpha_1, \alpha_2) : K(\alpha_1)] \cdot [K(\alpha_1) : K] \leq (n - 1)n.$$

Continuing this process, we obtain an extension $F = K(\alpha_1, \alpha_2, \ldots, \alpha_n)$ of K, where the polynomial $f(x)$ can be expressed as a product of linear factors and $[F : K] \leq n!$. It is obvious that F is a splitting field for the polynomial $f(x)$.
□

✳ Example 6.32.

The field $\mathbb{R}/\langle x^2 + 1 \rangle = \{\alpha a + b \ : \ a, b \in \mathbb{R}, \alpha^2 = -1\}$ is a splitting field for the polynomial $p(x) = x^2 + 1$, since $p(x)$ is irreducible over the field $\mathbb{R}$. Then, the homomorphism $f : \mathbb{R}(\alpha) \to \mathbb{R}(i) = \mathbb{C}$, given as follows

$$f(\alpha a + b) = ai + b$$

is an isomorphism of fields.

✳ Example 6.33.

Consider the extension $\mathbb{Q}(\sqrt{2}) = \{a + b\sqrt{2} \ : \ a, b \in \mathbb{Q}\}$ of the field $\mathbb{Q}$, which is a splitting field for the polynomial $p(x) = x^2 - 2 \in \mathbb{Q}[x]$. The number $\sqrt{2}$ is algebraic over $\mathbb{Q}$ with minimal polynomial $p(x) = x^2 - 2$. The degree of the algebraic number $\sqrt{2}$ is equal to 2.

6.6. Algebraically Closed Fields

It is well known that there are polynomials over the field of rational numbers or real numbers which have no roots in these fields. We also know that each polynomial over the field of complex numbers has roots which are complex numbers. The study of fields that have this property is the focus of this section.

✳ Definition 6.34.

A field K is called **algebraically closed** if each non-constant polynomial $P(x) \in K[x]$ has at least one root in K.

✳ Theorem 6.35.

A field K is algebraically closed if and only if each non-constant polynomial $P(x) \in K[x]$ splits over K.

Proof.

Suppose that K is an algebraically closed field, $P(x) \in K[x]$ and $\deg P(x) = n$. We will prove the theorem using the mathematical induction on the degree of $P(x)$. If $\deg P(x) = 1$, then $P(x) = a(x - b)$ and we are done. Suppose that $\deg P(x) = n > 1$ and each polynomial of degree $< n$ splits over K. Since K is an algebraically closed field, there is an element $a \in K$ such that $P(a) = 0$. By Theorem 5.40, $P(x)$ can be written in the form $P(x) = (x - a)P_1(x)$, where $P_1(x) \in K[x]$ and $\deg P_1(x) = n - 1$. By assumption of induction, $P_1(x)$ can be expressed as a product of linear factors belonging to $K[x]$. Therefore, the polynomial $P(x)$ can also be expressed in the same form, and so it splits over K.

The inverse statement is obvious. ☐

✳ **Corollary 6.36.**

If K is an algebraically closed field, then each non-constant polynomial $P(x) \in K[x]$ of degree n has exactly n roots in K, computing each root as many times as its multiplicity.

✳ **Corollary 6.37.**

A field K is algebraically closed if and only if each irreducible polynomial over K is linear.

✳ **Theorem 6.38.**

A field K is algebraically closed if and only if K has no proper algebraic extensions.

Proof.

$\Rightarrow$. Suppose that a field K is algebraically closed and a field F is an algebraic extension of K. Let $\alpha \in F$, then there is the minimal polynomial $p(x) \in K[x]$ of α, which is irreducible, and $p(\alpha) = 0$. Since, from Corollary 6.37, it follows that $p(x)$ is linear, we obtain that $p(x) = x - \alpha$, hence, $\alpha \in K$. In this way, $F = K$.

$\Leftarrow$. Consider a polynomial $p(x) \in K[x]$ of degree ≥ 1. If α is a root of $p(x)$, then we can consider an algebraic extension $F = K(\alpha)$. Since K has no proper algebraic extensions, $K(\alpha) = K$, so each root of $p(x)$ belongs to K. Therefore, the field K is algebraically closed. $\qquad \square$

The field of real numbers $\mathbb{R}$ is not algebraically closed, since, for example, the polynomial $x^2 + 1$ has no roots in $\mathbb{R}$. However, the field of complex numbers $\mathbb{C}$ is algebraically closed. Namely, we have the following theorem, which is called the **Fundamental Theorem of Algebra**. We omit the proof of this theorem as it exceeds the content of this book.

✳ **Theorem 6.39 (Fundamental Theorem of Algebra).**

The field of complex numbers $\mathbb{C}$ is algebraically closed.

As a corollary of this theorem and Corollary 6.36, we have the following corollary.

✳ **Corollary 6.40.**

Each polynomial $P(x)$ of degree $n > 0$ over the field of complex numbers has exactly n complex roots computing each root as many times as its multiplicity.

✴ **Definition 6.41.**

An extension F of a field K is called an **algebraic closure** of K if F is an algebraic extension of K and F is algebraically closed.

Note that, in this case, a field F is minimal among algebraically closed extensions of a field K. It can be proved that each field has only one algebraic closure.

✳ Example 6.42.

The algebraic closure of the field of real numbers $\mathbb{R}$ is the field of complex numbers $\mathbb{C}$.

6.7. Polynomials over Complex Numbers and Real Numbers

From the Fundamental Theorem of Algebra, we can obtain the number of corollaries, some of which will be considered in this section.

✳ Theorem 6.43.

Let a polynomial $P(z) = a_n z^n + a_{n-1} z^{n-1} + \cdots + a_z + a_0$ of degree $n > 0$ over the field of complex numbers have complex roots z_j of multiplicity $k_j \geq 1$ for $j = 1, 2, \ldots, m$ and $k_1 + k_2 + \cdots + k_m = n$. Then, $P(z)$ can be expressed as a product of linear factors of the following form:

$$P(z) = a_n (z - z_1)^{k_1} (z - z_2)^{k_2} \cdots (z - z_m)^{k_m} \qquad (6.7)$$

Proof.

Let $P(z) = a_n z^n + a_{n-1} z^{n-1} + \cdots + a_z + a_0 \in \mathbb{C}[z]$. Then, applying Theorem 6.39 n times, we obtain a sequence of equalities:

$$P(z) = (z - z_1) Q_1(z) = (z - z_1)(z - z_2) Q_2(z) = \cdots = (z - z_1)(z - z_2) \cdots (z - z_n) Q_n(z),$$

where $Q_i(z) \in \mathbb{C}[z]$ for $i = 1, 2 \ldots, n$, and $z_i \in \mathbb{C}$ are roots of the polynomial $P(z)$. It is obvious that $Q_n(z) = a_n$, i.e.,

$$P(z) = a_n (z - z_1)(z - z_2) \cdots (z - z_n).$$

Grouping the same roots, we obtain the required factorization of the polynomial $P(z)$:

$$P(z) = a_n (z - z_1)^{k_1} (z - z_2)^{k_2} \cdots (z - z_m)^{k_m},$$

where k_i is the multiplicity of the root z_i for $i = 1, 2, \ldots, m$ and $k_1 + k_2 + \cdots + k_m = n$. □

> **✳ Theorem 6.44 (Vieta's formulas).**
> Let $P(z) = a_n z^n + a_{n-1} z^{n-1} + \cdots + a_z + a_0$ be a polynomial of degree $n > 0$ over the field of complex numbers. Then, complex numbers $z_1, z_2, \ldots, z_n$ are roots of the polynomial $P(z)$ if and only if they satisfy the following equalities:
>
> $$\begin{cases} z_1 + z_2 + \cdots + z_n = -\dfrac{a_{n-1}}{a_n} \\[2mm] z_1 z_2 + z_1 z_3 + \cdots + z_{n-1} z_n = \dfrac{a_{n-2}}{a_n} \\[2mm] \cdots\cdots\cdots\cdots\cdots\cdots\cdots\cdots\cdots\cdots \\[2mm] z_1 z_2 \cdots z_n = (1)^n \dfrac{a_0}{a_n} \end{cases} \qquad (6.8)$$

Proof.
Let $P(z) = a_n z^n + a_{n-1} z^{n-1} + \cdots + a_z + a_0$ be a polynomial of degree n over the field of complex numbers and let complex numbers $z_1, z_2, \ldots, z_n$ be roots of $P(z)$. Then,

$$a_n z^n + a_{n-1} z^{n-1} + \cdots + a_z + a_0 = a_n(z - z_1)(z - z_2) \cdots (z - z_n).$$

Since $a_n \neq 0$, we obtain:

$$z^n + \frac{a_{n-1}}{a_n} z^{n-1} + \cdots + \frac{a_1}{a_n} z + \frac{a_0}{a_n} = (z - z_1)(z - z_2) \cdots (z - z_n).$$

Multiplying the factors on the right side of this equality, we obtain:

$$z^n + \frac{a_{n-1}}{a_n} z^{n-1} + \cdots + \frac{a_1}{a_n} z + \frac{a_0}{a_n} = z^n - (z_1 + z_2 + \cdots + z_n) z^{n-1} +$$

$$+ (z_1 z_2 + z_1 z_3 + \cdots + z_{n-1} z_n) z^{n-2} + \cdots + (-1)^n z_1 z_2 \cdots z_n.$$

From the equality of these polynomials, we obtain the required formulas (6.8):

$$\frac{a_{n-1}}{a_n} = -(z_1 + z_2 + \cdots + z_n)$$

$$\frac{a_{n-2}}{a_n} = z_1 z_2 + z_1 z_3 + \cdots + z_{n-1} z_n$$

$$\cdots\cdots\cdots\cdots\cdots\cdots\cdots\cdots\cdots\cdots$$

$$\frac{a_0}{a_n} = (-1)^n z_1 z_2 \cdots z_n$$

$\square$

The formulas (6.8) are called **Vieta's formulas** and they are some generalization of relations between roots of a quadratic trinomial and its coefficients.

✳ Examples 6.45.

Find a polynomial of degree 4 whose real roots $x_1 = 2$, $x_2 = -3$ are of multiplicity one and the real root $x_3 = x_4 = 5$ is of multiplicity two.

We can consider that the leading coefficient of this polynomial $a_4 = 1$. Then, from Vieta's formulas, we obtain:

$$a_3 = -(x_1 + x_2 + x_3 + x_4) = -(2 - 3 + 5 + 5) = -9$$

$$a_2 = x_1 x_2 + x_1 x_3 + x_1 x_4 + x_2 x_3 + x_2 x_4 + x_3 x_4 = 2 \cdot (-3) + 2 \cdot 5 + 2 \cdot 5 + (-3) \cdot 5 + (-3) \cdot 5 + 5 \cdot 5 = 9$$

$$a_1 = -(x_1 x_2 x_3 + x_1 x_2 x_4 + x_1 x_3 x_4 + x_2 x_3 x_4) = -[2 \cdot (-3) \cdot \cdot 5 + 2 \cdot (-3) \cdot 5 +$$

$$+ 2 \cdot 5 \cdot + (-3) \cdot 5 \cdot 5] = 85$$

$$a_0 = x_1 x_2 x_3 x_4 = 2 \cdot (-3) \cdot 5 \cdot 5 = -150$$

Therefore, the given polynomial has the following form: $f(x) = a(x^4 - 9x^3 + 9x^2 + 85x - 150)$, where a is an arbitrary real number.

✳ Remark 6.46.

Since the field of real numbers $\mathbb{R}$ is a subfield of the field of complex numbers, a polynomial $P(x) \in \mathbb{R}[x]$ can also be considered as a polynomial in the ring $\mathbb{C}[x]$. Then, for this polynomial, we can apply theorems for polynomials over the field $\mathbb{C}$. Then, we can obtain the following results.

✱ Theorem 6.47.

Let $P(x) = a_n x^n + a_{n-1} x^{n-1} + \cdots + a_x + a_0 \in \mathbb{R}[x]$ be a polynomial of degree $n > 0$. A complex number z_0 is a root of $P(x)$ if and only if the number $\overline{z_0}$ is also a root of this polynomial.

Proof.

From the definition of a conjugate complex number, we have that $\overline{P(x)} = P(\overline{x})$. Hence, if $P(z_0) = 0$, then $P(\overline{z_0}) = \overline{P(z_0)} = \overline{0} = 0$. □

As a corollary of this theorem, we get that if z_0 is a root of a polynomial $P(x) \in \mathbb{R}[x]$ and $\operatorname{Im} z_0 \neq 0$, then $P(x)$ is divisible by the polynomial $(x - z_0)(x - \overline{z_0}) = x^2 - (z_0 + \overline{z_0})x + z_0 \overline{z_0}$, i.e., by the quadratic trinomial $g(x) = x^2 + px + q \in \mathbb{R}[x]$, where $p = -(z_0 + \overline{z_0})$, $q = z_0 \overline{z_0} \in \mathbb{R}$. This trinomial is irreducible in $\mathbb{R}$ as it has no real roots. Dividing the polynomial $P(x)$ by the polynomial $g(x)$, we can again apply Theorem 6.43 and, in this way, we obtain that the multiplicity of roots z_0 and $\overline{z_0}$ is the same. Hence, we have the following corollary:

✱ Corollary 6.48.

Let $P(x) = a_n x^n + a_{n-1} x^{n-1} + \cdots + a_x + a_0 \in \mathbb{R}[x]$ be a polynomial of degree $n > 0$. A complex number z_0 is a root of multiplicity k of $P(x)$ if and only if the number $\overline{z_0}$ is a root of multiplicity k of $P(x)$.

Thus, if a polynomial $P(x) \in \mathbb{R}[x]$ has a real root x_i of multiplicity k_i, then $P(x)$ can be expressed in the following form: $P(x) = (x - x_i)^{k_i} Q(x)$, and if $P(x)$ has a complex root z_k of multiplicity m_k, then $P(x)$ can be expressed as follows: $P(x) = (g_k(x))^{m_k} R(x)$, where $g_k(x) = x^2 - (z_k + \overline{z_k})x + z_k\overline{z_k}$. Thus, the final factorization of the polynomial $P(x)$ with real coefficients has only two kinds of factors: $(x - x_i)^{k_i}$ and $(x^2 + p_k x + q_k)^{m_k}$. So, we obtain the following theorem:

❋ Theorem 6.49.

Let $P(x) = a_n x^n + a_{n-1}x^{n-1} + \cdots + a_x + a_0 \in \mathbb{R}[x]$ be a polynomial of degree $n > 0$. Let real numbers x_i be roots of multiplicity k_i of $P(x)$ for $i = 1, 2, \ldots, r$, and let complex numbers z_j, $\overline{z_j}$, where $\mathrm{Im}\, z_j > 0$, be roots of multiplicity m_j of $P(z)$ for $j = 1, 2, \ldots, s$. Moreover, $(k_1 + k_2 + \cdots + k_r) + 2(m_1 + m_2 + \cdots + m_s) = n$. Then, $P(x)$ can be expressed as a product of polynomials of the following form:

$$P(x) = a_n(x - x_1)^{k_1}(x - x_2)^{k_2} \cdots (x - x_r)^{k_r}.$$

$$\cdot(x^2 + p_1 x + q_1)^{m_1}(x^2 + p_2 x + q_2)^{m_2} \cdots (x^2 + p_s x + q_s)^{m_s},$$

where $p_j = -2\mathrm{Re}\, z_j$, $q_j = |z_j|^2$ for $j = 1, 2, \ldots, s$. In other words, each polynomial over the field of real numbers can be expressed as a product of real polynomials of degree ≤ 2.

From Theorem 6.47, it also follows that an arbitrary polynomial of odd degree with real coefficients has at least one real root.

❋ Examples 6.50.

Knowing that $x_1 = 2 + i$ is a root of the polynomial $P(x) = x^4 - x^3 - 5x^2 + 7x + 10 \in \mathbb{R}[x]$, find the rest roots of this polynomial.

Since the polynomial $P(x) \in \mathbb{R}[x]$, it possesses, along with the root $x_1 = 2 + i$, the conjugate root $x_2 = 2 - i$. Therefore, $P(x)$ is divisible by the polynomial $g(x) = x^2 + px + q$, where $p = -2 \cdot 2 = -4$, $q = 2^2 + 1 = 5$. Hence, $g(x) = x^2 - 4x + 5$. Therefore, $P(x) = (x^2 - 4x + 5)(x^2 + 3x + 2)$. Since the polynomial $x^2 + 3x + 2$ has two real roots $-1, -2$, all roots of polynomial $P(x)$ are: $x_1 = 2 + i$, $x_2 = 2 - i$, $x_3 = -1$, $x_4 = -2$.

6.8. Exercises

1. Show that if $F = \mathbb{Q}[x]/\langle x^3 - 2\rangle$ then $[F : \mathbb{Q}] = 3$.

2. Show that if $F = \mathbb{Q}[x]/\langle x^2 - 5\rangle$ then $[F : \mathbb{Q}] = 2$.

3. Prove that if a field K is algebraic extension of a field K and K is algebraic extension of a field L, then F is algebraic over L.

4. Show that the following numbers are algebraic over the field of rational numbers and find their minimal polynomials:

   (a) $\sqrt[3]{2} + \sqrt{5}$
   (b) $\sqrt{2 + \sqrt{3}}$
   (c) $\sqrt{3} + \sqrt{5}$
   (d) $\sqrt{\sqrt[3]{2} - i}$
   (e) $\cos(2\pi/5)$

5. Find the degree of the extensions:

   (a) $[\mathbb{Q}(\sqrt[3]{7} : \mathbb{Q}]$
   (b) $[\mathbb{Z}_3[x]/\langle x^2 + x + 2 \rangle : \mathbb{Z}_3]$
   (c) $[\mathbb{Q}(\sqrt{3} + \sqrt{5}) : \mathbb{Q}]$

6. Knowing that $x_1 = 1 - 2i$ is a root of the polynomial $P(x) = x^4 - x^3 + x^2 + 9x - 10 \in \mathbb{C}[x]$, find the rest roots of this polynomial.

7. Find all roots of the polynomial $P(z) = 2z^2 + (6 - 2i)z + 4 - 3i \in \mathbb{C}[z]$.

8. Find a real polynomial of minimal degree which has the real root $x_1 = 5$ of multiplicity 3 and the complex root $x_2 = 1 + i$ of multiplicity 2.

9. Write the polynomial $P(z) = z^4 + 1$ in $\mathbb{C}[z]$ as a product of irreducible polynomials over $\mathbb{C}$.

10. Write the polynomial $P(x) = x^4 + 16$ as a product of irreducible polynomials in $\mathbb{R}$.

11. Write the polynomial $P(x) = x^4 + x^2 + 1$ as a product of irreducible polynomials in $\mathbb{R}$.

References

[1] Ash, R.B. 2007. Basic Abstract Algebra, Dover Publications.

[2] Dummit, D.S. and R.M. Foote. 2004. Abstract Algebra (3rd Ed.), John Wiley & Sons.

[3] Jacobson, N. 1985. Basic Algebra I, Freeman, San Francisco, H. Freeman & Co.

[4] Judson, Th.W. 1994. Abstract Algebra. Theory and Applications, PWS Publishing Company, Michigan.

[5] Koblitz, N. 1998. Algebraic Aspects of Cryptography, Springer-Verlag Berlin-Heidelberg.

[6] Lang, S. 1984. Algebra, Addison-Wesley.

[7] Oggier, F. 2011. Algebraic Methods.
http://www1.spms.ntu.edu.sg/~frederique/AA11.pdf.

[8] Slinko, A. 2015. Algebra for Applications, Springer.

[9] Sterk, H. Algebra 3: Algorithms in Algebra.
http://www.win.tue.nl/~sterk/algebra3/hoofd.pdf.

Chapter 7

Examples of Applications

> *"We build too many walls and not enough bridges."*
> Isaac Newton

> *"It is not enough to have a good mind; the main thing is to use it well."*
> Rene Descartes

Throughout history, the theory of numbers was considered as a purely theoretical science with no practical applications and connections to the real world. Mathematicians considered number theory as a purely intellectual aspect. C.F. Gauss said that *"if mathematics is the queen of the sciences, then the theory of numbers is, because of its supreme uselessness, the queen of mathematics."*[1]

According to another outstanding English mathematician, G.H. Hardy, *"The 'real' mathematics of the 'real mathematicians', the mathematics of Fermat and Euler and Gauss and Abel and Riemann, is almost wholly 'useless' (and this is as true of 'applied' as of 'pure' mathematics)"*. However, in the same book G.H. Hardy wrote: *"If the theory of numbers could be employed for any practical and obviously honorable purpose, if it could be turned directly to the furtherance of human happiness or the relief of human suffering, as physiology and even chemistry can, then surely neither Gauss nor any other mathematician would have been so foolish as to decry or regret such applications."*

At last, in the 21th century, we can say that the theory of numbers as well as the theory of groups

G.H. Hardy
(1877-1947)

---

[1]G.H. Hardy wrote about this in his book, but he also noted that he was never able to find an exact quotation.

and rings really has a great influence on our life due to the digital revolution, because the results of these theories are the basis of different cryptographic systems providing security of transactions in the Internet.

The first important application of group theory was the classification of all possible structures of crystals, which is one of the basic problems in crystallography. The study of crystallographic groups was started by J.S. Fiedorow and continued by A. Schönflies at the end of the 19-th century. They showed that there are only 17 plane crystallographic groups and there are exactly 230 different three-dimensional crystallographic groups. It was the first example of the application of group theory to natural science. It is also interesting to note that the three-dimensional crystallographic groups were found mathematically before these 230 different types of crystals were actually discovered in nature.

Currently, the application of the theory of numbers, groups and rings covers such different branches of sciences and technics as cryptography and quantum mechanics, genetics and particle physics, biology and chemistry, even music and painting.

In this chapter, we consider some examples of the application of the theory of numbers, groups and rings.

7.1. Euler's φ-function and its Properties

One of the most important functions in number theory and its applications is Euler's totient function. It is an example of arithmetic and multiplicative functions.

※ **Definition 7.1.**

An **arithmetic function** is a function that maps the set of natural numbers to the set of complex numbers: $f : \mathbb{Z}^+ \to \mathbb{C}$.

An arithmetic function $f : \mathbb{Z}^+ \to \mathbb{R}$, which is not identically zero and satisfies the condition

$$f(mn) = f(m)f(n) \tag{7.1}$$

for all $m, n \in \mathbb{Z}^+$ such that $\gcd(m, n) = 1$, is called **multiplicative**.

An arithmetic function $f(x)$, which is not identically zero and satisfies the condition (7.1) for all $m, n \in \mathbb{Z}^+$, $f(x)$, is called **totally multiplicative**.

Recall that Euler's totient function $\varphi(m)$ is the number of positive integers less than m that are coprime to m. We assume that the number 1 is coprime to each positive integer m and $\varphi(1) = 1$. It is obvious that $\varphi(x)$ is an arithmetic function. We show that this function is also multiplicative.

> ⁕ **Theorem 7.2.**
> 1) If p is a prime then:
> $$\varphi(p) = p - 1 \tag{7.2}$$
> 2) If p is a prime then:
> $$\varphi(p^k) = p^k - p^{k-1} \tag{7.3}$$

Proof.

1. Since $\gcd(a, p) = 1$ for each $0 < a < p$, $\varphi(p) = p - 1$.

2. In the set of numbers $X = \{1, 2, \ldots, p-1, p, p+1, \ldots, p^2, p^2+1, \ldots, p^k-1, p^k\}$ the numbers that are not coprime to p are exactly the numbers that are multiples of p, i.e., the numbers of the set $Y = \{p, 2p, 3p, \ldots, (p-1)p, p^2, \ldots, p^{k-1}p = p^k\}$. Since $|X| = p^k$ and $|Y| = p^{k-1}$, we obtain that $\varphi(p^k) = p^k - p^{k-1}$. $\qquad\square$

⁕ **Examples 7.3.**

1. $\varphi(11) = 11 - 1 = 10$

2. $\varphi(31) = 31 - 1 = 30$

3. $\varphi(32) = \varphi(2^5) = 2^5 - 2^4 = 32 - 16 = 16$

4. $\varphi(81) = \varphi(3^4) = 3^4 - 3^3 = 81 - 27 = 54$

L. Euler
(1707-1783)

> ⁕ **Theorem 7.4.**
> Euler's totient function is an arithmetic multiplicative function, i.e., it satisfies the condition
> $$\varphi(m_1 m_2 \cdots m_k) = \varphi(m_1)\varphi(m_2) \cdots \varphi(m_k) \tag{7.4}$$
> for all $m_1, m_2, \ldots, m_k$ which are pairwise relatively prime positive integers.

Proof.

Consider the ring $\mathbb{Z}/m\mathbb{Z}$ where $m = m_1 m_2 \cdots m_k$ and $m_1, m_2, \ldots, m_k$ are pairwise relatively prime numbers which are ≥ 2. Then, by Generalized Chinese Remainder Theorem 4.97, we have an isomorphism of rings:

$$\mathbb{Z}/m\mathbb{Z} \cong \mathbb{Z}/m_1\mathbb{Z} \times \mathbb{Z}/m_2\mathbb{Z} \times \cdots \times \mathbb{Z}/m_k\mathbb{Z} \tag{7.5}$$

Let $(\mathbb{Z}/m\mathbb{Z})^*$ be a multiplicative group of invertible elements of the ring $\mathbb{Z}/m\mathbb{Z}$, and $(\mathbb{Z}/m_i\mathbb{Z})^*$ be a multiplicative group of invertible elements of the ring $\mathbb{Z}/m_i\mathbb{Z}$. Then, taking into account that an isomorphism of rings implies an isomorphism their multiplicative groups, from equality (7.5), we obtain an isomorphism of groups:

$$(\mathbb{Z}/m\mathbb{Z})^* \cong (\mathbb{Z}/m_1\mathbb{Z})^* \times (\mathbb{Z}/m_2\mathbb{Z})^* \times \cdots \times (\mathbb{Z}/m_k\mathbb{Z})^* \tag{7.6}$$

Since the number of elements of the group $(\mathbb{Z}/n\mathbb{Z})^*$ is equal to $\varphi(n)$, from equality (7.6), we obtain the equality (7.4). □

As an immediate corollary from Theorems 7.2 and 7.4, we obtain the following result.

✳ **Theorem 7.5.**
If $m = p_1^{k_1} p_2^{k_2} \cdots p_n^{k_n}$ is the canonic factorization of an integer $m \geq 2$ into prime factors, then:

$$\varphi(m) = (p_1^{k_1} - p_1^{k_1-1})(p_2^{k_2} - p_2^{k_2-1}) \cdots (p_n^{k_n} - p_n^{k_n-1}) \qquad (7.7)$$

The condition (7.7) can be rewritten as follows:

$$\varphi(m) = m(1 - \frac{1}{p_1})(1 - \frac{1}{p_2}) \cdots (1 - \frac{1}{p_n}) \qquad (7.8)$$

For $n = 2$, Theorem 7.5 has the following form which plays an important role in the cryptographic algorithm RSA.

✳ **Theorem 7.6.**
If $m = pq$, where p and q are relatively primes, then

$$\varphi(m) = (p - 1)(q - 1) \qquad (7.9)$$

Examples 7.7.

1. $\varphi(15) = \varphi(3)\varphi(5) = (3 - 1)(5 - 1) = 2 \cdot 4 = 8$

2. $\varphi(56) = \varphi(2^3 \cdot 7) = \varphi(2^3)\varphi(7) = (2^3 - 2^2)(7 - 1) = (8 - 4) \cdot 6 = 24$

✳ **Theorem 7.8.**
For any positive integer $m \geq 2$, the following equality holds:

$$\sum_{d|m} \varphi(d) = m \qquad (7.10)$$

Proof.
Consider the arithmetic function $F(m) = \sum_{d|m} \varphi(d)$. If $m = xy$, $\gcd(x, y) = 1$, and $d|xy$, then $d = d_1 d_2$ and $d_1|x$, $d_2|y$. Since $\gcd(d_1, d_2) = 1$, $\varphi(d) = \varphi(d_1)\varphi(d_2)$. Therefore,

$$F(xy) = \sum_{d_1|x \, d_2|y} \sum \varphi(d_1)\varphi(d_2) = \left(\sum_{d_1|x} \varphi(d_1)\right)\left(\sum_{d_2|y} \varphi(d_2)\right) = F(x)F(y),$$

implying that the function F is multiplicative. Note that

$$F(p^k) = \sum_{i=0}^{k} \varphi(p^i) = 1 + \sum_{i=0}^{k}(p^i - p^{i-1}) = p^k. \tag{7.11}$$

Let $m = p_1^{k_1} p_2^{k_2} \cdots p_s^{k_s}$ be the canonical prime factorizations of $m \geq 2$, where p_i are primes and k_i are positive integers. Then, from multiplicity of F and equality (7.11), we obtain:

$$F(m) = F(p_1^{k_1})F(p_2^{k_2}) \cdots F(p_s^{k_s}) = p_1^{k_1} p_2^{k_2} \cdots p_s^{k_s} = m.$$

So $\sum_{d|m} \varphi(d) = m.$ □

7.2. Euler's Theorem. Fermat's Little Theorem. Wilson's Theorem

In this section, we present three remarkable theorems which are considered to be the pearls of number theory.

✳ **Theorem 7.9 (Euler's Theorem).**
Let $m \in \mathbb{N}$. Then, for any integer a with $\gcd(a, m) = 1$, the number m divides $a^{\varphi(m)} - 1$, i.e.,
$$a^{\varphi(m)} \equiv 1 \,(\text{mod}\, m) \tag{7.12}$$

Proof.
Consider the multiplicative group $(\mathbb{Z}/m\mathbb{Z})^*$ of the ring $(\mathbb{Z}/m\mathbb{Z})$. The number of elements of this group is equal to $\varphi(m)$. Then, from Lagrange's theorem, it follows that, for an arbitrary element $[a] \in (\mathbb{Z}/m\mathbb{Z})^*$, the following holds:

$$[a]^{\varphi(m)} = [1],$$

which is equivalent to the congruence $a^{\varphi(m)} \equiv 1 \,(\text{mod}\, m)$, that required. □

If p is a prime, then, by Theorem 7.2(1), $\varphi(p) = p - 1$. Therefore, from Euler's theorem, another famous theorem, which is called **Fermat's Little Theorem**, immediately follows:

✳ **Corollary 7.10 (Fermat's Little Theorem).**
If p is a prime, then p divides $a^{p-1} - 1$ for any integer a with $\gcd(a, p) = 1$, i.e.,
$$a^{p-1} \equiv 1 \,(\text{mod}\, p) \tag{7.13}$$

✳ Corollary 7.11.

If p is a prime, then p divides $a^p - a$ for any integer a, i.e.,

$$a^p \equiv a \, (\text{mod} \, p) \tag{7.14}$$

Proof.

If $\gcd(a, p) = 1$, then, from Corollary 7.11, we have $a^{p-1} \equiv 1 \, (\text{mod} \, p)$. Multiplying both sides of this congruence by a, we obtain the equality (7.14).

If $\gcd(a, p) \neq 1$, then $\gcd(a, p) = p$. Hence, $a \equiv 0 \, (\text{mod} \, p)$ and $a^p \equiv 0 \, (\text{mod} \, p)$. Therefore, $a^p \equiv a \, (\text{mod} \, p)$. $\qquad\square$

✳ Corollary 7.12.

If $m > 1$ is a positive integer and $n \equiv k \, (\text{mod} \, \varphi(m))$, then

$$a^n \equiv a^k \, (\text{mod} \, m) \tag{7.15}$$

for all $a \in \mathbb{Z}$ with $\gcd(a, m) = 1$

Proof.

Since $n \equiv k \, (\text{mod} \, \varphi(m))$, there is an integer s such that $n = k + s\varphi(m)$. Then, from Euler's Theorem, it follows that $a^n = a^{k+s\varphi(m)} = a^k a^{s\varphi(m)} \equiv a^k \, (\text{mod} \, m)$. $\qquad\square$

✳ Corollary 7.13.

If p is a prime and $n \equiv k \, (\text{mod} \, p - 1)$, then

$$a^n \equiv a^k \, (\text{mod} \, p) \tag{7.16}$$

for all $a \in \mathbb{Z}$ with $\gcd(a, p) = 1$.

✳ Examples 7.14.

- Find the remainder on division of the number 2^{100} by 101.

  Here we have $a = 2$, $m = 101$, $\varphi(m) = \varphi(101) = 100$. Since $\gcd(2, 101) = 1$ and 101 is a prime, we can use Fermat's Little Theorem: $2^{100} \equiv 1 \, (\text{mod} \, 101)$.

- Find the remainder on division of the number 3^{102} by 101.

  Here we have $a = 3$, $m = 101$, $\varphi(m) = \varphi(101) = 100$. Since $\gcd(3, 101) = 1$ and 101 is a prime, by Fermat's Little Theorem: $3^{100} \equiv 1 \, (\text{mod} \, 101)$. Then, $3^{102} = 3^{100+2} \equiv 3^2 \equiv 9 \, (\text{mod} \, 101)$.

- Find the last digit of the number $2^{1000000}$ in the digital system with the base 7.

  Here we have $p = 7$, $p - 1 = 6$.

  Since $1000000 \equiv 4 \, (\text{mod} \, 6)$, by Corollary 7.12, we have: $2^{1000000} \equiv 2^4 \equiv 16 \equiv 2 \, (\text{mod} \, 7)$.

- Find the remainder on division of the number 731^{512} by 56.

  Taking into account that $731 \equiv 3 \, (\mathrm{mod}\, 56)$ we obtain $731^{512} \equiv 3^{512} \, (\mathrm{mod}\, 56)$.

  Here we have $a = 3$, $m = 56$, $\varphi(m) = \varphi(56) = 24$. Since $\gcd(3, 56) = 1$, by Euler's Theorem: $3^{24} \equiv 1 \, (\mathrm{mod}\, 56)$. Then $3^{512} = 3^{24 \cdot 21 + 8} \equiv 3^8 \equiv 3^4 \cdot 3^4 \equiv 81 \cdot 81 \equiv 25 \cdot 25 \equiv 125 \cdot 5 \equiv 13 \cdot 5 \equiv 65 \equiv 9 \, (\mathrm{mod}\, 56)$.

- Find the remainder on division of the number 37^{2011} by 7.

  Taking into account that $37 \equiv 2 \, (\mathrm{mod}\, 7)$, we obtain $37^{2011} \equiv 2^{2011} \, (\mathrm{mod}\, 7)$.

  Here we have $a = 2$, $m = 7$, $\varphi(m) = \varphi(7) = 6$. Since $2011 \equiv 1 \, (\mathrm{mod}\, 6)$ and $\gcd(2, 7) = 1$, by Corollary 7.13, we have $2^{2011} \equiv 2^1 \, (\mathrm{mod}\, 7)$. Therefore, $37^{2011} \equiv 2 \, (\mathrm{mod}\, 7)$.

- Find $2^{1000000}$ modulo 77.

  Since $\gcd(2, 77) = 1$ and $1000000 \equiv 40 \, (\mathrm{mod}\, 77)$, by Corollary 7.13 we have: $2^{1000000} \equiv 2^{40} \equiv 23^4 \equiv 23 \, (\mathrm{mod}\, 77)$.

The third pearl of number theory is the following theorem:

> ✳ **Theorem 7.15 (Wilson's Theorem).**
> If p is a prime then
> $$(p - 1)! \equiv -1 \, (\mathrm{mod}\, p). \tag{7.17}$$

Proof.

If $p = 2$ then $p - 1 = 1 \equiv -1 \, (\mathrm{mod}\, 2)$.

Let $p > 2$. Since p is a prime, $\mathbb{Z}_p$ is a field. Therefore, for any $a = 1, \ldots, p - 1$ a congruence $ax \equiv 1 \, (\mathrm{mod}\, p)$ has exactly one solution modulo p. At the time the congruence $x^2 \equiv 1 \, (\mathrm{mod}\, p)$ has two solutions modulo p: $x = 1$ and $x = p - 1$. Really, $p|(x - 1)(x + 1)$ if and only if $(p|(x - 1)) \vee (p|(x + 1))$. Since $0 \le x < p$ we obtain that $(x = 1) \vee (x = -1)$, i.e., $x = 1$ or $x = p - 1$.

Therefore, the congruence $ax \equiv 1 \, (\mathrm{mod}\, p)$ has exactly one solution $b \ne a$ modulo p, for any $a = 2, \ldots, p-1$. So, the set $\{2, \ldots, p-2\}$ can be divided into $(p-3)/2$ pairs of numbers (a, b) such that $a \ne b$ and $ab \equiv 1 \, (\mathrm{mod}\, p)$. Therefore, $2 \cdot 3 \cdots (p-2) \equiv 1 \, (\mathrm{mod}\, p)$. Multiplying both sides of this congruence by $p - 1$, we obtain:

$$(p - 1)! \equiv 2 \cdot 3 \cdots (p - 2)(p - 1) \equiv (p - 1) \equiv -1 \, (\mathrm{mod}\, p).$$

□

✳ **Examples 7.16.**
  1. $12! + 1 \equiv 0 \, (\mathrm{mod}\, 13)$
  2. $58! + 1 \equiv 0 \, (\mathrm{mod}\, 59)$

7.3. Solving Linear Congruences by Euler's Method

Consider a linear congruence $ax \equiv b \,(\mathrm{mod}\, m)$, where $a, b \in \mathbb{Z}$, $m \in \mathbb{Z}^+$. By Theorem 1.56, this congruence has a solution if and only if $d|b$, where $d = \gcd(a, m)$. If $d > 1$, we can divide the congruence by d and obtain an equivalent congruence of the form $a_1 x \equiv b_1 \,(\mathrm{mod}\, m_1)$ with $\gcd(a_1, m_1) = 1$.

Therefore, from the very beginning, we can assume that we have a congruence

$$ax \equiv b \,(\mathrm{mod}\, m) \tag{7.18}$$

with $\gcd(a, m) = 1$. In this case, we can use Euler's Theorem, which states that $a^{\varphi(m)} \equiv 1 \,(\mathrm{mod}\, m)$. Hence, $a^{\varphi(m)-1} \equiv a^{-1} \,(\mathrm{mod}\, m)$. Therefore, a solution of the congruence (7.18) can be given in the following form:

$$x \equiv a^{-1}b \equiv a^{\varphi(m)-1}b \,(\mathrm{mod}\, m) \tag{7.19}$$

If numbers a, m are large enough (that we have in practice in cryptographic algorithms), then the exponentiation operation of integers modulo natural number m can be very time-consuming. In this case, there is an effective method of quick modulo exponentiation. We describe this algorithm below.

Algorithm of quick modulo exponentiation

To compute $a^n \,(\mathrm{mod}\, m)$, we represent the number n in binary code, i.e., we write the number n in the following form:

$$n = \sum_{i=0}^{k} e_i 2^i,$$

where $e_i \in \{0, 1\}$. Then,

$$a^n = \prod_{i=0}^{k} (a^{2^i})^{e_i} = \prod_{0 \le i \le k,\ e_i=1} a^{2^i} \tag{7.20}$$

So the algorithm is performed as follows:

1. We compute step by step $a^{2^i} \,(\mathrm{mod}\, m)$ for all $0 \le i \le k$.
2. We compute a^n multiplying a^{2^i} for all $0 \le i \le k$ such that $e_i \ne 0$.

✳ Example 7.17.
Compute $6^{73} \,(\mathrm{mod}\, 100)$.
Here, $n = 73$. We write 73 in the binary code: $73 = 1 + 2^3 + 2^6$. Therefore, $6^{73} = 6 \cdot 6^{2^3} \cdot 6^{2^6}$.
Then, we compute step by step:
$6^2 = 36$,
$6^{2^2} = 36^2 \equiv -4 \,(\mathrm{mod}\, 100)$

$$6^{2^3} = (6^{2^2})^2 \equiv (-4)^2 \equiv 16 \,(\mathrm{mod}\,100)$$
$$6^{2^4} = (6^{2^3})^2 \equiv (16)^2 \equiv 56 \,(\mathrm{mod}\,100)$$
$$6^{2^5} = (6^{2^4})^2 \equiv (56)^2 \equiv 36 \,(\mathrm{mod}\,100)$$
$$6^{2^6} = (6^{2^5})^2 \equiv (36)^2 \equiv -4 \,(\mathrm{mod}\,100)$$

In this way:
$$6^{73} = 6 \cdot 6^{2^3} \cdot 6^{2^6} \equiv 6 \cdot 16 \cdot (-4) \equiv 16 \,(\mathrm{mod}\,100).$$

✳ Example 7.18.

Solve the congruence: $28x \equiv 33 \,(\mathrm{mod}\,43)$

Here, $a = 28$, $b = 33$ and $m = 43$. $\varphi(43) = 42$, $n = \varphi(43) - 1 = 41$.

Using the equality (7.19), we obtain the solution of the congruence in the form: $x \equiv 28^{41} \cdot 33 \,(\mathrm{mod}\,43)$.

We write $n = 41$ in the binary code: $41 = 1 + 2^3 + 2^5$. Therefore, $28^{41} = 28 \cdot 28^{2^3} \cdot 28^{2^5}$.

We compute step by step:
$$28^2 \equiv 10 \,(\mathrm{mod}\,43),$$
$$28^{2^2} = 10^2 \equiv 14 \,(\mathrm{mod}\,43)$$
$$28^{2^3} = (28^{2^2})^2 \equiv (14)^2 \equiv 24 \,(\mathrm{mod}\,43)$$
$$28^{2^4} = (28^{2^3})^2 \equiv (24)^2 \equiv 17 \,(\mathrm{mod}\,43)$$
$$28^{2^5} = (28^{2^4})^2 \equiv (17)^2 \equiv 31 \,(\mathrm{mod}\,43)$$
In this way:

$$28^{41} = 28 \cdot 28^{2^3} \cdot 28^{2^5} \equiv 28 \cdot 24 \cdot 31 \equiv 20 \,(\mathrm{mod}\,43).$$

Hence, $x \equiv 20 \cdot 33 \equiv 15 \,(\mathrm{mod}\,43)$ is the solution of the given congruence.

7.4. Solving Systems of Linear Congruences

Let K be a field. Consider a system of linear equations:

$$\mathbf{Ax} = \mathbf{y}, \tag{7.21}$$

where $\mathbf{A} \in M_n(K)$, $\mathbf{x}, \mathbf{y} \in K^n$.

Let $D = \det(\mathbf{A})$ be a determinant of the matrix $\mathbf{A}$. From linear algebra, it is well known that, if $D \neq 0$, then the matrix $\mathbf{A}$ is invertible and the system (7.21) has exactly one solution.

In particular, if p is a prime number and $K = \mathbb{Z}_p$, then the system (7.21) is equivalent to the system of linear congruences:

$$\mathbf{Ax} \equiv \mathbf{y} \,(\mathrm{mod}\,p), \tag{7.22}$$

where $\mathbf{A} \in M_n(\mathbb{Z}_p)$, $\mathbf{x}, \mathbf{y} \in (\mathbb{Z}_p)^n$.

In this case, the system (7.21) has one solution if $\gcd(D, p) = 1$, where $D = \det \mathbf{A}$.

✳ Example 7.19.

Solve the system of congruences:

$$\begin{cases} 17x + 11y \equiv 7 \,(\mathrm{mod}\,29) \\ 13x + 10y \equiv 8 \,(\mathrm{mod}\,29) \end{cases}$$

Here $\mathbf{A} = \begin{pmatrix} 17 & 11 \\ 13 & 10 \end{pmatrix}$, $D = \det(\mathbf{A}) = 27$ and $p = 29$.

Since $\gcd(27, 29) = 1$, there exists $D^{-1} = (27)^{-1} \equiv 14\,(\mathrm{mod}\,29)$. Hence, the inverse matrix: $\mathbf{A}^{-1} = D^{-1} \begin{pmatrix} 10 & -11 \\ -13 & 17 \end{pmatrix} \equiv D^{-1} \begin{pmatrix} 10 & 18 \\ 16 & 17 \end{pmatrix} \equiv \begin{pmatrix} 24 & 20 \\ 21 & 6 \end{pmatrix} (\mathrm{mod}\,29)$.

Therefore, the given system has one solution:

$\begin{pmatrix} x \\ y \end{pmatrix} = \begin{pmatrix} 24 & 20 \\ 21 & 6 \end{pmatrix} \begin{pmatrix} 7 \\ 8 \end{pmatrix} \equiv \begin{pmatrix} 9 \\ 21 \end{pmatrix} (\mathrm{mod}\,29)$, i.e., we obtain the solution in the following form:

$$\begin{cases} x \equiv 9\,(\mathrm{mod}\,29), \\ y \equiv 21\,(\mathrm{mod}\,29) \end{cases}$$

Now, let us consider a system of linear congruences in a general case:

$$\mathbf{A}\mathbf{x} \equiv \mathbf{y}\,(\mathrm{mod}\,m), \tag{7.23}$$

where $\mathbf{A} \in M_n(\mathbb{Z}_m)$, $\mathbf{x}, \mathbf{y} \in (\mathbb{Z}_m)^n$, and $m \in \mathbb{Z}^+$.

Then, the system (7.23) also has one solution if $\gcd(D, m) = 1$, where $D = \det \mathbf{A}$.

✳ Example 7.20.

Solve the system of congruences:

$$\begin{cases} 2x + 3y \equiv 1\,(\mathrm{mod}\,26) \\ 7x + 8y \equiv 2\,(\mathrm{mod}\,26) \end{cases}$$

Here, $\mathbf{A} = \begin{pmatrix} 2 & 3 \\ 7 & 8 \end{pmatrix}$, $D = \det(\mathbf{A}) = -5 \equiv 21\,(\mathrm{mod}\,26)$ and $m = 26$.

Since $\gcd(21, 29) = 1$, there exists $D^{-1} = (21)^{-1} \equiv 5\,(\mathrm{mod}\,26)$. Hence, the inverse matrix: $\mathbf{A}^{-1} = D^{-1} \begin{pmatrix} 8 & -3 \\ -7 & 2 \end{pmatrix} \equiv D^{-1} \begin{pmatrix} 8 & 23 \\ 19 & 2 \end{pmatrix} \equiv \begin{pmatrix} 14 & 11 \\ 17 & 10 \end{pmatrix} (\mathrm{mod}\,26)$.

Therefore, the given system has one solution:

$\begin{pmatrix} x \\ y \end{pmatrix} = \begin{pmatrix} 14 & 11 \\ 17 & 10 \end{pmatrix} \begin{pmatrix} 1 \\ 2 \end{pmatrix} \equiv \begin{pmatrix} 10 \\ 11 \end{pmatrix} (\mathrm{mod}\,26)$, i.e., we obtain the solution of the system in the following form:

$$\begin{cases} x \equiv 10\,(\mathrm{mod}\,26), \\ y \equiv 11\,(\mathrm{mod}\,26). \end{cases}$$

✳ Example 7.21.
Solve the system of congruences:

$$\begin{cases} x + 3y \equiv 1 \,(\mathrm{mod}\,26) \\ 7x + 9y \equiv 2 \,(\mathrm{mod}\,26) \end{cases}$$

Here, $\mathbf{A} = \begin{pmatrix} 1 & 3 \\ 7 & 9 \end{pmatrix}$, $D = \det(\mathbf{A}) = -12 \equiv 14 \,(\mathrm{mod}\,26)$ and $m = 26$.

Since $\gcd(14, 26) = 2 \neq 1$, there is no invertible matrix $\mathbf{A}^{-1} \,(\mathrm{mod}\,26)$, which means that the given system has no solutions.

Consider now solutions of a system of linear congruences of the following particular form:

$$x \equiv a_i \,(\mathrm{mod}\,m_i) \tag{7.24}$$

where $a_i \in \mathbb{Z}$, $m_i \in \mathbb{N}$ and $\gcd(m_i, m_j) = 1$ for all i, j with $i \neq j$. By the Generalized Chinese Remainder Theorem, this system has an integer solution.
The following theorem gives the formulas for obtaining this solution:

✳ Theorem 7.22.
Let $m = m_1 m_2 \cdots m_n$, where $m_i \in \mathbb{Z}^+$ and $\gcd(m_i, m_j) = 1$ for all i, j with $i \neq j$. Then, the system of congruences (7.24) has one solution in $\mathbb{Z}_m$. Moreover, if $x = b$ is a solution of this system, then the set of all solutions of this system coincides with the set of integers for which

$$x \equiv b \,(\mathrm{mod}\,m) \tag{7.25}$$

where

$$b = \sum_{i=1}^{n} a_i M_i N_i, \tag{7.26}$$

$M_i = m/m_i$, and $N_i = (M_i)^{-1}$ is the inverse element to M_i modulo m_i (i.e., the inverse element in the multiplicative group $\mathbb{Z}_{m_i}^*$).

Proof.
The uniqueness of the solution follows from the Generalized Chinese Remainder Theorem. If $m = m_1 m_2 \cdots m_n$, then $m = m_i M_i$, where $M_i = m/m_i$ for $i = 1, 2, \ldots, n$. It is obvious that $\gcd(M_i, m_i) = 1$. Therefore, for any $i = 1, 2, \ldots, n$ there exists an element $N_i = (M_i)^{-1}$ which is the unique solution of the congruence: $M_i N_i \equiv 1 \,(\mathrm{mod}\,m_i)$. Then, the element (7.26) is a solution of the system (7.24), since $M_i N_i \equiv 1 \,(\mathrm{mod}\,m_i)$ and $M_i \equiv 0 \,(\mathrm{mod}\,m_j)$ for all i, j with $i \neq j$. So

$$b \equiv a_i \,(\mathrm{mod}\,m_i), \quad i = 1, 2, \ldots, n.$$

□

Using Theorem 7.22, we obtain the following algorithm.

Algorithm of solving systems of linear congruences

1. Compute $m = m_1 m_2 \cdots m_n$.

2. Verify that $\gcd(m_i, m_j) = 1$ for all i, j with $i \neq j$.

3. Compute $M_i = m/m_i$, for all i

4. Compute $N_i = (M_i)^{-1} \pmod{m_i}$, for all i.

5. Compute $a_i M_i N_i$, for all i.

6. Compute $b = \sum_{i=1}^{n} a_i M_i N_i$.

✳ Example 7.23.
Solve the system of congruences:

$$\begin{cases} x \equiv 36 \pmod{41} \\ x \equiv 5 \pmod{17} \end{cases}$$

Here, $m_1 = 41$, $m_2 = 17$, $m = 41 \cdot 17 = 697$.
Since $\gcd(41, 17) = 1$, we can use Theorem 7.22.
For $i = 1, 2$, we compute:
$M_1 = m/m_1 = 17$, $N_1 = (17)^{-1} \equiv 29 \pmod{41}$
$M_2 = m/m_2 = 41$, $N_2 = (41)^{-1} \equiv 5 \pmod{17}$
Therefore, by (7.23), (7.24), we obtain that

$$x \equiv 36 \cdot 17 \cdot 29 + 5 \cdot 41 \cdot 5 \equiv 17748 \equiv 651 \pmod{697}$$

is the solution of the given system of congruences.

✳ Example 7.24.
Find the least positive integer which gives remainders 4, 3, 1 on division, respectively, by 5, 7, 9.
The required number is the solution of the system of congruences:

$$\begin{cases} x \equiv 4 \pmod{5} \\ x \equiv 3 \pmod{7} \\ x \equiv 1 \pmod{9} \end{cases}$$

Here, $m_1 = 5$, $m_2 = 7$, $m_3 = 9$, $m = 5 \cdot 7 \cdot 9 = 315$.
Since $\gcd(m_i, m_j) = 1$ for i, j with $i \neq j$, we can use Theorem 7.22.
For $i = 1, 2, 3$, we compute:
$M_1 = m/m_1 = 63$, $N_1 = (63)^{-1} \equiv (3)^{-1} \equiv 2 \pmod{5}$
$M_2 = m/m_2 = 45$, $N_2 = (45)^{-1} \equiv (3)^{-1} \equiv 5 \pmod{7}$
$M_3 = m/m_3 = 35$, $N_3 = (35)^{-1} \equiv (8)^{-1} \equiv 8 \pmod{7}$

So $x \equiv 63 \cdot 2 \cdot 4 + 45 \cdot 5 \cdot 3 + 35 \cdot 8 \cdot 1 \equiv 1459 \equiv 199 \,(\mathrm{mod}\,315)$ is the solution of the given system of congruences. Therefore, the least positive integer satisfying the given conditions is $x = 199$.

7.5. Lagrange's Interpolation Polynomials

In this section, we consider the problem of constructing a special polynomial that passes through a given set of data points.

Consider the following problem:

Let K be a field. Given a set of n elements $\{(x_i, y_i)\}_{i=0}^n$ in K^2 with distinct x-coordinates, i.e., $x_i \neq x_j$ for all i, j with $i \neq j$, find a polynomial $f_n(x) \in K[x]$ of degree at most n such that $f_n(x_i) = y_i$ for all $i = 0, 1, 2, \ldots, n$.

The method of solving this problem is called **polynomial interpolation**, the elements $x_0, x_1, \ldots, x_n \in K$ are called **interpolation points** and the polynomial $f_n(x) \in K[x]$ is called the **interpolation polynomial**. This method is based on the following theorem.

✳ **Theorem 7.25.**

Let $u_0, u_1, \ldots, u_n$ be distinct elements of a field K and let $y_0, y_1, \ldots, y_n$ be arbitrary elements of K. Then, there is a unique polynomial $f(x) \in K[x]$ of degree at most n such that $f(u_i) = y_i$ for all $i = 0, 1, \ldots, n$.

Proof.

Consider a polynomial $f(x) = a_0 + a_1 x + a_2 x^2 + \ldots + a_n x^n \in K[x]$. Since $f(u_i) = y_i$ for $i = 0, 1, \ldots, n$, in order to find the coefficients a_i, we obtain the following system of linear equations:

$$\begin{cases} y_0 = a_0 + a_1 u_0 + a_2 u_0^2 + \cdots + a_n u_0^n \\ y_1 = a_0 + a_1 u_1 + a_2 u_1^2 + \cdots + a_n u_1^n \\ \cdots\cdots\cdots\cdots\cdots\cdots\cdots\cdots\cdots\cdots\cdots\cdots\cdots\cdots \\ y_n = a_0 + a_1 u_n + a_2 u_n^2 + \cdots + a_n u_n^n \end{cases} \tag{7.27}$$

or in the matrix form:

$$\begin{pmatrix} y_0 \\ y_1 \\ \vdots \\ y_n \end{pmatrix} = \begin{pmatrix} 1 & u_0 & u_0^2 & \cdots & u_0^n \\ 1 & u_1 & u_1^2 & \cdots & u_1^n \\ \vdots & \vdots & \vdots & \ddots & \vdots \\ 1 & u_n & u_n^2 & \cdots & u_n^n \end{pmatrix} \begin{pmatrix} a_0 \\ a_1 \\ \vdots \\ a_n \end{pmatrix}$$

The matrix of coefficients

$$\mathbf{A} = \begin{pmatrix} 1 & u_0 & u_0^2 & \cdots & u_0^n \\ 1 & u_1 & u_1^2 & \cdots & u_1^n \\ \vdots & \vdots & \vdots & \ddots & \vdots \\ 1 & u_n & u_n^2 & \cdots & u_n^n \end{pmatrix}$$

of this system is the **Vandermonde matrix** with determinant:

$$\det \mathbf{A} = \prod_{1 \leq i < j \leq n} (u_j - u_i). \tag{7.28}$$

This determinant is not equal to zero because by assumption $u_i \neq u_j$ for all i, j with $i \neq j$. Therefore, the system (7.27) has a solution and $f(x) = a_0 + a_1 x + a_2 x^2 + \ldots + a_n x^n \in K[x]$ is a required polynomial.

Suppose we have two distinct polynomials $f(x)$ and $g(x)$ of degree n such that $f(u_i) = g(u_i) = y_i$ for all $i = 0, 1, \ldots, n$. Hence, the polynomial $h(x) = f(x) - g(x) \in K[x]$ is of degree n and has $n + 1$ distinct roots $u_0, u_1, \ldots, u_n$ in K, that is impossible, by Theorem 5.43. $\qquad\square$

✳ **Theorem 7.26 (Lagrange's interpolation).**

Let K be a field. Then, any polynomial $f(x)$ of degree n over a field K is uniquely determined by its values $y_0, y_1, y_2, \ldots, y_n$ in arbitrary distinct $n + 1$ points $c_0, c_1, c_2, \ldots, c_n \in K$. In this case, $f(x) = p_n(x)$, where $p_n(x)$ is the **Lagrange interpolation polynomial** of the form:

$$p_n(x) = \sum_{i=0}^{n} y_i L_{n,i} \tag{7.29}$$

where

$$L_{n,i} = \prod_{k=0, k \neq i}^{n} (x - c_k)(c_i - c_k)^{-1} \in K[x] \tag{7.30}$$

If the number of points is less than n, then the polynomial $f(x)$ is not unique.

Proof.

Consider a polynomial $f(x) = a_0 + \sum_{j=1}^{n-1} a_j x^j \in K[x]$. Assume that the values of this polynomial are known in the points c_i: $f(c_i) = y_i$ for $i = 0, 1, 2, \ldots, n$. Then, by Theorem 7.25, there exists a unique polynomial of degree at most n that satisfies these conditions. Since the polynomials $L_{n,i}(x) = \prod_{k=0, \; k \neq i}^{n} (x - c_k)(c_i - c_k)^{-1}$ have the property:

$$L_{n,i}(x_j) = \begin{cases} 1 & \text{if } i = j \\ 0 & \text{if } i \neq j \end{cases}$$

the polynomial $p_n(x) = \sum_{i=0}^{n} y_i L_{n,i}$ has the property: $p_n(c_i) = y_i$ for $i = 0, 1, 2, \ldots, n$, i.e., $p_n(x)$ satisfies the conditions of Theorem 7.25. Therefore, $p_n(x)$ is the unique polynomial of degree n that determined by $n + 1$ values $y_0, y_1, y_2, \ldots, y_n$ in distinct $n + 1$ points $c_0, c_1, c_2, \ldots, c_n \in K$.

If the number of points k is less than $n+1$, then the system of equations (7.27) has k equations with $n+1 > k$ unknowns and, therefore, has more than $n+1-k$ different solutions. □

The polynomials $L_{n,i}$ defined by (7.30) are called **Lagrange's polynomials**.

✳ Example 7.27.

Use Lagrange's interpolation to find the polynomial $f(x) \in \mathbb{R}[x]$ satisfying the following conditions:

$$f(c_0) = 12 \quad \text{if} \quad c_0 = 2$$
$$f(c_1) = 6 \quad \text{if} \quad c_1 = -1$$
$$f(c_2) = 5 \quad \text{if} \quad c_2 = 3$$
$$f(c_3) = -10 \quad \text{if} \quad c_3 = -2$$

In this case, $n = 4$, so the given polynomial $f(x)$ has the degree 3 and it is interpolated by Lagrange's interpolation polynomial $p_3(x)$ constructed by the Lagrange's polynomials $L_{3,i}$ for $i = 0, 1, 2, 3$:

$$L_{3,0}(x) = (x - c_1)(x - c_2)(x - c_3)[(c_0 - c_1)(c_0 - c_2)(c_0 - c_3)]^{-1} =$$

$$= (x+1)(x-3)(x+2)[(2+1)(2-3)(2+2)]^{-1} = -\frac{1}{12}(x^3 - 7x - 6)$$

$$L_{3,1}(x) = (x - c_0)(x - c_2)(x - c_3)[(c_1 - c_0)(c_1 - c_2)(c_1 - c_3)]^{-1} =$$

$$= (x-2)(x-3)(x+2)[(-1-2)(-1-3)(-1+2)]^{-1} = \frac{1}{12}(x^3 - 3x^2 - 4x + 12)$$

$$L_{3,2}(x) = (x - c_0)(x - c_1)(x - c_3)[(c_2 - c_0)(c_2 - c_1)(c_2 - c_3)]^{-1} =$$

$$= (x-2)(x+1)(x+2)[(3-2)(3+1)(3+2)]^{-1} = \frac{1}{20}(x^3 + x^2 - 4x - 4)$$

$$L_{3,3}(x) = (x - c_0)(x - c_1)(x - c_2)[(c_3 - c_0)(c_3 - c_1)(c_3 - c_2)]^{-1} =$$

$$= (x-2)(x+1)(x-3)[(-2-2)(-2+1)(-2-3)]^{-1} = -\frac{1}{20}(x^3 - 4x^2 + x + 6)$$

Therefore, Lagrange's interpolation polynomial $p_3(x)$ is given by:

$$p_3(x) = \sum_{i=0}^{3} f(c_i)L_{n,i} = f(c_0)L_{3,0}(x) + f(c_1)L_{3,1}(x) + f(c_2)L_{3,2}(x) + f(c_3)L_{3,3}(x) =$$

$$= 12 \cdot [-\frac{1}{12}(x^3 - 7x - 6)] + 6 \cdot [\frac{1}{12}(x^3 - 3x^2 - 4x + 12)] + 5 \cdot [\frac{1}{20}(x^3 + x^2 - 4x - 4)] -$$

$$-10 \cdot [-\frac{1}{20}(x^3 - 4x^2 + x + 6)] = \frac{1}{4}(x^3 - 13x^2 + 18x + 56)$$

✳ Example 7.28.

Use Lagrange's interpolation to construct the polynomial $f(x) \in \mathbb{Z}_p[x]$, where $p = 31$, that satisfies the following properties:

$$f(c_0) = 16 \quad \text{if} \quad c_0 = 1$$
$$f(c_1) = 5 \quad \text{if} \quad c_1 = 3$$
$$f(c_2) = 7 \quad \text{if} \quad c_2 = 5$$

In this case, $n = 3$, so the given polynomial $f(x)$ has the degree 2 and it is interpolated by Lagrange's interpolation polynomial $p_2(x)$ constructed by Lagrange's polynomials $L_{2,i}$ for $i = 0, 1, 2$.

$$L_{2,0}(x) = (x-c_1)(x-c_2)[(c_0-c_1)(c_0-c_2)]^{-1} = (x-3)(x-5)[(1-3)(1-5)]^{-1}$$

$$L_{2,1}(x) = (x-c_0)(x-c_2)[(c_1-c_0)(c_1-c_2)]^{-1} = (x-1)(x-5)[(3-1)(3-5)]^{-1}$$

$$L_{2,0}(x) = (x-c_0)(x-c_1)[(c_2-c_0)(c_2-c_1)]^{-1} = (x-1)(x-3)[(5-1)(5-3)]^{-1}$$

Taking into account the arithmetic in $\mathbb{Z}_{31}$, we get:

$$L_{2,0}(x) = 4(x^2 + 23x + 15)$$

$$L_{2,1}(x) = 23(x^2 + 25x + 5)$$

$$L_{2,2}(x) = 4(x^2 + 27x + 3)$$

Therefore,

$$f(x) = 16 \cdot 4(x^2+23x+15)+5 \cdot 23(x^2+25x+5)+7 \cdot 4(x^2+27x+3) = 21x^2+19x+7$$

7.6. Secret Sharing

In this section, we consider the well known problem of secret sharing, which consists of recovering a secret S from a set of shares. At the beginning, we consider the simplest variant of this problem.

Problem:

1. Adam wants to send to Alice and Bob a message $M \in \mathbb{Z}_m$.

2. Adam want to be sure that only Alice and Bob together can read this message if they both agree to do this.

An easy way to solve this problem leads to the following simple algorithm:

Secret Sharing Scheme

1. Adam divides the message M into two parts. For this aim, he chooses randomly a number $r \in \mathbb{Z}_m$.

2. Alice receives from Adam the number r.

3. Bob receives from Adam the number $(M - r) \,(\text{mod}\, m)$.

In what way can we generalize this problem for n persons? Similarly to the previous variant for two persons, we have the following simple variant of this problem for n persons.

Problem:

- Adam wants to "share" a secret $M \in \mathbb{Z}_m$ between n persons.

- Each person receives his "share".

- If an arbitrary $t \leq n$ persons agree to recover the secret, then they can do this.

- If only $k < t$ persons agree to recover the secret, then they cannot get any information about this secret.

In the general case, a **secret sharing** is a cryptographic protocol in which some **secret** is "divided" into "shares" which are distributed to users in such a way that this secret can be recovered only by subgroups of users that satisfy certain criteria.

The aim of protocols is the security of a message before the loss of its safeguarding key. The implementation of more copies of the key reduces security of the system, and the implementation of less copies of the key increases the risk of losing all of them. Protocols of secret sharing realize the first case, which improves system reliability without increasing risk.

In a secret sharing protocol, the following persons take part: The person named a **dialer**, who prepares and the shares a secret in fragments, and the shareholders who keep these fragments.

There are a several types of secret sharing schemes. The most known types are the so-called threshold schemes. A protocol is called (t, n)-**threshold** if the number of shares is n, and to recover a secret it is necessary to put together t such shares. At the same time, an arbitrary set of $t - 1$ random shares can not be allowed to recover even a part of a secret.

✳ **Definition 7.29.**

A **threshold scheme** (t, n) is a method that enables a secret S to be shared for n persons $P_1, P_2, \ldots, P_n$ in such a way that the following conditions hold:

- $t \leq n$,

- each of the persons P_i receives some piece y_i of the secret S,

- any t persons can find the secret S from their pieces,

- no group of less than t users can get any information about the secret S from their pieces.

A secret sharing threshold scheme was first developed independently by Adi Shamir and George Blakley in 1979. Both protocols are theoretically secure, which means that even an enemy possessing infinite computing power can not recover a secret. Shamir's threshold secret sharing scheme is based on polynomial interpolation, and George Blakley's secret sharing scheme is based on geometric methods.

In this section, we consider Shamir's threshold access algorithm. Existence of Lagrange's interpolation polynomials are given by Theorem 7.26.

Shamir's threshold algorithm of secret sharing works as follows:

- The secret S is chosen as a value of some polynomial $P(x)$ of degree $t-1$ at the point 0, i.e., $S = P(0)$.

- The shares are chosen as the values of this polynomial in distinct t points.

- The set of values in t distinct points allows the interpolation polynomial $P(x)$ to be constructed and $P(0)$ to be computed.

Using polynomials over real numbers is not practical, since due to the large randomness of their coefficients, the values of polynomials in distinct points can take a lot of space. Therefore, we usually consider polynomials over finite fields such as $\mathbb{Z}_p$, where p is a prime. In this case, each share takes the same space as a secret. The number of random bits required in order to construct a polynomial is also minimal in this case.

Shamir's Threshold Access Algorithm

Suppose that a secret S needs to be shared among a group of $n \geq 2$ users and the number of persons which can recover the secret S is equal to $2 \leq t \leq n$, i.e., $t \leq n$ is the threshold of the scheme. The dealer does the following:

- Choose a prime p which is greater than an arbitrary possible message S and choose n random distinct elements $c_1, c_2, \ldots, c_n \in \mathbb{Z}_p$ such that $c_1 < c_2 < \cdots < c_n$.

- Choose $t-1$ random numbers $a_j \in \mathbb{Z}_p$ and construct a polynomial of degree $t-1$ over the finite field $\mathbb{Z}_p$:

$$f(x) = f(0) + \sum_{j=1}^{t-1} a_j x^j \in \mathbb{Z}_p[x], \qquad (7.31)$$

- Choose $f(0) = S$.

- Choose the shares as $y_i = f(c_i)$ for all $i = 1, 2, \ldots, n$.

- Send (c_i, y_i) to each i-th shareholder for all $i = 1, 2, \ldots, n$ using a secure communication channel.

If t arbitrary shareholders want to recover a secret S from a set $\{(c_i', y_i')\}_{i=1}^{t}$, where $c_i' \in \{c_1, \ldots, c_n\}$, $y_i' = f(c_i')$ and $f(x)$ is the Lagrange interpolation polynomial in the form (7.29), then they compute $S = f(0)$, i.e.,

$$S = f(0) = \sum_{i=1}^{t} y_i' \prod_{j=1, i \neq j}^{t} a_{ij} \,(\mathrm{mod}\, p), \tag{7.32}$$

where a_{ij} are solutions of congruences $(c_j' - c_i')a_{ij} \equiv c_j' \,(\mathrm{mod}\, p)$.

✳ **Example 7.30.**

Consider Shamir's threshold algorithm, when $n = 5$, $t = 3$, and the secret $S = 3 \in \mathbb{Z}_{17}$.

We choose randomly the elements $a_1 = 14$, $a_2 = 15$. Then, the polynomial has the form: $f(x) = 15x^2 + 14x + 3$. We choose the elements $c_i = i$ for $1 \leq i \leq 5$ and compute shares:

$y_1 = f(1) = 15$,
$y_2 = f(2) = 6$,
$y_3 = f(3) = 10$,
$y_4 = f(4) = 10$,
$y_5 = f(5) = 6$.

To recover the secret from the first three shares, we compute:

$$f(0) = \sum_{i=1}^{t} y_i \prod_{j=1, i \neq j}^{t} a_{ij} \,(\mathrm{mod}\, p) \equiv 15 \cdot 3 + 6 \cdot (-3) + 10 \cdot 1 \equiv 3 \,(\mathrm{mod}\, 17),$$

where

$a_{12} \equiv 2 \,(\mathrm{mod}\, 17)$, $2a_{13} \equiv 3 \,(\mathrm{mod}\, 17)$, $a_{13} \equiv 10 \,(\mathrm{mod}\, 17)$,
$a_{12}a_{13} \equiv 3 \,(\mathrm{mod}\, 17)$
$a_{21} \equiv -1 \,(\mathrm{mod}\, 17)$, $a_{23} \equiv 3 \,(\mathrm{mod}\, 17)$, $a_{21}a_{23} \equiv -3 \,(\mathrm{mod}\, 17)$
$2a_{31} \equiv -1 \,(\mathrm{mod}\, 17)$, $a_{32} \equiv -2 \,(\mathrm{mod}\, 17)$, $a_{31} \equiv 8 \,(\mathrm{mod}\, 17)$,
$a_{31}a_{32} \equiv 1 \,(\mathrm{mod}\, 17)$
Therefore, the secret $S = f(0) = 3$.

Other types of threshold secret sharing algorithms are based on the Chinese Remainder Theorem. In this section, we consider the two most important such algorithms: Mignotte's threshold secret sharing scheme and Asmuth-Bloom's threshold secret sharing scheme.

In Mignotte's threshold secret sharing scheme, we choose randomly a strictly increasing sequence of n positive integers $m_1 < m_2 < \ldots < m_n$ which are pairwise relatively prime, i.e., $\gcd(m_i, m_j) = 1$ for all i, j with $i \neq j$, and such that the product of the smallest t of them is greater than the secret S and the product of the $t - 1$ biggest ones is greater than S.

These conditions enable the Chinese Remainder Theorem to be used for an arbitrary choice of t numbers among the numbers $m_1, m_2, \ldots, m_n$, however, any choice of less than t numbers does not satisfy the uniqueness of the Chinese Remainder Theorem.

Mignotte's Threshold Secret Sharing Scheme

Suppose that a secret S needs to be shared among a group of $n \geq 2$ users and the number of persons which can recover the secret S is equal to $2 \leq t \leq n$, i.e., $t \leq n$ is the threshold of the scheme. The dealer does the following:

- Choose randomly a set of n pairwise relatively prime integers $m_1, m_2, \ldots, m_n$, i.e., $\gcd(m_i, m_j) = 1$ for all i, j with $i \neq j$ satisfying conditions:

$$m_1 < m_2 < \cdots < m_n \tag{7.33},$$

$$\prod_{i=0}^{t-2} m_{n-i} < \prod_{i=1}^{t} m_i \tag{7.34}$$

$$\prod_{i=0}^{t-2} m_{n-i} < S < \prod_{i=1}^{t} m_i \tag{7.35}$$

- Choose the shares as: $y_i \equiv S \,(\mathrm{mod}\, m_i)$ for all $i = 1, 2, \ldots, n$.

- Send (y_i, m_i) to each i-th shareholder using a secure communication channel.

Given t shareholders possessing t shares $y_{i_1}, y_{i_2}, \ldots, y_{i_t}$, the secret S can be recovered using the Chinese Remainder Theorem, as the unique solution modulo $m_{i_1} m_{i_2} \ldots m_{i_t}$ of the system of congruences:

$$\begin{cases} x \equiv y_{i_1} \,(\mathrm{mod}\, m_{i_1}) \\ x \equiv y_{i_2} \,(\mathrm{mod}\, m_{i_2}) \\ \cdots\cdots\cdots\cdots \\ x \equiv y_{i_t} \,(\mathrm{mod}\, m_{i_t}) \end{cases} \tag{7.36}$$

Note that the inequality (7.34) is equivalent to the inequality with:

$$\max_{1 \leq i_1 < \cdots < i_{t-1} \leq n} (p_{i_1} p_{i_2} \cdots p_{i_{t-1}}) < \min_{1 \leq i_1 < \cdots < i_{t-1} \leq n} (p_{i_1} p_{i_2} \cdots p_{i_t}), \tag{7.37}$$

i.e., the product of arbitrary $t-1$ integers is less than the product of arbitrary t integers among the integers $m_1, m_2, \ldots, m_n$.

Suppose $m_{i_1} m_{i_2} \ldots m_{i_t} = M_t$. Then, from this inequality, it follows that $\prod_{i=1}^{t} m_i < M_t$. Taking this inequality and (7.35) into account, we obtain that

$$0 < S < \prod_{i=1}^{t} m_i < M_t,$$

i.e., S lies in $\mathbb{Z}_{M_t}$, and so it is the unique solution modulo M_t of the system of congruences (7.36). On the other hand, if we have only $t-1$ distinct shares $y_{i_1}, y_{i_2}, \ldots, y_{i_{t-1}}$, we obtain only that $S \equiv y_0 \pmod{(y_{i_1} y_{i_2} \ldots y_{i_{t-1}})}$, where y_0 is the unique solution modulo $y_{i_1} y_{i_2} \ldots y_{i_{t-1}}$ of the resulted system, since, in this case, the secret S satisfies the inequalities

$$S > \prod_{i=0}^{t-2} m_{n-i} \geq y_{i_1} y_{i_2} \ldots y_{i_{t-1}} > y_0.$$

To solve system of congruences (7.36), we can use formulas (7.22)–(7.23).

✳ **Example 7.31.**

Consider Mignotte's Threshold Secret Sharing Scheme, when $n = 5$, $t = 3$, and the secret $S = 671875$.

We choose pairwise prime integers $m_1 = 97$, $m_2 = 98$, $m_3 = 99$, $m_4 = 101$, $m_5 = 103$ for which

$m = m_1 m_2 m_3 m_4 m_5 = 9790200882$, $m_1 m_2 m_3 = 941094$, $m_4 m_5 = 10403$ and

$10403 < S = 671875 < 941094$

Then, we compute the shares:

$y_1 \equiv S \pmod{97} = 53$

$y_2 \equiv S \pmod{98} = 85$

$y_3 \equiv S \pmod{99} = 61$

$y_4 \equiv S \pmod{101} = 23$

$y_5 \equiv S \pmod{103} = 6$

Assume that the group of shareholders P_1, P_2, P_3 want to recover the secret S. They compute $m = m_1 m_2 m_3 = 941094$ and:

$M_1 = m/m_1 = 9702$, $M_2 = m/m_2 = 9603$, $M_3 = m/m_3 = 9506$

$M_1 N_1 \equiv 1 \pmod{m_1} \Rightarrow N_1 = 49$

$M_2 N_2 \equiv 1 \pmod{m_2} \Rightarrow N_2 = 97$

$M_3 N_3 \equiv 1 \pmod{m_3} \Rightarrow N_3 = 50$

Then, the secret is:

$S \equiv y_1 M_1 N_1 + y_2 M_2 N_2 + y_3 M_3 N_3 \pmod{m} \equiv 133366129 \pmod{941094} = 671875$.

Assume that the group of shareholders P_1, P_4 want to recover the secret. They compute $m = m_1 m_4 = 9797$ and

$M_1 = m/m_1 = 101$, $M_4 = m/m_4 = 97$.

$$M_1N_1 \equiv 1 \,(\mathrm{mod}\,m_1) \;\Rightarrow\; N_1 = 49$$
$$M_4N_4 \equiv 1 \,(\mathrm{mod}\,m_4) \;\Rightarrow\; N_4 = 25$$

In this case,

$$S^* \equiv y_1 M_1 N_1 + y_4 M_4 N_4 \,(\mathrm{mod}\,m) \equiv 332622 \,(\mathrm{mod}\,9797) = 9321 \neq S,$$

so they can not recover the secret.

Asmuth-Bloom's Threshold Secret Sharing Scheme

Suppose that a secret S needs to be shared among a group of $n \geq 2$ users and the number of persons which can recover the secret S is equal to $2 \leq t \leq n$, i.e., $t \leq n$ is the threshold of the scheme. Suppose $1 \leq S < m_0$. The dealer does the following:

- Choose randomly n pairwise relatively prime integers $m_1, m_2, \ldots, m_n$, satisfying the conditions: $\gcd(m_0, m_i) = 1 \;\forall(i = 1, \ldots, n)$ and

$$m_1 < m_2 < \cdots < m_n \tag{7.38}$$

$$m_0 \cdot \prod_{i=0}^{t-2} m_{n-i} < \prod_{i=1}^{t} m_i \tag{7.39}$$

- Choose a random integer α such that $S + \alpha m_0 < m_1 m_2 \cdots m_t$.

- Choose the shares as: $y_i \equiv S + \alpha m_0 \,(\mathrm{mod}\,p_i)$ for all $i = 1, 2, \ldots, n$.

- A dialer send (y_i, m_i, m_0) to each i-th shareholder using a secure communication channel.

When t shareholders, possessing t shares $y_{i_1}, y_{i_2}, \ldots, y_{i_t}$, want to recover a secret, they must solve the system of congruences (7.36), which, by assumption on the choice of elements $m_1, m_2, \ldots, m_n$ and the Chinese Remainder Theorem, has the unique solution S_0 modulo $m_t = m_{i_1} m_{i_2} \cdots m_{i_t}$. Then, the secret is $S = S_0 \,(\mathrm{mod}\,m_0)$.

✳ Example 7.32.

Consider Asmuth-Bloom's Threshold Secret Sharing Scheme, when $n = 5$, $t = 3$.

Consider pairwise prime integers

$m_0 = 3$, $m_1 = 7$, $m_2 = 11$, $m_3 = 13$, $m_4 = 17$, $m_5 = 19$

These numbers satisfy the inequalities (7.38) and (7.39), since

$969 = 3 \cdot 17 \cdot 19 < 7 \cdot 11 \cdot 13 = 1001$.

Let the secret be $S = 2$. We choose $\alpha = 251$ such that:

$S_0 = S + \alpha m_0 = 2 + 251 \cdot 3 = 755 < 7 \cdot 11 \cdot 13 = 1001$.

A dialer computes the shares:

$y_1 \equiv S_0 \,(\mathrm{mod}\,7) = 6$

$y_2 \equiv S_0 \,(\mathrm{mod}\,11) = 7$

$y_3 \equiv S_0 \,(\mathrm{mod}\,13) = 1$

$y_4 \equiv S_0 \,(\text{mod}\,17) = 7$
$y_5 \equiv S_0 \,(\text{mod}\,19) = 16$
To recover the secret, we choose three shares: $(y_1, y_3, y_4) = (6, 1, 7)$.
$m = m_1 m_3 m_4 = 7 \cdot 13 \cdot 17 = 1547,$
The shareholders must solve the system:

$$\begin{cases} x \equiv 6 \,(\text{mod}\,7) \\ x \equiv 1 \,(\text{mod}\,13) \\ x \equiv 7 \,(\text{mod}\,17) \end{cases} \tag{7.40}$$

$M_1 = m/m_1 = 13 \cdot 17 = 221, \; M_3 = m/m_3 = 7 \cdot 17 = 119, \; M_4 = m/m_4 = 7 \cdot 13 = 91$

$(M_1)^{-1} = (221)^{-1} \equiv 4^{-1} \equiv 2 \,(\text{mod}\,7)$
$(M_3)^{-1} = (119)^{-1} \equiv 2^{-1} \equiv 7 \,(\text{mod}\,13)$
$(M_4)^{-1} = (91)^{-1} \equiv 6^{-1} \equiv 3 \,(\text{mod}\,17)$
In this way, the system of congruences (7.40) has the solution:
$x \equiv 6 \cdot 221 \cdot 2 + 1 \cdot 119 \cdot 7 + 7 \cdot 91 \cdot 3 \equiv 2302 \equiv 755 \,(\text{mod}\,1547)$, i.e., $S_0 = 755$.
Therefore, $S \equiv 755 \equiv 2 \,(\text{mod}\,3)$, so $S = 2$ is the given secret.

7.7. Cryptographic Algorithm RSA

In general, cryptography is the method of hiding a secret by encryption. Participants of interchanging information are a sender and a recipient who want to have an assurance that their correspondence will not be intercepted and read by an adversary.

A security process of transformation of an original message or information into secret code to hide its content is called **encryption**, and an encrypting message is called a **ciphertext**. The message or information in its original form is called the **plaintext** or the **original text**. The reverse process of finding the original message from the ciphertext is called **decryption**. Data or plaintext is encrypted into a ciphertext with **encryption algorithm** or **encryption function**, while the reverse algorithm of decoding a ciphertext into an original message is called an **decryption algorithm** or **decryption function**. The pair of an encryption algorithm and a decryption algorithm is often called a **cryptosystem**, or a **cryptographic algorithm**, or a **cipher**.

In modern cryptographic algorithms, the security is provided thanks to using encryption and decryption keys which are some parameters of a cipher. A key is a secret and, ideally, it must be known only for confidential correspondents. There are two types of keys: **Private** and **public keys**. Depending on using keys, cryptosystems are divided by two different types.

1. **Symmetric cryptosystems** use keys which are identical or easily obtained from each other, to encrypt a plaintext and to decrypt a ciphertext.

2. **Asymmetric cryptosystems**, which are often called **public-key cryptosystems**, use two different types of keys for encrypting and decrypting data. They use the public key, which can be available everywhere, to encrypt

data. While, for decrypting data, they use the private key, which must be protected and kept in secret, and it must be ideally known only to one user.

The historian David Kah describes cryptography with public key as "the most revolutionary thought from the beginning of polyalphabet encryption during the Renaissance".

Algorithms with public keys are often based on computational complexity of "difficult" mathematical problems. For example, the algorithm RSA is based on difficulty of factorization of large integers into products of primes, and the Diffie-Hellmann algorithm is based on difficulty of computing discrete logarithms. The new elliptic-curve cryptographic algorithms are based on algebraic structure of elliptic curves over finite fields.

The algorithm RSA is one of the first and now the most popular asymmetric cryptographic algorithm. The algorithm was invented in 1977 by Ron Rivest, Adi Shamir and Leonard Adleman and was the first among algorithms which could be used for both encryption and for digital signatures. The name of the algorithm RSA comes from the first letters of the names of its discoverers: **R**ivest, **S**hamir, **A**dleman. The security of encryption of this algorithm is based on the difficulty of factorization of large numbers.

Consider the general problem of a private communication between two correspondents, Alice and Bob, using public-key cryptosystem.

Alice creates her public key, which can be accessible everywhere, in particular, it will be known to Bob. Bob also creates his public key. Alice and Bob also generate their own private keys, which they keep in secret.

If Alice wants to send a private message to Bob using the public-key cryptosystem, she does the following:

1. Using Bob's public key, Alice encrypts her message and sends it to Bob.

2. Bob, using his private key, decrypts the received ciphertext from Alice.

3. If Bob wants to send Alice his private message, he does the same as Alice, i.e., he encrypts his message using Alice's public key and sends it to Alice.

4. Alice, using her private key, decrypts the received ciphertext from Bob.

So, in general, the public-key cryptosystem consists of three steps:

1. Generation of public and private keys.

2. Encryption of a message.

3. Decryption of a message

Consider each of these steps more exactly in mathematical notation for the algorithm RSA.

1. Algorithm of generation of keys

In order to generate her public and private keys, Alice does the following:
a) generate randomly two large distinct primes p, q.
b) calculate their product $m = p \cdot q$ and $\varphi(m) = (p-1)(q-1)$.
c) pick randomly an integer k such that $1 < k < \varphi(m)$ and

$$\gcd(k, \varphi(m)) = 1.$$

d) find an integer h such that $1 < h < \varphi(m)$ and

$$kh \equiv 1 \pmod{\varphi(m)}.$$

The pair of numbers (m, k) is Alice's public key and the pair of numbers (m, h) is her private key.

2. Encryption algorithm

For encrypting original message M, satisfying the inequalities $1 \le M \le m - 1$, using the public key (m, k) of Alice, Bob does the following:

Compute the ciphertext: $C \equiv M^k \pmod{m}$ and send this text to Alice.

2. Decryption algorithm

Alice, receiving the ciphertext C from Bob and using her private key (m, h), does the following:

Compute $M \equiv C^h \pmod{m}$, which is the original text from Bob.

Now, we will show the correctness of the algorithm RSA. Consider two different cases.

1. Let $\gcd(M, m) = 1$.

   Then, by Euler's Theorem, $M^{\varphi(m)} \equiv 1 \pmod{m}$. From the encryption algorithm, we have $C \equiv M^k \pmod{m}$.

   Since, from the algorithm of generation of keys $\gcd(k, \varphi(m)) = 1$, there are integers h, s such that $kh = 1 + s\varphi(m)$. Therefore, from the decryption algorithm, we obtain:

   $$C^h \equiv M^{kh} \equiv M^{1+s\varphi(m)} \equiv M \cdot (M^{\varphi(m)})^s \equiv M \pmod{m},$$

   i.e., C^h is the original message.

2. Let $\gcd(M, m) \ne 1$.

   Since $m = pq$, where p, q are distinct primes, and $\gcd(k, \varphi(m)) = 1$, by the algorithm of generation of keys, there are integers h, s such that $kh = 1 + s\varphi(m) = 1 + s(p-1)(q-1)$.

(a) If $\gcd(M,p) = 1$, then, from Fermat's Little Theorem, we have that $M^{p-1} \equiv 1 \,(\mathrm{mod}\,p)$. Therefore, from the decryption algorithm, we obtain:

$$C^h \equiv M^{kh} \equiv M^{1+s(p-1)(q-1)} \equiv M \cdot (M^{p-1})^{s(q-1)} \equiv M \,(\mathrm{mod}\,p).$$

(b) If $\gcd(M,p) = p$, then we obtain that $M \equiv 0 \,(\mathrm{mod}\,p)$ and $M^{kh} \equiv 0 \,(\mathrm{mod}\,p)$, so $M^{kh} \equiv M \,(\mathrm{mod}\,p)$.

Therefore, $M^{kh} \equiv M \,(\mathrm{mod}\,p)$ for any M.

Analogously, we get that $M^{kh} \equiv M \,(\mathrm{mod}\,q)$ for any M.

Thus, $p|(M^{kh} - M)$ and $q|(M^{kh} - M)$. Since p and q are distinct primes, $pq|(M^{kh} - M)$, i.e., $m|(M^{kh} - M)$, which means that $M^{kh} \equiv M \,(\mathrm{mod}\,m)$ for any M. Therefore, from the decryption algorithm we have:

$$C^h \equiv M^{kh} \equiv M \,(\mathrm{mod}\,m).$$

since $M < m$, $C^h = M$ is the original message.

* **Example 7.33.**
Suppose that Alice and Bob want to correspond. Both of them create the pair of keys: Public keys and private keys.

Bob create his keys as follows. He chooses two primes: $p = 13$, $q = 19$. Then, $m = pq = 247$ and $\varphi(m) = (p-1)(q-1) = 216$.

He randomly picks a number $1 < k < 216$ such that $\gcd(k, 216) = 1$. Let $k = 5$. Then, the pair of numbers $(253, 5)$ is Bob's public key.

To obtain the private key, he solves the congruence: $kh \equiv 1 \,(\mathrm{mod}\,216)$, i.e., $5h \equiv 1 \,(\mathrm{mod}\,216)$. The solution of this congruence is the number $h = 173$. So, the pair of numbers $(247, 173)$ is Bob's private key.

If Alice wants to sent the message $M = 165$ to Bob, she encrypts this message using Bob's public key $(253, 5)$. Then, $C \equiv 165^5 \equiv 185 \,(\mathrm{mod}\,247)$ is the encryption message which is obtained by Bob. To decrypt this message, he uses his private key $(247, 173)$ doing the following: $M = C^{173} \equiv 185^{173} \,(\mathrm{mod}\,247)$. To compute $185^{173} \,(\mathrm{mod}\,247)$, the algorithm of quick modulo exponentiation can be used, writing the exponent 173 in its binary representation

$$173 = 1 + 0 \cdot 2 + 1 \cdot 2^2 + 1 \cdot 2^3 + 0 \cdot 2^4 + 1 \cdot 2^5 + 0 \cdot 2^6 + 1 \cdot 2^7$$

and then computing

$$185^{173} = 185 \cdot 185^4 \cdot 185^8 \cdot 185^{32} \cdot 185^{128}.$$

Since
$$185^2 \equiv 139 \,(\mathrm{mod}\,247),$$
$$185^4 \equiv 139^2 \equiv 55 \,(\mathrm{mod}\,247),$$

$$185^8 \equiv 55^2 \equiv 61 \,(\mathrm{mod}\,247),$$
$$185^{16} \equiv 61^2 \equiv 16 \,(\mathrm{mod}\,247),$$
$$185^{32} \equiv 16^2 \equiv 9 \,(\mathrm{mod}\,247),$$
$$185^{64} \equiv 9^2 \equiv 81 \,(\mathrm{mod}\,247),$$
$$185^{128} \equiv 81^2 \equiv 139 \,(\mathrm{mod}\,247),$$

Bob obtains that

$$185^{173} \equiv 185 \cdot 55 \cdot 61 \cdot 9 \cdot 139 \equiv 165 \,(\mathrm{mod}\,247).$$

So, $M = 165 < 247$ is the original text.

* **Example 7.34.**

Suppose that Alice and Bob want to correspond. Both of them create the pair of keys: Public keys and private keys.

Alice creates her public and private keys as follows. She chooses two primes $p = 43$, $q = 59$. Then, $m = 43 \cdot 59 = 2537$ and $\varphi(m) = (p-1)(q-1) = 42 \cdot 58 = 2436$. She also picks the number $k = 13$, such that $1 < k < 2436$ and $\gcd(k, 2436) = 1$. So, $(2537, 13)$ is Alice's public key. To find the private key, she needs to solve the congruence: $13h \equiv 1 \,(\mathrm{mod}\,2436)$, which has the solution $h = 937 \,(\mathrm{mod}\,2436)$. So, $(2537, 937)$ is Alice's private key.

Assume that Bob wants to send to Alice the text: HELP.

To encrypt this text, he changes each letter in it by a natural number accordingly with ASCII table. In this way, the text HELP is presented as: HELP=07041115. Then, Bob divides the encryption text into blocks in such a way that each block will be less than $m = 2537$: $M = (M_1 \, M_2)$ where $M_1 = 0704$, $M_2 = 1115$.

To find the ciphertext, Bob uses Alice's public key, doing the following: $C_1 = M_1^{13} = 0704^{13} = 0981$, $C_2 = M_2^{13} = 0461$. Then, he sends Alice the ciphertext as the pair of two integers: $C = (C_1, C_2)$, where $C_1 = 0981$ and $C_2 = 0461$.

Obtaining this encryption text Alice decrypts it using her private key. Therefore, she computes:

$M_1 = C_1^{937} = 0981^{937} \equiv 0704 \,(\mathrm{mod}\,2537)$ and $M_1 = C_2^{937} = 0461^{937} \equiv 1115 \,(\mathrm{mod}\,2537)$.

So, she read the original message: 0704 1115. In the ASCII code, it corresponds to the text: HELP.

7.8. Exercises

1. Find Euler's function $\varphi(m)$ if

   (a) $m = 25$,

   (b) $m = 60$,

   (c) $m = 1000000$.

2. Find the remainder on division:

   (a) 19^{71} by 12
   (b) 5^{135} by 11
   (c) 54^{59} by 17
   (d) 6^{100} by 7
   (e) 2^{100} by 101

3. Find a secret S as an integer in $\mathbb{Z}_{17}$ using Shamir's algorithm, if it is known that the secret is shared between 5 persons, $t = 3$ persons can recover this secret, provided that $y_1 = 8$, $y_3 = 10$ and $y_5 = 11$ are given shares.

4. Find a secret S as an integer in $\mathbb{Z}_{31}$ using Shamir's algorithm, if it is known that the secret is shared between 7 persons, $t = 3$ persons can recover this secret, provided that $y_1 = 16$, $y_3 = 5$ and $y_7 = 22$ are given shares.

5. Find a secret S as an integer in $\mathbb{Z}_{1613}$ by Shamir's algorithm, if it is known that the secret is shared between 6 persons, $t = 3$ persons can recover this secret, provided that $y_2 = 329$, $y_3 = 176$ and $y_5 = 1188$ are given shares.

6. Find a secret S using Mignotte's $(5, 3)$ Threshold Secret Sharing Scheme if it is known that 3 shareholders received information about their shares as follows:

   1) $(53, 97)$, $(61, 99)$ and $(6, 103)$;
   2) $(85, 98)$, $(61, 99)$ and $(6, 103)$.

7. Find a secret S using Asmuth-Bloom's $(4, 3)$ Threshold Secret Sharing Scheme if it is known that 3 shareholders received information about their shares as follows:

   1) $(1, 11, 3)$, $(12, 13, 3)$ and $(2, 17, 3)$;
   2) $(1, 11, 3)$, $(2, 17, 3)$ and $(3, 19, 3)$.

8. Encrypt the message $M = 10$ and decrypt it using the RSA algorithm, if $p = 17$, $q = 13$. Generate the private key, knowing that (91, 5) is the public key.

9. Encrypt the message $M = 2$ and decrypt it using the RSA algorithm, if $p = 3$, $q = 11$. Generate the private key, knowing that (33,7) is the public key.

10. Encrypt the message $M = 42$ and decrypt it using the RSA algorithm, if $p = 5$, $q = 17$. Generate the private key, knowing that (85, 11) is the public key.

11. Encrypt the message $M = 439$ and decrypt it using the RSA algorithm, if $p = 31$, $q = 23$. Generate the private key, knowing that $(713, 223)$ is the public key.

12. Using the RSA algorithm with Alice's public key $(143, 7)$:

    (a) Encrypt the message $M = 100$ and send it to Alice.

    (b) Find the private key h of Alice.

    (c) Decrypt the message received by Alice.

13. The ciphertext $C = 5859$ is obtained using the RSA algorithm with the public key $(11413, 7467)$. Use the factorization $11413 = 101 \cdot 113$ to decrypt this text.

14. The ciphertext $C = 6868$ is encrypted using the RSA algorithm with the public key $(7597, 4947)$. Knowing that $\varphi(7597) = 7420$, decrypt this message.

References

[1] Ash, R.B. 2007. Basic Abstract Algebra, Dover Publications.

[2] Asmuth, C.A. and J. Bloom. 1983. A modular approach to key safeguarding. IEEE Transactions on Information Theory, IT 29(2): 208–210.

[3] Blakley, G.R. 1979. Safeguarding cryptographic keys. *In*: Proc. AFIPS 1979 National Computer Conference, AFIPS, 313–317.

[4] Birkhoff, G. and T.C. Bartee. 1970. Modern Applied Algebra, New York, McGraw-Hill.

[5] Buchmann J.A. 2004. Introduction to Cryptography, Springer-Verlag, New York.

[6] Chou T.-W.J. and G.E. Collins. 1982. Algorithms for the solution of systems of Linear Diophantine equations. SIAM J. Computing 11: 687–708.

[7] Coutinho S.C. 1999. The Mathematics of Ciphers: Number Theory and RSA Cryptography, A K Peters/CRC Press.

[8] Ding, C., D. Pei and A. Salomaa. 1996. Chinese Remainder Theorem: Applications in Computing, Coding and Cryptography, World Scientific Publishing.

[9] Dummit, D.S. and R.M. Foote. 2004. Abstract Algebra (3rd Ed.), John Wiley & Sons.

[10] Gilbert, W.J. and W.K. Nicholson. 2004. Modern Algebra with Applications, John Wiley & Sons, New Jersey.

[11] Goldreich, O., D. Ron and and M. Sudan. 2000. Chinese remaindering with errors. IEEE Transactions on Information Theory 46(4): 1330–1338.

[12] Joyner, D., R. Kreminski and J. Turisco. 2004. Applied Abstract Algebra, The Johns Hopkins University Press.

[13] Judson, Th.W. 1994. Abstract Algebra. Theory and Applications, PWS Publishing Company, Michigan.

[14] Koblitz, N. 1994. A Course in Number Theory and Cryptography, Springer-Verlag, New-York.

[15] Koblitz, N. 1998. Algebraic Aspects of Cryptography, Springer-Verlag, Berlin-Heidelberg.

[16] Lazard, D. 1993. Systems of algebraic equations (algorithms and complexity). *In*: Eisenbud, D. and L. Robbiano [eds.] 1993. Computational Algebraic Geometry and Commutative Algebra, Cambridge University Press, London.

[17] Lazebnik, F. 1996. On systems of Linear Diophantine equations. Mathematics Magazine 69(4): 261–266.

[18] Lidl, R. and G. Pilz. 1997. Applied Abstract Algebra, (2nd Ed.), Springer-Verlag, New York.

[19] Mignotte, M. 1983. How to Share a Secret, Cryptography, Workshop Proceedings, Lecture Notes in Computer Science, 149, Springer-Verlag, 371–375.

[20] Rivest, R., A. Shamir and L.N. Adelman. 1978. A method for obtaining digital signatures and public key cryptosystems. Communications of the ACM 21(2): 120–126.

[21] Shamir, A. 1979. How to share a secret. Communications of the ACM 22(11): 612–613.

[22] Slinko, A. 2015. Algebra for Applications, Springer.

[23] Smith, H.J.S. 1961. On systems of linear indeterminate equations and congruences. Phil. Trans. London 151: 293–326.

[24] Sorin, I. 2007. General secret sharing based on the chinese remainder theorem with applications in E-Voting. Electronic Notes in Theoretical Computer Science 186: 67–84.

[25] Sterk, H. Algebra 3: Algorithms in Algebra.
http://www.win.tue.nl/~sterk/algebra3/hoofd.pdf

[26] Stinson, D. and R. Wei. 1998. Bibliography on Secret Sharing Schemes.
http://www.cacr.math.uwaterloo.ca/~dstinson/ssbib.html

Chapter 8

Polynomials in Several Variables

"Science is built up of facts, as a house is with stones. But a collection of facts is no more a science than a heap of stones is a house."

Henri Poincare

"To solve math problems, you need to know the basic mathematics before you can start applying it."

Catherine Asaro

Polynomials are basic objects in mathematics. They have many applications in various fields of mathematics. One of the most important statements in the theory of polynomials is the Hilbert Basis Theorem, proved by David Hilbert (1862-1943) in 1888, which states that every polynomial ring in several variables over a Noetherian ring is also Noetherian. Noetherian rings were named in honor of Emmy Noether (1882-1935) who systematically studied these rings and made a significant contribution to the development of ring theory. Emmy Noether played a major role in the development of algebra as a consistent, abstract and axiomatic field of mathematics.

In particular, when polynomials are considered over a field, the Hilbert Basis Theorem states that each ideal of the polynomial ring in several variables is finitely generated. Unfortunately, the proof of this theorem was nonconstructive and showed only the existence of such a finite set of generators. About the proof of this theorem, Paul Gordan (1837-1912), one of the leading mathematicians in the theory of invariants, said: "Das ist nicht Mathematik. Das ist Theologie". ("This is not mathematics. This is theology.)"

At the present time, the constructive method of finding a finite set of generators of ideals of the polynomial rings, which is based on the method of Gröbner bases, is known. The theory of Gröbner bases for polynomials in several variables was developed by Bruno Buchberger in 1965, who named it on the honor of his adviser, Wolfgang Gröbner. The notion of Gröbner basis, their main properties and some of their applications were introduced and developed by Bruno Buchberger in his Ph.D. thesis.

The importance of Gröbner bases shows the facts that many fundamental problems in mathematics (e.g., exact and complete solutions of systems of algebraic equations, analysis of affine varieties, calculations of Hilbert functions and Hilbert polynomials, solutions of systems of diophantine equations with polynomial coefficients, solutions of various problems of theory of invariants and integer programming) can be reduced by simple algorithms for constructing Gröbner bases. It should also be emphasized that these algorithms can be applied not only to polynomials over fields but to polynomials over some rings and also for noncommutative polynomials and differential algebras. It is also important that the algorithms for constructing Gröbner bases and many of their applications are now available in the form of programs for modern mathematical programming systems, such as MATHEMATICS, MAPLE and others.

8.1. Polynomial Rings in Several Variables

Similar to polynomials in one variable, we can introduce polynomials in two variables with coefficients in a commutative ring K.

> ✳ **Definition 8.1.**
>   A polynomial in two variables x_1, x_2 over a commutative ring K is defined as
> $$P(x_1, x_2) = \sum_{i=0}^{n} \sum_{j=0}^{m} a_{ij} x_1^i x_2^j, \qquad (8.1)$$
> where $a_{ij} \in K$ for $i = 0, 1, \ldots, n$; $j = 0, 1, \ldots, m$. Elements a_{ij} are called **coefficients of a polynomial**. We assume that $x_i^0 = 1$ and $1 \cdot x_i = x_i^1 = x_i$ for $i = 1, 2$.

We consider two polynomials

$$P(x_1, x_2) = \sum_{i=0}^{n} \sum_{j=0}^{m} a_{ij} x_1^i x_2^j,$$

$$Q(x_1, x_2) = \sum_{i=0}^{k} \sum_{j=0}^{t} b_{ij} x_1^i x_2^j$$

to be equal if $n = k$, $m = t$ and $a_{ij} = b_{ij}$ for all $i = 0, 1, \ldots, n$ and all $j = 0, 1, \ldots, m$.

The polynomial in which all coefficients a_{ij} are equal to zero is called the **zero polynomial** and is written as 0.

The set of all polynomials in two variables is written as $K[x_1, x_2]$. In a natural way, we can introduce the operations of addition and multiplication on the set of polynomials as follows:

$$P(x_1, x_2) + Q(x_1, x_2) = \sum_{i=0}^{n}\sum_{j=0}^{m} a_{ij}x_1^i x_2^j + \sum_{i=0}^{k}\sum_{j=0}^{t} b_{ij}x_1^i x_2^j =$$

$$\sum_{i=0}^{\max(n,k)}\sum_{j=0}^{\max(m,t)} (a_{ij} + b_{ij})x_1^i x_2^j,$$

$$P(x_1, x_2) \cdot Q(x_1, x_2) = \sum_{p,q}\left(\sum_{\substack{i+k=p\\ j+r=q}} a_{ij}b_{kr}\right)x_1^p x_2^q.$$

It is easy to show that all axioms of a commutative ring hold.

Each element

$$P(x_1, x_2) = \sum_{i=0}^{n}\sum_{j=0}^{m} a_{ij}x_1^i x_2^j$$

of the set $K[x_1, x_2]$ corresponds to the element $\sum_{j=0}^{m}(\sum_{i=0}^{n} a_{ij}x_1^i)x_2^j$ of the ring $(K[x_1])[x_2]$ and vice versa. Analogously, each element $P(x_1, x_2) \in K[x_1, x_2]$ corresponds to the element $\sum_{i=0}^{n}(\sum_{j=0}^{m} a_{ij}x_2^j)x_1^i$ of the ring $(K[x_2])[x_1]$ and vice versa. It is easy to show that these maps are ring isomorphisms, and so we have the following theorem:

> ✳ **Theorem 8.2.**
> For a commutative ring with identity K, a set $K[x_1, x_2]$ is a commutative ring which is isomorphic to $(K[x_1])[x_2]$ and $(K[x_2])[x_1]$:
>
> $$(K[x_1])[x_2] \simeq (K[x_2])[x_1] \simeq K[x_1, x_2].$$
>
> In other words, polynomials in two variables can be considered as polynomials in one variable x_2 over the ring $K[x_1]$ or as polynomials in one variable x_1 over the ring $K[x_2]$.

Analogously, we can define polynomials in n variables.

A polynomial in n variables $x_1, x_2, \ldots, x_n$ is the formal expression of the form:

$$P(x_1, x_2, \ldots, x_n) = \sum_{i_1=0}^{m_1}\sum_{i_2=0}^{m_2}\cdots\sum_{i_n=0}^{m_n} a_{i_1 i_2 \ldots i_n} x_1^{i_1} x_2^{i_2} \ldots x_n^{i_n}, \qquad (8.2)$$

where $a_{i_1 i_2 \ldots i_n} \in K$ for $i_k = 0, 1, \ldots, m_k$; $k = 1, \ldots, n$. Elements $a_{i_1 i_2 \ldots i_n}$ are called the coefficients of the polynomial $P(x_1, x_2, \ldots, x_n)$. We assume that $x_i^0 = 1$ and $1 \cdot x_i = x_i^1 = x_i$ for $i = 1, 2, \ldots, n$.

We consider two polynomials

$$P(x_1, x_2, \ldots, x_n) = \sum_{i_1=0}^{m_1} \sum_{i_2=0}^{m_2} \cdots \sum_{i_n=0}^{m_n} a_{i_1 i_2 \ldots i_n} x_1^{i_1} x_2^{i_2} \ldots x_n^{i_n},$$

$$Q(x_1, x_2, \ldots, x_n) = \sum_{i_1=0}^{t_1} \sum_{i_2=0}^{t_2} \cdots \sum_{i_n=0}^{t_n} b_{i_1 i_2 \ldots i_n} x_1^{i_1} x_2^{i_2} \ldots x_n^{i_n},$$

to be equal if $m_i = t_i$ for all $i = 1, 2, \ldots, n$ and $a_{i_1 i_2 \ldots i_n} = b_{i_1 i_2 \ldots i_n}$ for all sequences $(i_1, i_2, \ldots, i_n)$, where $0 \leq i_k \leq m_k$, $k = 1, 2, \ldots, n$.

The polynomial which all coefficients a_{ij} are equal to zero is called the **zero polynomial** and is written as 0.

The set of all polynomials in n variables is written as $K[x_1, x_2, \ldots, x_n]$. In a natural way, we can define the operations of addition and multiplication on this set and show that all axioms of a commutative ring hold.

✳ **Theorem 8.3.**

For a commutative ring with identity K, the set $K[x_1, x_2, \ldots, x_n]$ is a commutative ring which is isomorphic to $(K[x_1, x_2, \ldots, x_{n-1}])[x_n]$. In other words, polynomials in n variables can be considered as polynomials in one variable x_n over the ring $K[x_1, x_2, \ldots, x_{n-1}]$.

✳ **Definition 8.4.**

A polynomial in n variables of the form $x_1^{i_1} x_2^{i_2} \ldots x_n^{i_n} \in K[x_1, x_2, \ldots, x_n]$, where all $i_k \geq 0$, is called a **monomial**.

We will write a monomial $x_1^{\alpha_1} x_2^{\alpha_2} \cdots x_n^{\alpha_n}$ in the form u^α, where $\alpha = (\alpha_1, \alpha_2, \ldots, \alpha_n) \in \mathbb{N}^n$. In particular, $x^0 = x_1^0 x_2^0 \cdots x_n^0 = 1$. Moreover, $u^\alpha \cdot u^\beta = u^{\alpha+\beta}$ for all $\alpha, \beta \in \mathbb{N}^n$. The set of all monomials of the form $x_1^{\alpha_1} x_2^{\alpha_2} \cdots x_n^{\alpha_n}$ will be denoted by T_n. Therefore, each polynomial in n variables $P(x_1, x_2, \ldots, x_n) \in K[x_1, x_2, \ldots, x_n]$ can be written in the form of a linear combination of monomials:

$$P(u) = \sum_\alpha c_\alpha u^\alpha, \tag{8.3}$$

where $c_\alpha \in K$, $\alpha \in \mathbb{N}^n$, $u^\alpha \in T_n$.

The set of monomials $\operatorname{supp}(P) = \{u^\alpha : c_\alpha \neq 0\}$ is called the **support** of a polynomial $P(u)$, written as (8.3).

✳ **Definition 8.5.**

An element $c_\alpha u^\alpha$ is called a **term** of the polynomial $P(u)$, written as (8.3), and an element c_α is called its **coefficient**.

The **degree** of a monomial $u^\alpha = x_1^{\alpha_1} x_2^{\alpha_2} \cdots x_n^{\alpha_n}$, where $\alpha = (\alpha_1, \alpha_2, \ldots, \alpha_n) \in \mathbb{N}^n$, is the sum

$$|\alpha| = \alpha_1 + \alpha_2 + \ldots + \alpha_n \in \mathbb{N}. \tag{8.4}$$

The non-zero elements of K are regarded as monomials of degree 0. We also assume that $\deg(0) = -\infty$.

The greatest degree of all terms occurring in a polynomial $P(x_1, x_2, \ldots, x_n)$ is called the **total degree** of this polynomial and is denoted by $\deg(P)$.

Note that the notions of a monomial and a term of a polynomial often do not differ and they are considered as the same.

✳ Example 8.6.

Let $P(x, y, z) = 2x^2 y^3 z + 3x^2 y^2 z^2 - 5x^3 \in \mathbb{Q}[x, y, z]$. Then, we write
$P(u) = 2u^\alpha + 3u^\beta - 5u^\gamma$, where $u = (x, y, z)$, and $\alpha = (2, 3, 1)$, $\beta = (2, 2, 2)$, $\gamma = (3, 0, 0)$,
$|\alpha| = |\beta| = 6$, $|\gamma| = 3$.

Similar to polynomials in one variable (see Theorem 5.6), it can be shown that for arbitrary polynomials in n variables $P(x_1, x_2, \ldots, x_n), Q(x_1, x_2, \ldots, x_n) \in K[x_1, x_2, \ldots, x_n]$, where K is an integral domain, the following hold:

$$\deg(P + Q) \leq \max(\deg(P), \deg(Q)),$$

$$\deg(PQ) = \deg(P) + \deg(Q).$$

Using the mathematical induction, it is easy to obtain from Theorem 5.7 the following theorem:

✳ Theorem 8.7.

If K is an integral domain, then the polynomial ring $K[x_1, x_2, \ldots, x_n]$ is also an integral domain.

However, unlike polynomials in one variable, the ring $K[x_1, x_2, \ldots, x_n]$ of all polynomials in n variables over a field K for $n > 1$ is not a principal ideal ring. For example, the ideal $\mathcal{I} = \{xP(x, y) + yQ(x, y) \mid P(x, y), Q(x, y) \in K[x, y]\}$ is not a principal ideal in the ring $K[x, y]$.

For polynomials in n variables, we can define the notion of polynomial value at a point $c \in K^n$.

✳ Definition 8.8.

A value of a polynomial $P(x_1, x_2, \ldots, x_n) = \sum_{i_1=0}^{m_1} \sum_{i_2=0}^{m_2} \cdots \sum_{i_n=0}^{m_n} a_{i_1 i_2 \ldots i_n} x_1^{i_1} x_2^{i_2} \ldots x_n^{i_n}$ at a point $c = (c_1, c_2, \ldots, c_n) \in K^n$
is defined as an element $\sum_{i_1=0}^{m_1} \sum_{i_2=0}^{m_2} \cdots \sum_{i_n=0}^{m_n} a_{i_1 i_2 \ldots i_n} c_1^{i_1} c_2^{i_2} \ldots c_n^{i_n} \in K$ and
denoted by $P(c) = P(c_1, c_2, \ldots, c_n)$.

For any polynomials $P(x_1, x_2, \ldots, x_n), Q(x_1, x_2, \ldots, x_n) \in K[x_1, x_2, \ldots, x_n]$ and any element $(c_1, c_2, \ldots, c_n) \in K^n$ the following equalities hold:

$$P(c_1, c_2, \ldots, c_n) + Q(c_1, c_2, \ldots, c_n) = (P + Q)(c_1, c_2, \ldots, c_n),$$

$$P(c_1, c_2, \ldots, c_n) \cdot Q(c_1, c_2, \ldots, c_n) = (P \cdot Q)(c_1, c_2, \ldots, c_n).$$

This means that the mapping $\varphi : K[x_1, x_2, \ldots, x_n] \to K$ such that $\varphi(P(x_1, x_2, \ldots, x_n)) = P(c_1, c_2, \ldots, c_n)$ is a homomorphism of the ring $K[x_1, x_2, \ldots, x_n]$ to the ring K.

> ✳ **Definition 8.9.**
> We say that a non-zero polynomial $P \in K[x_1, x_2, \ldots, x_n]$ is called **uniform** (or a **form**) of degree k if it can be written as a sum of monomials of degree k. In particular, a form of the first degree is called **linear**, and a form of the second degree is called **quadratic**.

It can be shown that a sum and a product of uniform polynomials are also a uniform polynomial.

Any polynomial in n variables can be written as a sum of uniform polynomials.

8.2. Symmetric Polynomials

In this chapter, we consider the special subring of the ring of all polynomials in n variables which form symmetric polynomials. Symmetric polynomials have an interesting structure and have several applications in various areas of mathematics, e.g., in combinatorics and representation theory.

Let $P(x_1, x_2, \ldots, x_n)$ be a polynomial in the ring $K[x_1, x_2, \ldots, x_n]$ and let $\sigma = \begin{pmatrix} 1 & 2 & \cdots & n \\ \sigma(1) & \sigma(2) & \cdots & \sigma(n) \end{pmatrix}$ be a permutation of the symmetric group S_n. Using a permutation σ for variables $x_1, x_2, \ldots, x_n$, we get a polynomial $Q(x_1, x_2, \ldots, x_n) = P(x_{\sigma(1)}, x_{\sigma(2)}, \ldots, x_{\sigma(n)})$ which is, in general, distinct from the polynomial $P(x_1, x_2, \ldots, x_n)$.

If $P(x_{\sigma(1)}, x_{\sigma(2)}, \ldots, x_{\sigma(n)}) = P(x_1, x_2, \ldots, x_n)$, we say that a permutation σ does not change the polynomial $P(x_1, x_2, \ldots, x_n)$.

✳ **Examples 8.10.**

1. If $P(x_1, x_2, x_3) = 2x_1 + x_2 - 4x_3$ and $\sigma = \begin{pmatrix} 1 & 2 & 3 \\ 2 & 3 & 1 \end{pmatrix}$ then, using this permutation to the polynomial $P(x_1, x_2, x_3)$, we get the new polynomial $Q(x_1, x_2, x_3) = -4x_1 + 2x_2 + x_3$, which is not equal to $P(x_1, x_2, x_3)$.

2. If $P(x_1, x_2, x_3) = 2x_1 x_2 + 4x_3$, then the permutation $\sigma = \begin{pmatrix} 1 & 2 & 3 \\ 2 & 1 & 3 \end{pmatrix}$ does not change the polynomial $2x_1 x_2 + 4x_3$.

It is obvious that if permutations $\sigma, \tau \in S_n$ do not change a polynomial $P(x_1, x_2, \ldots, x_n)$, then their product $\sigma\tau$ also does not change this polynomial. Also, the permutation σ^{-1} which is inverse to a permutation σ does not change the polynomial $P(x_1, x_2, \ldots, x_n)$. Thus, the set of all permutations which do not change the polynomial $P(x_1, x_2, \ldots, x_n)$ forms a subgroup of the group S_n.

> ✳ **Definition 8.11.**
> The **group of symmetries of a polynomial** $P(x_1, x_2, \ldots, x_n) \in K[x_1, x_2, \ldots, x_n]$ is a subgroup of S_n consisting of all permutations $\sigma \in S_n$ which do not change this polynomial, i.e.,
> $$P(x_{\sigma(1)}, x_{\sigma(2)}, \ldots, x_{\sigma(n)}) = P(x_1, x_2, \ldots, x_n). \tag{8.5}$$

✳ **Examples 8.12.**

- The group of symmetries of the polynomial $x_1 x_2 + 4 x_3 x_4$ consists of four elements:

$$\sigma_1 = e = \begin{pmatrix} 1 & 2 & 3 & 4 \\ 1 & 2 & 3 & 4 \end{pmatrix}; \quad \sigma_1 = \begin{pmatrix} 1 & 2 & 3 & 4 \\ 2 & 1 & 3 & 4 \end{pmatrix};$$

$$\sigma_3 = \begin{pmatrix} 1 & 2 & 3 & 4 \\ 1 & 2 & 4 & 3 \end{pmatrix}; \quad \sigma_4 = \begin{pmatrix} 1 & 2 & 3 & 4 \\ 2 & 1 & 4 & 3 \end{pmatrix}.$$

- The group of symmetries of the polynomial of n variables:

$$(x_1 - x_2)(x_1 - x_3) \cdots (x_1 - x_n)(x_2 - x_3) \cdots (x_2 - x_n) \cdots (x_{n-1} - x_n)$$

  is the alternative group A_n.

- The group of symmetries of the polynomial of n variables: $x_1 + x_2 + \cdots + x_n$ is the symmetric group S_n.

> ✳ **Definition 8.13.**
> A polynomial $P(x_1, x_2, \ldots, x_n) \in K[x_1, x_2, \ldots, x_n]$ is called **symmetric** if its symmetric group is equal to S_n. In other words, $P(x_1, x_2, \ldots, x_n)$ is symmetric if and only if it is unchanged by any permutation of its variables, i.e.,
> $$P(x_1, x_2, \ldots, x_n) = P(x_{\sigma(1)}, x_{\sigma(2)}, \ldots, x_{\sigma(n)})$$
> for all $\sigma \in S_n$.

✳ Examples 8.14.

- The polynomial $P(x_1, x_2) = 3x_1^3 + 4x_1x_2 + 3x_2^3 - 2x_1^2 - 2x_2^2$ is a symmetric polynomial in two variables.

- The polynomial $P(x_1, x_2, x_3) = 3(x_1 + x_2 + x_3)x_1x_2x_3 + 2(x_1x_2 + x_2x_3 + x_1x_3)$ is a symmetric polynomial in three variables.

It is easy to verify that a sum, a difference and a product of symmetric polynomials are also symmetric polynomials. Therefore, the set of all symmetric polynomials in n variables form a subring in the ring $K[x_1, x_2, \ldots, x_n]$ and it is called the **ring of symmetric polynomials** in n variables.

✳ Definition 8.15.

The **elementary symmetric polynomials** in n variables are polynomials $\tau_1, \tau_2, \ldots, \tau_n$, defined as follows:

$$\tau_1 = \tau_1(x_1, x_2, \ldots, x_n) = x_1 + x_2 + \cdots + x_n$$

$$\tau_2 = \tau_2(x_1, x_2, \ldots, x_n) = \sum_{i<j} x_i x_j$$

$$\tau_3 = \tau_3(x_1, x_2, \ldots, x_n) = \sum_{i<j<k} x_i x_j x_k$$

$$\ldots\ldots\ldots\ldots\ldots\ldots\ldots\ldots\ldots\ldots\ldots\ldots$$

$$\tau_n = \tau_n(x_1, x_2, \ldots, x_n) = x_1 x_2 \cdots x_n$$

The role of elementary symmetric polynomials in the theory of symmetric polynomials shows the Fundamental Theorem on Symmetric Polynomials which states that any symmetric polynomial can be expressed in terms of the elementary symmetric polynomials. In this sense, the elementary symmetric polynomials play the role of basic building blocks for symmetric polynomials.

✳ Theorem 8.16 (Fundamental Theorem on Symmetric Polynomials)

For any symmetric polynomial $P \in K[x_1, x_2, \ldots, x_n]$, where K is a commutative ring with identity, there exists a unique polynomial $F \in K[x_1, x_2, \ldots, x_n]$ such that $P(x_1, x_2, \ldots, x_n) = F(\tau_1, \tau_2, \ldots, \tau_n)$, where $\tau_1, \tau_2, \ldots, \tau_n \in K[x_1, x_2, \ldots, x_n]$ are elementary symmetric polynomials.

Proof.

We prove the theorem by mathematical induction on the number of variables n. If $n = 1$, then the statement is trivial because, in this case, $\tau_1 = \tau_1(x) = x$, and so $P(x) = P(\tau_1)$ for any symmetric polynomial P.

Assume that the statement is true for all symmetric polynomials in $n-1$ variables. Let $P(x_1, x_2, \ldots, x_n)$ be a symmetric polynomial in n variables and $d = \deg(P)$. In this case, we prove by mathematical induction on degree d. If $d = 0$, then the statement is obvious. Assume that $d > 0$ and for all symmetric polynomials in n variables and degree $< d$ the statement is true.

Since $P(x_1, x_2, \ldots, x_{n-1}, 0)$ is a symmetric polynomial in $n-1$ variables, by induction hypothesis, there exists a polynomial $G(\tau_1', \tau_2', \ldots, \tau_{n-1}')$ such that

$$P(x_1, x_2, \ldots, x_{n-1}, 0) = G(\tau_1', \tau_2', \ldots, \tau_{n-1}'),$$

where $\tau_1', \tau_2', \ldots, \tau_{n-1}'$ are elementary symmetric polynomials in $n-1$ variables $x_1, x_2, \ldots, x_{n-1}$. It is obvious that $\deg(G(\tau_1', \tau_2', \ldots, \tau_{n-1}')) \leq d$. Consider a polynomial $G(\tau_1, \tau_2, \ldots, \tau_{n-1})$ which contains variables $x_1, x_2, \ldots, x_n$. It is easy to see that $\deg(G(\tau_1, \tau_2, \ldots, \tau_{n-1})) \leq d$. Consider a polynomial

$$P_1(x_1, x_2, \ldots, x_n) = P(x_1, x_2, \ldots, x_n) - G(\tau_1, \tau_2, \ldots, \tau_{n-1}) \qquad (8.6)$$

The polynomial $P_1(x_1, x_2, \ldots, x_n)$ is a symmetric polynomial in n variables with degree $\leq d$. We have $P_1(x_1, x_2, \ldots, x_{n-1}, 0) = 0$, which means that this polynomial is divided by x_n. Since $P_1(x_1, x_2, \ldots, x_n)$ is a symmetric polynomial in n variables, it is also divided by $x_1 x_2 \cdots x_n = \tau_n$, i.e.,

$$P_1(x_1, x_2, \ldots, x_n) = \tau_n P_2(x_1, x_2, \ldots, x_n) \qquad (8.7)$$

Since $\deg(P_2(x_1, x_2, \ldots, x_n)) \leq d - n < d$, by induction assumption there exists a polynomial G_1 in n variables which satisfies the equality:

$$P_2(x_1, x_2, \ldots, x_n) = G_1(\tau_1, \tau_2, \ldots, \tau_n). \qquad (8.8)$$

Then, from equalities (8.6)-(8.8), we obtain:

$$P(x_1, x_2, \ldots, x_n) = G(\tau_1, \tau_2, \ldots, \tau_{n-1}) + \tau_n G_1(\tau_1, \tau_2, \ldots, \tau_n) = F(\tau_1, \tau_2, \ldots, \tau_n),$$

where $\tau_1, \tau_2, \ldots, \tau_n$ are elementary symmetric polynomials in n variables $x_1, x_2, \ldots, x_n$, which proves the existence of a polynomial $F(\tau_1, \tau_2, \ldots, \tau_n)$.

We show that such a polynomial is unique. Suppose there are two polynomials $F(\tau_1, \tau_2, \ldots, \tau_n)$, $F_1(\tau_1, \tau_2, \ldots, \tau_n)$ such that

$$P(x_1, x_2, \ldots, x_n) = F(\tau_1, \tau_2, \ldots, \tau_n) = F_1(\tau_1, \tau_2, \ldots, \tau_n).$$

Then, a polynomial $H(\tau_1, \tau_2, \ldots, \tau_n) = F(\tau_1, \tau_2, \ldots, \tau_n) - F_1(\tau_1, \tau_2, \ldots, \tau_n)$ is the zero polynomial. We show that, in this case, $P(x_1, x_2, \ldots, x_n)$ is also the zero polynomial. We write the polynomial $H(\tau_1, \tau_2, \ldots, \tau_n)$ as a polynomial in one variable x_n with coefficients in the ring $K[x_1, x_2, \ldots, x_{n-1}]$:

$$H(x_1, x_2, \ldots, x_n) = H_0(x_1, x_2, \ldots, x_{n-1}) + x_n H_1(x_1, x_2, \ldots, x_{n-1}) + \cdots +$$

$$+ x_n^m H_m(x_1, x_2, \ldots, x_{n-1}) = H_0(x_1, x_2, \ldots, x_{n-1}) + x_n R(x_1, x_2, \ldots, x_{n-1}),$$

where $\deg R(x_1, x_2, \ldots, x_{n-1}) < \deg H(x_1, x_2, \ldots, x_{n-1}, x_n) \le d$.

If $H_0(x_1, x_2, \ldots, x_{n-1}) = 0$, then $H(x_1, x_2, \ldots, x_{n-1}, x_n) = x_n R(x_1, x_2, \ldots, x_{n-1}) = 0$, implying that $H(\tau_1, \tau_2, \ldots, \tau_n) = \tau_n S(\tau_1, \tau_2, \ldots, \tau_n) = 0$; and so $S(\tau_1, \tau_2, \ldots, \tau_n) = 0$. Since $\deg S(\tau_1, \tau_2, \ldots, \tau_n) < d$, by induction hypothesis, we get that $H(x_1, x_2, \ldots, x_n)$ is the zero polynomial.

If $H_0(x_1, x_2, \ldots, x_{n-1}) \ne 0$, then we have:

$$H(x_1, x_2, \ldots, x_n) = H_0(x_1, x_2, \ldots, x_{n-1}) + x_n R(x_1, x_2, \ldots, x_{n-1}).$$

Since $H(x_1, x_2, \ldots, x_n)$ is a symmetric polynomial, we get:

$$H(\tau_1, \tau_2, \ldots, \tau_n) = H_0(\tau_1, \tau_2, \ldots, \tau_{n-1}) + \tau_n S(\tau_1, \tau_2, \ldots, \tau_{n-1}).$$

Substituting $\tau_n = 0$ into this equality, we obtain: $H(\tau_1, \tau_2, \ldots, \tau_n) = H_0(\tau_1, \tau_2, \ldots, \tau_{n-1}) = 0$ which, by induction hypothesis, means that $H(x_1, x_2, \ldots, x_n)$ is the zero polynomial. $\qquad\square$

Note that the proof of this theorem also gives an algorithm to express symmetric polynomials in terms of elementary symmetric polynomials.

✳ Examples 8.17.

1. Write the polynomial $P(x, y, z) = x^2 + y^2 + z^2$ in terms of the elementary symmetric polynomials.

We have: $P(x, y, z) = x^2 + y^2 + z^2 = (x+y+z)^2 - 2(xy+xz+yz) = \tau_1^2 - 2\tau_2$

2. Write the polynomial $P(x, y) = x^3 + y^3$ in terms of the elementary symmetric polynomials.

We have $P(x, 0) = x^3 = \tau_1(x)^3$. Then, $P(x, y) - \tau_1(x, y)^3 = (x^3 + y^3) - (x + y)^3 = -3x^2 y - 3xy^2 = -3xy(x + y) = -3\tau_1(x, y)\tau_2(x, y)$. Therefore, $P(x, y) = \tau_1^3 - 3\tau_1(x, y)\tau_2(x, y)$.

Issac Newton extensively studied the symmetric polynomials and he not only proved the Fundamental Theorem on Symmetric Polynomials but he also found the identities which express the symmetric polynomials $w_k = x_1^k + x_2^k + \cdots + x_n^k$ in terms of elementary symmetric polynomials $\tau_1, \tau_2, \ldots, \tau_n$. We give this theorem without proof.

✳ Theorem 8.18 (Newton's identities)
Let K be a commutative ring and $w_k = x_1^k + x_2^k + \cdots + x_n^k \in K[x_1, x_2, \ldots, x_n]$. If $1 \le k \le n$, then:

$$k\tau_k = \sum_{i=1}^{k} (-1)^{i-1} \tau_{k-i} w_i$$

and if $k > n$, then:

$$0 = \sum_{i=k-n}^{k} (-1)^{i-1} \tau_{k-i} w_i,$$

where $\tau_0 = 1$

8.3. Noetherian Rings. Hilbert Basis Theorem

Throughout this section, we will assume that R is a commutative ring with identity.

In this section, we consider Noetherian rings that have fundamental importance in ring theory and Hilbert's Basis Theorem which hold for polynomials over such rings. Noetherian rings are characterized by the property that every ideal in them is finitely generated. These rings are named after outstanding German mathematician Emmy Noether (1882-1935). She was the first to discover the true importance of this property. Emmy Noether made a great contribution to the development of a new field of mathematics, abstract algebra, and, in particular, ring theory. Hilbert's Basis Theorem states that the polynomial ring in n variables over a Noetherian ring is also Noetherian. The theorem is named in honor of David Hilbert, who first stated and proved this theorem in 1888 for the case of a field.

Emmy Noether
(1882-1935)

> ✳ **Definition 8.19.**
> An ideal $\mathcal{I}$ of a ring R is called **finitely generated** if there is a finite subset $X = \{x_1, x_2, \ldots, x_k\} \subset \mathcal{I}$ such that each element $x \in \mathcal{I}$ can be written in the form:
>
> $$x = a_1 x_1 + a_2 x_2 + \cdots + a_k x_k, \quad \text{where all} \ \ a_i \in R.$$
>
> Elements $x_1, x_2, \ldots, x_k$ are called **generators** of $\mathcal{I}$.

We write a finitely generated ideal in the form:

$$\mathcal{I} = \langle x_1, x_2, \ldots, x_k \rangle.$$

Note that if $X = \{a\}$, then $I = \langle a \rangle$ is a principal ideal.

> ✳ **Definition 8.20.**
> A ring R is called **Noetherian** if each its ideal is finitely generated.

✳ **Examples 8.21.**

1. Any field is a Noetherian ring.

2. Each principal ideal ring is Noetherian, e.g., $\mathbb{Z}$, $\mathbb{Z}[i]$, $\mathbb{R}[x]$.

3. The ring $\mathbb{Z}[x]$ is not a principal ideal ring but it is a Noetherian ring.

✴ **Theorem 8.22.**

The following conditions are equivalent for a ring R:
1. A ring R is Noetherian.
2. Every infinite strictly ascending chain of ideals in R

$$\mathcal{I}_1 \subset \mathcal{I}_2 \subset \cdots \subset \mathcal{I}_m \subset \cdots \qquad (8.9)$$

stabilizes, i.e., there exists an integer n such that $\mathcal{I}_n = \mathcal{I}_{n+1} = \cdots$.

Proof.

$1 \Rightarrow 2$. Let R be a Noetherian ring. Consider an ascending chain of ideals

$$\mathcal{I}_1 \subset \mathcal{I}_2 \subset \cdots \subset \mathcal{I}_m \subset \cdots$$

Suppose that this chain does not stabilize. Then, we consider $\bigcup_{k=1}^{\infty} \mathcal{I}_k$, which is a proper ideal, by Theorem 4.32. Since R is a Noetherian ring, the ideal $\mathcal{I}$ is finitely generated. Let $\mathcal{I} = \langle x_1, x_2, \ldots, x_k \rangle$. Therefore, there exists some index t such that $x_i \in \mathcal{I}_t$ for all $i = 1, 2, \ldots, k$. So, we obtain $\mathcal{I} \subseteq \mathcal{I}_t \subset \mathcal{I}_{t+1} \subseteq \cdots \subseteq \mathcal{I}$, where $\mathcal{I}_t \neq \mathcal{I}_{t+1}$. This contradiction shows that $\mathcal{I}_t = \mathcal{I}_{t+1} = \cdots$, i.e., the chain of ideals stabilizes.

$2 \Rightarrow 1$. Let $\mathcal{I}$ be an ideal of a ring R. If $\mathcal{I} \neq 0$, then there is a non-zero element $x_1 \in \mathcal{I}$. Consider an ideal $\mathcal{I}_1 = \langle x_1 \rangle$. If $\mathcal{I}_1 = \mathcal{I}$, then $\mathcal{I}$ is a finitely generated ideal, and we are done. Otherwise, there exists an element $x_2 \in I$ such that $x_2 \notin \mathcal{I}_1$. If $\mathcal{I}_2 = \langle x_1, x_2 \rangle = \mathcal{I}$, then the ideal $\mathcal{I}$ is finitely generated, and the theorem is proved. Otherwise, we obtain that $\mathcal{I}_1 \subset \mathcal{I}_2$. Continuing this process and taking into account that each strictly ascending chain of ideals $\langle x_1 \rangle \subset \langle x_1, x_2 \rangle \subset \langle x_1, x_2, x_3 \rangle \subset \cdots$ must stabilize, we get that $I = \langle x_1, x_2, \ldots, x_k \rangle$, i.e., the ideal $\mathcal{I}$ is finitely generated. Since the ideal $\mathcal{I}$ is an arbitrary one, R is a Noetherian ring. $\qquad \square$

✴ **Theorem 8.23.**

The following conditions are equivalent for a ring R:
1. A ring R is Noetherian.
2. Each non-empty family of ideals of a ring R has a maximal element with respect to inclusion.

Proof.

$1 \Rightarrow 2$. Let S be a non-empty family of ideals of a Noetherian ring R. We choose an ideal $\mathcal{I}_1 \in S$. If $\mathcal{I}_1$ is maximal in S, then the theorem is proved. Otherwise, there exists an ideal $\mathcal{I}_2 \in S$ such that $\mathcal{I}_1 \subset \mathcal{I}_2$. If $\mathcal{I}_2$ is maximal in S, then the theorem is proved. Otherwise, there exists an ideal $\mathcal{I}_3 \in S$ such that $\mathcal{I}_2 \subset \mathcal{I}_3$. Continuing this process, we either get a maximal ideal or obtain an infinite strictly ascending chain of ideals

$$\mathcal{I}_1 \subset \mathcal{I}_2 \subset \cdots \subset \mathcal{I}_m \subset \cdots$$

that contradicts Theorem 8.22.

2 $\Rightarrow$ 1. Suppose that R is not a Noetherian ring. Then, there is an infinite ascending sequence of ideals $\mathcal{I}_1 \subset \mathcal{I}_2 \subset \cdots \subset \mathcal{I}_m \subset \cdots$, which does not stabilize. Consider the set $S = \{\mathcal{I}_1, \mathcal{I}_2, \ldots, \mathcal{I}_n, \ldots\}$ which consists of ideals of this sequence. Then, S has no a maximal element with respect to inclusion. However, this contradicts condition 2 of the theorem. $\square$

Recall that an element a of an integral domain R is decomposable in R if it can be written as a product $a = xy$, where elements $x, y \in R$ are not invertible in R. A non-zero non-invertible element $x \in R$ is called **irreducible** (or **indecomposable**) in R if it cannot be written as a product of non-invertible elements of R.

> ✳ **Definition 8.24.**
> An integral domain R is called a **ring with decomposition** if each non-zero element $a \in R$ can be decomposed into a finite product:
>
> $$a = p_1 p_2 \cdots p_n,$$
>
> where $n \geq 1$ and elements $p_1, p_2, \ldots, p_n$ are irreducible in R.

> ✳ **Theorem 8.25.**
> If R is a Noetherian integral domain, then R is a ring with decomposition, i.e., each its non-zero element can be decomposed into a finite product of irreducible elements.

Proof.

The proof of this theorem is very similar to the proof of theorem 4.80.

Suppose there is a non-zero element $a \in R$ such that it cannot be written as a finite product of irreducible elements. Then, there are elements $a_1, a_2 \in R$ such that $a = a_1 b_1$ and elements a_1, b_1 are non-invertible. Moreover, at least one of these elements also cannot be written as a finite product of irreducible elements. Without loss of generality, we can consider that such an element is a_1. Since $a = a_1 b_1$ and b_1 is not invertible, from Lemma 4.53, it follows that $\langle a \rangle \subset \langle a_1 \rangle$. Since a_1 is a non-invertible element, there are elements $a_2, b_2 \in R$ such that $a_1 = a_2 b_2$ and elements a_2, b_2 are non-invertible. Again, at least one of these elements also cannot be written as a finite product of irreducible elements. Without loss of generality, we can consider that such an element is a_2. Since b_2 is a non-invertible element, from Lemma 4.53, it follows that $\langle a_1 \rangle \subset \langle a_2 \rangle$. Continuing this process, we get a sequence of elements $a_0 = a$, $a_1, a_2, \ldots$ satisfying the condition: $a_{n+1} | a_n$ for $n = 0, 1, 2, \ldots$. Therefore, we get an infinite strictly ascending chain of ideals:

$$\mathcal{I}_1 \subset \mathcal{I}_2 \subset \cdots \subset \mathcal{I}_m \subset \mathcal{I}_{m+1} \subset \cdots,$$

that contradicts the fact that R is a Noetherian ring. $\square$

Recall that an element a of an integral domain R is **prime** if a is non-invertible in R and the condition $(a|bc) \wedge (b, c \in P)$ implies that $(a|b) \wedge (b|c)$.

So, from Theorems 4.81 and 8.25, we immediately get the following statement:

✳ Theorem 8.26.

If R is a Noetherian integral domain, then R is a ring with unique decomposition if and only if each irreducible element $a \in R$ is prime.

✳ Theorem 8.27.

If R is a Noetherian integral domain and each principal ideal generated by an irreducible element $a \in$ is prime, then R is a ring with unique decomposition.

Proof.

$\Longrightarrow$. Let $\mathcal{I} = \langle a \rangle$ be a principal ideal of a Noetherian integral domain R, where $a \in R$ is an irreducible element. Suppose that $uv \in \mathcal{I}$, then $uv = ax$, where $x \in P$. Therefore, $a|uv$. Since, from Theorem 8.23, it follows that a is a prime element, we get that $(a|u) \vee (a|v)$, i.e., $(u = au_1) \vee (v = av_1)$, which means that $(u \in \mathcal{I}) \vee (v \in \mathcal{I})$. Thus, $\mathcal{I}$ is a prime ideal.

$\Longleftarrow$. Let $a \in R$ be an irreducible element. Then, the principal ideal $\mathcal{I} = \langle a \rangle$ is prime. Suppose that $a = uv$. Since $a \in \mathcal{I}$ and $\mathcal{I}$ is a prime ideal, we have that $(u \in \mathcal{I}) \vee (v \in \mathcal{I})$, and, therefore, $(u = au_1) \vee (v = av_1)$, which means that $(a|u) \vee (a|v)$. Therefore, a is a prime element and the statement follows from Theorem 8.26. $\qquad\square$

✳ Theorem 8.28 (Hilbert Basis Theorem).

If R is a Noetherian ring, then the polynomial ring $R[x]$ is also Noetherian.

Proof.

Let R be a Noetherian ring and let $\mathcal{I}$ be an arbitrary ideal of the ring $R[x]$. Clearly, the set

$$M = \{a_n \in R \ : \ a_0 + a_1 x + ... + a_n x^n \in \mathcal{I}, \ a_n \neq 0\} \cup \{0\} \qquad (8.10)$$

forms an ideal of R. Since R is a Noetherian ring, the ideal M is finitely generated. Therefore, there is a finite set of generators $b_1, ..., b_s$ such that $M = \langle b_1, ..., b_s \rangle$. Let $f_i(x) \in \mathcal{I}$ be a polynomial with the leading coefficient b_i: $f_i(x) = b_i x^{n_i} + ...$ $(i = 1, ..., s)$ and $n = \max\{n_1, n_2, \ldots, n_s\}$.

Suppose $f(x) \in \mathcal{I}$ is a polynomial with the leading coefficient a and $\deg(f) = m$. We will show that $f(x)$ can be written as follows:

$$f(x) = f_1(x) g_1(x) + ... + f_s(x) g_s(x) + h(x), \qquad (8.11)$$

where $f_i(x) \in \mathcal{I}$ and $\deg(h(x)) \leq n - 1$. If $m < n$, then everything is proved. Let $m \geq n$. We have $a = \sum_{i=1}^{s} b_i c_i$, where $c_1, ..., c_s \in R$. Consider the polynomial

$$t_1(x) = f(x) - \sum_{i=1}^{s} c_i f_i(x) x^{m - n_i}.$$

Obviously, $t_1(x) \in \mathcal{I}$ and $\deg(t_1(x)) < m$. If $\deg(t_1(x)) > n - 1$, we can apply to it the construction considered above and obtain a polynomial $t_2(x)$ such that $\deg(t_2(x)) < \deg(t_1(x))$. Continuing this process, we obtain the need form.

The coefficients at x^{n-i} in the polynomials of the ideal $\mathcal{I}$ with degrees $\leq n - i$, $(i = 1, ..., n)$, form an ideal L_i of the ring R. Denote by $d_1^i, ..., d_{s_i}^i$ the system of generators of the ideal L_i and by $f_j^i(x) \in \mathcal{I}$ a polynomial of degree $n - i$ with the leading coefficient d_j^i $(i = 1, ..., n; j = 1, ..., s_i)$. It is easy to verify that the polynomials $h(x) \in \mathcal{I}$ with degree $\leq n - 1$ can be expressed by the polynomials $f_j^i(x)$. Therefore, a system of generators of the ideal $\mathcal{I}$ is formed by the polynomials $f_1(x), ..., f_s(x)$ and $f_j^i(x)$ $(i = 1, ..., n; j = 1, ..., s_i)$. Thus, the theorem is proved. $\qquad\square$

David Hilbert
(1862-1943)

✳ Corollary 8.29.
If R is a Noetherian ring, then the polynomial ring $R[x_1, ..., x_n]$ is also Noetherian.

Proof. The proof follows immediately from the previous theorem by mathematical induction on the number of variables n, taking into account that $R[x_1, x_2, \ldots, x_n] = R[x_1, x_2, \ldots, x_{n-1}][x_n]$. $\qquad\square$

✳ Definition 8.30.
An ideal $\mathcal{I}$ generated by a set S of monomials in $K[x_1, x_2, \ldots, x_n]$, where K is a field, is called a **monomial ideal**.

Note, that this definition does not say that a set of monomials $S \subset \mathcal{I}$ is finite. Using the Hilbert Basis Theorem, we can say that each monomial ideal is generated by a finite number of polynomials. However, in this case, we can prove much more. Namely, each monomial ideal $\mathcal{I}$ is generated by a finite number of monomials from $\mathcal{I}$.

✳ Theorem 8.31.
Let $\mathcal{I}$ be a monomial ideal in $K[x_1, x_2, \ldots, x_n]$, where K is a field. Then,
   1) A polynomial $f \in \mathcal{I}$ if and only if each term of f belongs to $\mathcal{I}$, i.e., $\text{supp}(f) \subset \mathcal{I}$.
   2) There is a finite number of monomials $g_1, g_2, \ldots, g_m \in \mathcal{I}$ which generates $\mathcal{I}$, i.e.,
$$\mathcal{I} = \langle g_1, g_2, \ldots, g_m \rangle$$

Proof.
   1) $\Leftarrow$. Since $\mathcal{I}$ is a monomial ideal, it is generated by a set $S \subset \mathcal{I}$ of monomials u^β. Therefore, if $f \in \mathcal{I}$, then f can be written in the form

$f = \sum\limits_{j=1}^{m} g_j u^{\beta_j}$, where $u^{\beta_j} \in S$ and $g_j \in K[x_1, x_2, \ldots, x_n]$. On the other hand,

the polynomial f can be written in the form $f = \sum\limits_{i=1}^{k} f_i$, where $f_i = c_i u^{\alpha_i} \in$ supp(f). Since supp$(f) = \bigcup\limits_j$ supp$(g_j u^{\beta_j})$, for each $f_i \in$ supp(f) there exists an index j_i such that $f_i \in$ supp$(g_{j_i} u^{\beta_{j_i}})$. So, there is a monomial $v_i \in T_n$ such that $f_i = v_i u^{\beta_{j_i}}$, which means that $f_i \in \mathcal{I}$.

$\Rightarrow$. It is obvious.

2) From the Hilbert Basis Theorem, it follows that $\mathcal{I}$ is a finitely generated ideal of $K[x_1, x_2, \ldots, x_n]$, i.e., there exist a finite number of polynomials $f_1, f_2, \ldots, f_k \in K[x_1, x_2, \ldots, x_n]$ that generates $\mathcal{I}$, i.e., $\mathcal{I} = \langle f_1, f_2, \ldots, f_k \rangle$. So, each polynomial $f \in \mathcal{I}$ can be written in the form $f = \sum\limits_{i=1}^{k} h_i f_i$, where $h_i \in K[x_1, x_2, \ldots, x_n]$. By part 1 of this theorem, supp$(f_i) \subset \mathcal{I}$ for all $i = 1, 2, \ldots, k$. So, $S = \bigcup\limits_{i=1}^{k}$ supp$(f_i) \subset \mathcal{I}$ is a finite set of monomials which generates $\mathcal{I}$. $\qquad \square$

8.4. Monomial Order

In this section, we consider different ways to order monomials in n variables of the polynomial ring $K[x_1, x_2, \ldots, x_n]$, where K is a field.

From the definition of polynomials, we can write terms of a polynomial in an arbitrary order. For example, the expressions:

$$x_1^3 x_2 + x_1^2 x_3 + x_1 x_2 x_3 \quad \text{and} \quad x_1^2 x_3 + x_1 x_2 x_3 + x_1^3 x_2$$

represent the same polynomial.

In the case of polynomials in one variable x, we define the leading term of a polynomial and there are two natural ways to write polynomials, namely: In order of increasing degrees and in order of decreasing degrees of x, i.e., in the following forms:

$$a_0 + a_1 x^1 + \cdots + a_{n-1} x^{n-1} + a_n x^n \quad \text{or} \quad a_n x^n + a_{n-1} x^{n-1} + \cdots + a_1 x^1 + a_0.$$

In the case of polynomials in $n > 1$ variables, there are no natural ways to write polynomials. Polynomials are often written in order of increasing or decreasing degrees of monomials, but this writing is not unique. For example, polynomials

$$x_1^3 x_2 + x_1 x_3^2 + x_1 x_2^2 x_3 + x_1 x_3 + x_2 \quad \text{and} \quad x_1^2 x_3^2 + x_1 x_2^2 x_3 + x_1^3 x_2 + x_1 x_3 + x_2$$

are written in order of increasing degrees of monomials.

Nevertheless, in the case of polynomials in n variables, there are also unambiguous orders to write polynomials.

Recall that a binary relation $\leq$ on a set X is called a **partial order** if it is reflexive, antisymmetric and transitive, i.e., for all $a, b, c \in X$ we have:

1) $a \leq a$ (reflexivity)
2) if $(a \leq b) \wedge (b \leq a)$ then $a = b$ (antisymmetry)
3) if $(a \leq b) \wedge (b \leq c)$ then $a \leq c$ (transitivity)

A partial order $\leq$ on X is called a **total order** (or a **linear order**) if, in addition, for all $a, b \in X$:

4) either $a \leq b$ or $b \leq a$.

Moreover, we write $a < b$ if $a \leq b$ and $a \neq b$. We also write $a \geq b$ $(a > b)$ if $b \leq a$ $(b < a)$.

A total order $\leq$ on X is called a **well-ordering** if every subset $Y \subseteq X$ contains a least element, i.e., there is an element $d \in Y$ such that $d \leq y$ for all $y \in Y$.

> **✳ Definition 8.32.**
>   We say that a well-ordering $\leq$ on the set of monomials $T_n \subset K[x_1, x_2, \ldots, x_n]$ is called a **monomial order** (or **monomial ordering**) if it satisfies the following property:
>   if $u^\alpha \leq u^\beta$ then $u^{\alpha+\gamma} \leq u^{\beta+\gamma}$ for all monomials $u^\alpha, u^\beta, u^\gamma \in T_n$.

Consider some examples of monomial orderings on the set T_n.

1. Pure Lexicographic Order

> **✳ Definition 8.33.**
>   Let $\alpha = (\alpha_1, \alpha_2, \ldots, \alpha_n)$, $\beta = (\beta_1, \beta_2, \ldots, \beta_n) \in \mathbb{N}^n$. The **lexicographic order** $\leq_{\text{lex}}$ on the set $\mathbb{N}^n$ is defined as follows: $\alpha \leq_{\text{lex}} \beta$ if and only if the left-most non-zero entry in $\alpha - \beta$ is negative, i.e., there is a number $1 \leq k \leq n$ such that
>
> $$\alpha_1 = \beta_1, \alpha_2 = \beta_2, \ldots, \alpha_{k-1} = \beta_{k-1}, \text{ and } \alpha_k \leq \beta_k.$$

It is easy to show that the lexicographic order $\leq_{\text{lex}}$ is an example of total order. Moreover, it is a well-ordering, since $1 = u^0 \leq u^\alpha$, for every $\alpha \in \mathbb{N}^n$ and $0 = (0, 0, \ldots, 0) \in \mathbb{N}^n$.

✳ Examples 8.34.
1. $(1, 0, 0) <_{\text{lex}} (2, 0, 0)$
2. $(1, 2, 0) <_{\text{lex}} (1, 2, 1)$
3. $(0, 1, 0) <_{\text{lex}} (8, 0, 0)$

We can transfer the lexicographic order to the set of all monomials in n variables.

✻ **Definition 8.35.**

Let $u^\alpha, u^\beta \in T_n$. The relation of monomials $\leq_{\text{lex}}$ such that $u^\alpha \leq_{\text{lex}} u^\beta$ if and only if $\alpha \leq_{\text{lex}} \beta$ is called the **pure lexicographic order** of monomials.

✴ **Examples 8.36.**

1. Determine the pure lexicographic order for monomials: $u^\alpha = 5x^2yz^4$ and $u^\beta = -2x^3z^3$

   In this case, $\alpha = (2,1,4)$, $\beta = (3,0,3)$. Since $\alpha - \beta = (-1,1,-1)$, we have that $\alpha <_{\text{lex}} \beta$. So that, by definition, $u^\alpha <_{\text{lex}} u^\beta$, i.e., $5x^2yz^4 <_{\text{lex}} -2x^3z^3$.

2. Determine the pure lexicographic order for monomials: $u^\alpha = 2x^2y^3z$, $u^\beta = 3x^2y^2z^2$, and $u^\gamma = -5x^3$

   In this case, $\alpha = (2,3,1)$, $\beta = (2,2,2)$, $\gamma = (3,0,0)$. Since $\beta - \alpha = (0,-1,1)$, we have that $\beta <_{\text{lex}} \alpha$. So that, by definition, $u^\beta <_{\text{lex}} u^\alpha$, i.e., $3x^2y^2z^2 <_{\text{lex}} 2x^2y^3z$.

   Since $\alpha - \gamma = (-1,3,1)$, we get that $\alpha <_{\text{lex}} \gamma$, so that $u^\alpha <_{\text{lex}} u^\gamma$ and $u^\beta <_{\text{lex}} u^\alpha <_{\text{lex}} u^\gamma$, i.e., $3x^2y^2z^2 <_{\text{lex}} 2x^2y^3z <_{\text{lex}} -5x^3$.

2. Reverse Pure Lexicographic Order

✻ **Definition 8.37.**

Let $u^\alpha, u^\beta \in T_n$. The relation of monomials $\leq_{\text{revlex}}$ such that $u^\alpha \leq_{\text{revlex}} u^\beta$ if and only if $\beta \leq_{\text{lex}} \alpha$ is called the **reverse pure lexicographic order** of monomials.

✴ **Example 8.38.**

Determine the reverse pure lexicographic order for monomials: $u^\alpha = 7xy^2z^4$ and $u^\beta = -2x^2z^3$

Since $\alpha - \beta = (1,2,4) - (2,0,3) = (-1,2,1)$, $\alpha <_{\text{lex}} \beta$, therefore, $u^\beta <_{\text{revlex}} u^\alpha$, i.e., $-2x^2z^3 <_{\text{revlex}} 7xy^2z^4$.

3. Graded Lexicographic Order

✻ **Definition 8.39.**

Let $\alpha = (\alpha_1, \alpha_2, \ldots, \alpha_n)$, $\beta = (\beta_1, \beta_2, \ldots, \beta_n) \in \mathbb{N}^n$ and $|\alpha| = \sum_{i=1}^n \alpha_i$, $|\beta| = \sum_{i=1}^n \beta_i$. The **graded lexicographic order** $\leq_{\text{grlex}}$ on the set $\mathbb{N}^n$ is defined as follows:

$$\alpha \leq_{\text{grlex}} \beta \text{ if and only if } (|\alpha| < |\beta|) \vee \{(|\alpha| = |\beta|) \wedge (\alpha \leq_{\text{lex}} \beta)\}.$$

✳ **Examples 8.40.**

1. $\alpha = (1,2,0) <_{\text{grlex}} \beta = (1,2,3)$, since $|\alpha| = 3 < 4 = |\beta|$.

2. $\alpha = (1,2,6) <_{\text{grlex}} \beta = (1,8,0)$, since $|\alpha| = 9 = |\beta|$ and $(1,2,6) <_{\text{lex}}$ $(1,8,0)$.

❋ **Definition 8.41.**
    Let $u^\alpha, u^\beta \in T_n$. The relation of monomials $\leq_{\text{grlex}}$ such that $u^\alpha \leq_{\text{grlex}}$ u^β if and only if $\alpha \leq_{\text{grlex}} \beta$ is called the **graded lexicographic order** of monomials.

✳ **Examples 8.42.**

1. Determine the graded lexicographic order for monomials $u^\alpha = -3x^2yz^4$ and $u^\beta = 4x^3z^3$.

   Here we have: $\alpha = (2,1,4)$, $\beta = (3,0,3)$, and $|\alpha| = 7$, $|\beta| = 6$.

   Since $|\beta| = 6 < 7 = |\alpha|$, $\beta <_{\text{grlex}} \alpha$ and so $4x^3z^3 <_{\text{grlex}} -3x^2yz^4$.

2. Determine the graded lexicographic order for monomials: $u^\alpha = -7x^2yz^3$ and $u^\beta = 4x^3z^3$.

   Here we have: $\alpha = (2,1,3)$, $\beta = (3,0,3)$ and $|\alpha| = |\beta| = 6$.

   Since $\alpha - \beta = (-1,1,0)$, $\alpha <_{\text{grlex}} \beta$ and so $-7x^2yz^3 <_{\text{grlex}} 4x^3z^3$.

3. The lexicographic order on the set of monomials T_2 in $K[x,y]$ is:

   $1 <_{\text{lex}} y <_{\text{lex}} y^2 <_{\text{lex}} \cdots <_{\text{lex}} x <_{\text{lex}} xy <_{\text{lex}} xy^2 <_{\text{lex}} \cdots <_{\text{lex}} x^2 <_{\text{lex}} x^2y <_{\text{lex}} x^2y^2 <_{\text{lex}} \cdots$

4. The graded lexicographic order on the set of monomials T_2 in $K[x,y]$ is:

   $1 <_{\text{grlex}} y <_{\text{grlex}} x <_{\text{grlex}} y^2 <_{\text{grlex}} xy <_{\text{grlex}} x^2 <_{\text{grlex}} y^3 <_{\text{grlex}} xy^2 <_{\text{grlex}} x^2y <_{\text{grlex}} x^3 <_{\text{grlex}} \cdots$

3. Graded Reverse Lexicographic Order

❋ **Definition 8.43.**
    Let $u^\alpha, u^\beta \in T_n$. The relation of monomials $\leq_{\text{grevlex}}$ such that $u^\alpha \leq_{\text{grevlex}} u^\beta$ if and only if $\beta \leq_{\text{grlex}} \alpha$ is called the **graded reverse lexicographic order** of monomials.

✳ **Examples 8.44.**
    Define the graded reverse lexicographic order for monomials $u^\alpha = 5x^3y^2z^4$ and $u^\beta = 4x^3z^5$.

Here we have: $\alpha = (3, 2, 4)$, $\beta = (3, 0, 5)$, and $|\alpha| = 9$, $|\beta| = 8$.
Since $|\beta| = 8 < 9 = |\alpha|$, $\beta <_{\text{grevlex}} \alpha$ and so $5x^3y^2z^4 <_{\text{grevlex}} 4x^3z^5$.

Note that all the relations considered above are monomial orders and for all of them we assume that:

$$x_1 < x_2 < \cdots < x_n$$

and, in particular, we assume that $x < y < z$.

✳ Proposition 8.45.

The pure lexicographic order, reverse lexicographic order, graded lexicographic order and graded reverse lexicographic order are monomial orders.

The proof of this proposition can be done easily by verifying all conditions of Definition 8.32 of a monomial order, and we leave its proof to the Reader.

Suppose we have a monomial order $\leq$ on the set T_n. Then, each polynomial in n variables $f(x_1, x_2, \ldots, x_n) \in K[x_1, x_2, \ldots, x_n]$ can be written in the form:

$$f(u) = c_1 u^{\alpha_1} + c_2 u^{\alpha_2} + \cdots + c_s u^{\alpha_s}, \qquad (8.12)$$

where $0 \neq c_i \in K$, $u^{\alpha_i} \in T_n$ and $u^{\alpha_s} < u^{\alpha_{s-1}} < \cdots < u^{\alpha_1}$.

✳ Example 8.46.

The polynomial:

$$P(x_1, x_2, x_3, x_4) = x_1^4 + 2x_1^2 x_2^3 x_3 - x_1^2 x_2^3 x_4^3 + +3x_1 x_3 x_4^2 + x_1 x_3 + 4x_1 + x_2 x_4^2$$

is written in the lexicographic order.

✳ Definition 8.47.

Given a monomial order $\leq$ on the set T_n, if $f(u) \in K[x_1, x_2, \ldots, x_n]$ is a non-zero polynomial written in the form (8.12), we define:

- the **leading term** of $f(u)$ is $\text{lt}(f) = c_1 u^{\alpha_1}$;

- the **initial monomial** of $f(u)$ is $\text{in}(f) = u^{\alpha_1}$;

- the **leading coefficient** of $f(u)$ is $\text{lc}(f) = c_1$, the coefficient of the leading term of f;

- the **multidegree** of $f(u)$ is $\text{multideg}(f) = \alpha_1$, the maximal degree of monomials occurring in f.

Moreover, for convenience, we also define that $\text{in}(0) = 0$, $\text{lc}(0) = 0$, and $\text{multideg}(0) = -\infty$.

It is easy to prove the following lemma, which we leave for the Reader as an exercise.

❋ **Lemma 8.48.**

If K is a field, then

1) $\text{in}(f_1 f_2 \ldots f_m) = \text{in}(f_1)\text{in}(f_2) \cdots \text{in}(f_m)$
2) $\text{lc}(f_1 f_2 \ldots f_m) = \text{lc}(f_1)\text{lc}(f_2) \cdots \text{lc}(f_m)$
3) $\text{in}(f_1 + f_2 + \cdots + f_m) \leq \max_{1 \leq i \leq n} [\text{in}(f_i)]$.

   Moreover, $\text{in}(f_1 + f_2 + \cdots + f_m) < \max_{1 < i \leq n} [\text{in}(f_i)]$ if and only if $\sum_i \text{lc}(f_i) = 0$,

where the sum is taken only over those i for which $\text{in}(f_i) \geq \text{in}(f_j)$ for all j.

4) $\text{multideg}(f_1 f_2 \ldots f_m) = \text{multideg}(f_1) + \text{multideg}(f_2) + \cdots \text{multideg}(f_m)$

❋ **Examples 8.49.**

1. $f_1(x, y) = xy - 2y$;

   With respect to the monomial order $\leq_{\text{lex}}$ or $\leq_{\text{grlex}}$, we have:

   $\text{in}(f_1) = xy$, $\text{multideg}(f_1) = (1, 1)$

2. $f_2(x, y) = 2y^2 - x^2 + x$;

   With respect to the monomial order $\leq_{\text{lex}}$ or $\leq_{\text{grlex}}$, we have:

   $\text{in}(f_2) = -x^2$, $\text{multideg}(f_2) = (2, 0)$

3. $f_3(x, y) = 2y^3 - xy$;

   With respect to the monomial order $\leq_{\text{lex}}$, we have:

   $\text{in}(f_3) = -xy$, $\text{multideg}(f_3) = (1, 1)$

   With respect to the monomial order $\leq_{\text{grlex}}$, we have:

   $\text{in}(f_3) = 2y^3$, $\text{multideg}(f_3) = (0, 3)$.

8.5. Division Algorithm for Polynomials

According to Theorem 5.9, in the ring of polynomials in one variable $K[x]$ there is division with remainder (Remainder Theorem). Namely, for any polynomials $P(x), Q(x) \in K[x]$, $Q(x) \neq 0$ and $\deg(P) \geq \deg(Q)$ there exist polynomials $S(x), R(x) \in K[x]$ such that $P(x) = Q(x)S(x) + R(x)$ and either $R(x) = 0$ or $0 \leq \deg(R) < \deg(Q)$.

For the ring of polynomials in n variables $R = K[x_1, x_2, \ldots, x_n]$ over a field K, there is some generalization of this algorithm which allows a polynomial $f \in R$ to be divided by a finite set of polynomials. Given a monomial order $\leq$ on the ring R, then each polynomial $f \in R$ can be divided by a finite set F of non-zero polynomials $g_1, g_2, \ldots, g_m \in R$ according to the following definition:

> **✳ Definition 8.50.**
>
> Let $\leq$ be a monomial order on $R = K[x_1, x_2, \ldots, x_n]$. We say that a polynomial $f \in R$ is divided by a finite set $F = \{g_1, g_2, \ldots, g_m\}$ of non-zero polynomials with respect to $\leq$ if there exist polynomials $p_1, p_2, \ldots, p_m, r \in R$ such that
>
> $$f = p_1 g_1 + p_2 g_2 + \cdots + p_m g_m + r, \tag{8.13}$$
>
> where either $r = 0$ or no term of the polynomial r is divisible by any of the initial monomials $\text{in}(g_i)$. Moreover, $\text{in}(p_i g_i) \leq \text{in}(f)$, for all $i = 1, 2, \ldots, m$, and $\text{in}(f) = \max_{1 \leq i \leq n} [\text{in}(p_i g_i), \text{in}(r)]$. The polynomial r is called the **remainder** of f with respect to the set F, and the form (8.13) is called the **standard expression** of f.

✳ Example 8.51.

Divide the polynomial $f(x, y) = xy^2 - 2xy + x$ by the polynomial $g(x, y) = xy - 4x$ with respect to the lexicographical order: $\leq_{\text{lex}}$ and $y \leq_{\text{lex}} x$.

Here, we have $\text{in}(f) = xy^2$ and $\text{in}(g) = xy$. Then, $f(x, y) = p(x, y)g(x, y) + r(x, y) = (y + 2)(xy - 4x) + 9x$ is the standard expression of $f(x, y)$, where $p(x, y) = y + 2$ and the polynomial $r(x, y) = 9x$ is the remainder, because $\text{in}(pg) = xy^2 = \text{in}(f)$ and $9x$ is not divisible by $\text{in}(g) = xy$.

✳ Example 8.52.

Divide the polynomial $f(x, y) = x^3 y^2 + x^2 y + x^2 + xy^2$ by the polynomial $g(x, y) = xy^2 + 1$ with respect to the lexicographical order: $\leq_{\text{lex}}$ and $y <_{\text{lex}} x$.

Here, we have $\text{in}(f) = x^3 y^2$, $\text{multideg}(f) = (3, 2)$, $\text{in}(g) = xy^2$, $\text{multideg}(g) = (1, 2)$.

Therefore, $f(x, y) = p(x, y)g(x, y) + r(x, y)$, where $p(x, y) = x^2 + 1$, is the standard expression of f, and the polynomial $r(x, y) = x^2 y - 1$ is the remainder, since $\text{in}(pg) = x^3 y^2 = \text{in}(f)$ and neither $x^2 y$ nor 1 is divisible by $\text{in}(g) = xy^2$.

✳ Example 8.53.

Divide the polynomial $f(x, y) = x^2 y + xy^2 + y^2$ by the set of polynomials $F = \{g_1 = xy - 1, g_2 = y^2 - 1\}$ with respect to the lexicographical order: $\leq_{\text{lex}}$ and $y <_{\text{lex}} x$.

Here, $\text{in}(f) = x^2 y$, $\text{in}(g_1) = xy$ and $\text{in}(g_2) = y^2$. Then, $f(x, y) = (x + y)g_1(x, y) + h(x, y) = (x + y)(xy - 1) + h(x, y)$, where $h(x, y) = x + y^2 + y$. Since $h(x, y) = 1 \cdot g_2 + r_1(x, y) = 1 \cdot (y^2 - 1) + (x + y + 1)$, we obtain that:

$$f(x, y) = (x + y)g_1 + 1 \cdot g_2 + r_1(x, y) =$$

$$= (x + y)(xy - 1) + 1 \cdot (y^2 - 1) + (x + y + 1)$$

is the standard expression of f and $r_1(x, y) = x + y + 1$ is the remainder, since $\text{in}((x + y)g_1) = x^2 y = \text{in}(f)$, $\text{in}(g_2) = y^2 <_{\text{lex}} \text{in}(f) = x^2 y$ and no term of $r_1(x, y)$ is divisible by $\text{in}(g_1) = xy$ or $\text{in}(g_2) = y^2$.

Since the set $F = \{g_1 = xy - 1, g_2 = y^2 - 1\} = \{g_2 = y^2 - 1, g_1 = xy - 1\}$ does not depend on the order of writing its elements, we can change the choice of polynomials by which we divide. Then, we obtain:

$$f(x, y) = (x + y)g_2 + xg_1 + r(x, y) = (x + y)(y^2 - 1) + x(xy - 1) + r_2(x, y)$$

where $r_2(x, y) = 2x + 1$ is the remainder, since no its term is divisible by $\text{in}(g_1) = xy$ or $\text{in}(g_2) = y^2$.

Note, that this remainder is not the same as the previous one.

So, this example shows that the result (i.e., the expression of a polynomial f in the form (8.13)) on dividing by a set of polynomials $F = \{g_1, g_2, \ldots, g_m\}$ is not unambiguous. It depends not only on a set F but also can depend on the order of elements $g_1, g_2, \ldots, g_m$ by which we divide.

The following theorem shows that, for each set of polynomials $F = \{g_1, g_2, \ldots, g_m\} \subset K[x_1, x_2, \ldots, x_n] = P$, any polynomial $f \in F$ can be divided by F, i.e., can be written in the form (8.13).

✳ Theorem 8.54.

Given a monomial order $\leq$ on the ring of polynomials $R = K[x_1, x_2, \ldots, x_n]$, a non-zero polynomial $f \in R$ can be divided by a finite set F of non-zero polynomials $g_1, g_2, \ldots, g_m \in R$, i.e., f can be written in the form (8.13).

Proof.

Note that every strictly descending chain of monomials $v_0 > v_1 > v_2 > \cdots$ of the ring $R = K[x_1, x_2, \ldots, x_n]$ terminates. If $f \in R$ and either $f = 0$ or $f = c \in K$, then the theorem is trivial.

We prove this theorem by mathematical induction on monomials, $v_0, v_1, v_2, \ldots$, which are initial monomials of polynomials, such that they form a strictly descending chain $v_0 > v_1 > v_2 > \cdots$.

1. If $\text{in}(f)$ is not divisible by any $\text{in}(g_i)$ for all $i = 1, \ldots, m$, we set $f_1 = f - \text{lt}(f)$ and $v_0 = \text{in}(f)$. Then, $v_1 = \text{in}(f_1) < \text{in}(f) = v_0$. Therefore, by induction hypothesis, the polynomial f_1 has the standard expression:

$$f_1 = s_1 g_1 + s_2 g_2 + \cdots + s_m g_m + r_1, \qquad (8.14)$$

where $r_1 = 0$ or no term of r_1 is divisible by any of the initial monomials $\text{in}(g_i)$. Moreover, $\text{in}(s_i g_i) \leq \text{in}(f_1)$, for all $i = 1, 2, \ldots, m$, and $\text{in}(f_1) = \max_{1 \leq i \leq m} [\text{in}(s_i g_i), \text{in}(r_1)]$. Then, $\text{in}(s_i g_i) \leq \text{in}(f_1) < \text{in}(f)$ for all $i = 1, \ldots, m$. If $r = r_1 + \text{lt}(f) \neq 0$, then $\text{in}(r) = \text{in}(r_1 + \text{lt}(f)) = \text{in}(f)$, since $\text{in}(r_1) \leq \text{in}(f_1) < \text{in}(f)$. Therefore, $\max_{1 \leq i \leq m} [\text{in}(s_i g_i), \text{in}(r)] = \max_{1 \leq i \leq m} [\text{in}(s_i g_i), \text{in}(f)] = \text{in}(f)$, since $\text{in}(s_i g_i) \leq \text{in}(f_1) < \text{in}(f)$. Thus,

$$f = s_1 g_1 + s_2 g_2 + \cdots + s_m g_m + r,$$

where $r = r_1 + \text{lt}(f)$, is the standard expression of f.

2. Suppose that $\text{in}(f)$ is divisible by $\text{in}(g_i)$ for some $i = 1, \ldots, m$, i.e., there exists a monomial $h_i \in T_n$ such that $\text{in}(f) = h_i\text{in}(g_i)$. Then, $\text{lt}(f) = m_i\text{lt}(g_i)$, where $m_i = c_ih_i$ and $c_i = \text{lc}(f)[\text{lc}(g_i)]^{-1}$. We set $f_1 = f - m_ig_i$. Then, $v_1 = \text{in}(f_1) = \text{in}(f - m_ig_i) < \text{in}(f) = v_0$. Therefore, we can apply the induction hypothesis for the polynomial f_1. So that f_1 has the standard expression (8.14). Therefore,

$$f = p_1g_1 + p_2g_2 + \cdots + p_ig_i + \cdots + p_mg_m + r_1,$$

where $p_j = s_j$ for all i, j with $i \neq j$ and $p_i = s_i + m_i$, is the standard expression of f. Indeed, r_1 is a remainder on dividing f by $g_1, g_2, \ldots, g_m$, since either $r_1 = 0$ or no term of the polynomial r_1 is divisible by any of the initial monomials $\text{in}(g_i)$ for all $i = 1, \ldots, m$. Moreover, $\text{in}(p_jg_j) \leq \text{in}(f_1) < \text{in}(f)$ for all $j \neq i$. For $j = i$, we have:

$$\text{in}(p_ig_i) = \text{in}((s_i + m_i)g_i) = \max[\text{in}(s_ig_i), \text{in}(m_ig_i)] = \text{in}(f),$$

since $s_i + m_i \neq 0$, otherwise we have a contradiction: $\text{in}(f_1) \geq \text{in}(s_ig_i) = \text{in}(m_ig_i) = \text{in}(f) > \text{in}(f_1)$. Therefore, $\max_{1 \leq i \leq m} [\text{in}(p_ig_i), \text{in}(r_1)] = \text{in}(f)$. $\square$

From the proof of this theorem, we obtain the the following division algorithm with remainder.

Division Algorithm with Remainder.

INPUT: $f \in R = K[x_1, x_2, \ldots, x_n]$;
$F = \{g_1, g_2, \ldots, g_m : 0 \neq g_i \in R; i = 1, 2, \ldots, m\}$.
OUTPUT: $p_1, p_2, \ldots, p_m, r$ such that $f = p_1g_1 + p_2g_2 + \cdots + p_mg_m + r$
INITIALIZATION: $p_i := 0$ for each i; $r := 0$; $h := f$

WHILE $h \neq 0$ DO
    IF $\exists i \in \{1, 2, \ldots, m\}$ such that $\text{in}(g_i)|\text{in}(h)$ THEN
        $m_i := \text{lt}(h)/\text{lt}(g_i)$
        $p_i := p_i + m_i$
        $h := h - m_ig_i$
    ELSE
        $r := r + \text{lt}(h)$
        $h := h - \text{lt}(h)$
END DO
RETURN $p_1, p_2, \ldots, p_m, r$

✳ Example 8.55.
    Divide the polynomial $f(x, y) = xy^2 - 2xy + x$ by the polynomial $g(x, y) = xy - 4x$ with respect to the monomial order $\leq_{\text{lex}}$ and $y <_{\text{lex}} x$.

We have: $1 <_{\text{lex}} x <_{\text{lex}} xy <_{\text{lex}} xy^2$ and
$\text{in}(f) = \text{lt}(f) = xy^2$; $\text{in}(g) = \text{lt}(g) = xy$, $\text{in}(g) <_{\text{lex}} \text{in}(f)$

1. INITIALIZATION: $h := f = xy^2 - 2xy + x$; $r := 0$; $p := 0$

2. $m := \text{lt}(h)/\text{lt}(g) = y$

   $p := p + m = 0 + y = y$

   $h := h - mg = (xy^2 - 2xy + x) - y(xy - 4x) = 2xy + x$

   $\text{in}(h) = xy$; $\text{lt}(h) = 2xy$; $(xy)|(2xy)$;

3. $m := \text{lt}(h)/\text{lt}(g) = 2$

   $p := p + m = y + 2$

   $h := h - mg = (2xy + x) - 2(xy - 4x) = 9x \neq 0$

   $\text{lt}(h) = 9x$

   $\text{in}(g) = xy$ does not divide $\text{in}(h) = x$.

4. $r = 0 + \text{lt}(h) = 9x$

   $h := h - \text{lt}(h) = 9x - 9x = 0$

   Therefore, $\quad p = y + 2, \quad r = 9x$.

$$f = pg + r = (y + 2)(xy - 4x) + 9x$$

where $r = 9x$ is the remainder on dividing f by g.

✳ Example 8.56.

Divide the polynomial $f = x^2y^2z + xz$ by the set $F = \{g_1, g_2\}$, where $g_1 = xy - z$, $g_2 = z$ with respect to the monomial order $\leq_{\text{lex}}$ and $z <_{\text{lex}} y <_{\text{lex}} x$.

We have: $\text{lt}(f) = \text{in}(f) = x^2y^2z$, $\text{lt}(g_1) = \text{in}(g_1) = xy$, $\text{lt}(g_2) = \text{in}(g_2) = z$

1. INITIALIZATION: $\quad h := f$; $\ p_1 = p_2 = 0$; $\ r = 0$

2. $m_1 = \text{lt}(h)/\text{lt}(g_1) = xyz$

   $p_1 := 0 + m_1 = xyz$;

   $h := h - m_1 g_1 = xyz^2 + xz \neq 0$;

   $\text{lt}(h) = \text{in}(h) = xyz^2$;

3. $m_1 = \text{lt}(h)/\text{lt}(g_1) = z^2$

   $p_1 := p_1 + m_1 = xyz + z^2$

   $h := h - m_1 g_1 = xz + z^3 \neq 0$;

   $\text{lt}(h) = \text{in}(h) = xz$;

4. $m_2 = \text{lt}(h)/\text{lt}(g_2) = xz/z = x$

   $p_2 := 0 + m_2 = x$

   $h := h - m_2 g_2 = z^3 \neq 0$;

   $\text{lt}(h) = \text{in}(h) = z^3$

5. $m_2 = \text{in}(h)/\text{in}(g_2) = z^2$

$p_2 = p_2 + m_2 = x + z^2;$

$h := z^3 - z^3 = 0.$

Therefore, $p_1 = xyz + z^2;$ $p_2 = x + z^2,$ $r = 0.$

$$f = p_1 g_1 + p_2 g_2 + r = (xyz + z^2)g_1 + (x + z^2)g_2,$$

where $r = 0$ is the remainder on dividing f by the set F.

Moreover, $\text{in}(f) = \text{in}(p_1 g_1) = x^2 y^2 z.$

8.6. Initial Ideals. Gröbner Basis

As shown in the previous section, we can get the different remainders for the same polynomials in several variables by applying the division algorithm, that is, in general, the standard expression (8.13) and the reminder r for a polynomial $f \in K[x_1, x_2, \ldots, x_n]$ are not unique.

Our main question is: "What does a set F need to be so that the remainder on dividing every polynomial $f \in K[x_1, x_2, \ldots, x_n]$ by F will be unique?"

Let $F = \{g_1, g_2, \ldots, g_k\}$ be a set of polynomials of $R = K[x_1, x_2, \ldots, x_n]$ and let $\mathcal{I} = \langle g_1, g_2, \ldots, g_k \rangle$ be a non-zero ideal of R. Then, the remainder r of the division of a polynomial f by the set F can be given as $f \equiv r \pmod{\mathcal{I}}$. Since the generators of I can be chosen in a different way, there is a natural question: "Does a set of polynomials generated the ideal $\mathcal{I}$ exist such that for any polynomial $f \in R$ its reminder of the division of f by this set is unique? The answer is "Yes". It turns out that such a set of generators exists for any ideal $\mathcal{I}$ of R and it is called a **Gröbner basis**.

First, a Gröbner basis for polynomial rings was introduced by Bruno Buchberger in his PhD thesis in 1965. Bruno Buchberger named it a **Gröbner basis** in 1976 to honor his advisor, professor Wolfgang Gröbner.

It should be noted that the main idea that this theory is based on can been traced from the earlier sources, for example, papers of Paul Gordon in 1900, David Hilbert in 1890 or G.Hermann in 1926. Nevertheless, Bruno Buchberger was the first who gave the simple algorithm for computing Gröbner bases.

In this section, we introduce the main terminology and notions regarding the Gröbner basis theory, while Buchberger's algorithm, which gives an efficient method for computing Gröbner bases, will be given in the next section.

Towards the end of this chapter, we will assume that a monomial order $\leq$ on the ring $R = K[x_1, x_2, \ldots, x_n]$ is specified.

> ✳ **Definition 8.57.**
>
> Let S be a non-empty subset of $R = K[x_1, x_2, \ldots, x_n]$. A monomial ideal generated by the set of initial monomials of all polynomials of S:
>
> $$\text{in}(S) = \{\text{in}(f) \mid 0 \neq f \in S\} \tag{8.15}$$
>
> is called the **initial ideal** of S and is denoted by $\langle \text{in}(S) \rangle$.

Note, that in this definition we can replace "initial monomials" by "leading terms", since they differ only by constants, which are elements of a field K.

Though the initial ideal is generated by initial monomials of all polynomials of S, by Theorem 8.31, we can choose a finite number of them. Hence, we obtain the following corollary.

✳ Corollary 8.58.

If S is a non-empty subset of the ring $K[x_1, x_2, \ldots, x_n]$, then there exists a finite number of polynomials $f_1, f_2, \ldots, f_k \in S$ such that

$$\langle \text{in}(S) \rangle = \langle \text{in}(f_1), \text{in}(f_2), \ldots, \text{in}(f_k) \rangle \qquad (8.16)$$

This corollary gives a possibility to introduce the notion of a Gröbner basis.

✳ Definition 8.59.

Let $\mathcal{I}$ be a non-zero ideal in the ring $R = K[x_1, x_2, \ldots, x_n]$, and let $\leq$ be a monomial ordering on R. A finite subset $G = \{g_1, g_2, \ldots, g_m\}$ of $\mathcal{I}$ is called a **Gröbner basis** of $\mathcal{I}$ with respect to $\leq$ if the ideal $\langle \text{in}(\mathcal{I}) \rangle$ is generated by the initial monomials of polynomials $g_1, g_2, \ldots, g_m$, i.e.,

$$\langle \text{in}(\mathcal{I}) \rangle = \langle \text{in}(g_1), \text{in}(g_1), \ldots, \text{in}(g_m) \rangle = \langle \text{in}(G) \rangle \qquad (8.17)$$

From Corollary 8.58, the important statement follows:

✳ Theorem 8.60.

For each ideal of the ring $K[x_1, x_2, \ldots, x_n]$, there exists its Gröbner basis.

Thought not each system of generators of an ideal $\mathcal{I}$ needs to be a Gröbner basis of $\mathcal{I}$, the elements of a Gröbner basis form a system of generators of $\mathcal{I}$.

✳ Theorem 8.61.

Let $G = \{g_1, g_2, \ldots, g_m\}$ be a Gröbner basis of an ideal $\mathcal{I}$ of $R = K[x_1, x_2, \ldots, x_n]$ with respect to a monomial order $\leq$. Then, G is a system of generators of $\mathcal{I}$, i.e., $\mathcal{I} = \langle G \rangle$.

Proof.

Let $f \in \mathcal{I}$. Consider the standard expression of f on dividing f by the set G: $f = p_1 g_1 + p_2 g_2 + \cdots + p_m g_m + r$, where r is the remainder. Then, $r \in \mathcal{I}$, since $g_i \in \mathcal{I}$ for all $i = 1, 2, \ldots, m$. Suppose that $r \neq 0$. Then, $\text{in}(r) \in \langle \text{in}(\mathcal{I}) \rangle$. Since G is a Gröbner basis of $\mathcal{I}$, $\langle \text{in}(\mathcal{I}) \rangle = \langle \text{in}(g_1), \text{in}(g_2), \ldots, \text{in}(g_m) \rangle$. Thus, $\text{in}(r) = \sum\limits_{i=1}^{m} t_i \text{in}(g_i)$, where $t_i \in R$. So $\text{in}(r) \in \bigcup\limits_{i=1}^{m} \text{supp}(t_i \text{in}(g_i))$. Therefore,

$\text{in}(r) \in \text{supp}(t_i \text{in}(g_i))$ for some i. This implies that $\text{in}(g_i)|\text{in}(r)$, which contradicts the definition of the remainder. Thus, $r = 0$ and G is a system of generators of $\mathcal{I}$. $\qquad \square$

✻ Theorem 8.62.

Let $G = \{g_1, g_2, \ldots, g_m\}$ be a Gröbner basis of an ideal $\mathcal{I}$ of $R = K[x_1, x_2, \ldots, x_n]$ with respect to a monomial order $\leq$. Then,

1. For any polynomial $f \in R$, there exists the unique remainder $r \in P$ on dividing f by G such that

   (a) either $r = 0$ or no term of r is divisible by $\text{in}(g_i)$ for all $i = 1, 2, \ldots, m$;

   (b) $f \equiv r \pmod{\mathcal{I}}$, i.e., $f - r \in \mathcal{I}$.

2. A polynomial $f \in \mathcal{I}$ if and only if the reminder r on dividing f by G is equal to 0.

Proof.

1. Let $G = \{g_1, g_2, \ldots, g_m\}$ be a Gröbner basis of an ideal $\mathcal{I}$ of $P = K[x_1, x_2, \ldots, x_n]$. Then, G is a system of generators of $\mathcal{I}$, by Theorem 8.61.

Let $f \in R$. Assume that there are two standard expressions of f with regard to the set G:

$$f = p_1 g_1 + p_2 g_2 + \cdots + p_m g_m + r$$

and

$$f = t_1 g_1 + t_2 g_2 + \cdots + t_m g_m + s$$

with remainders r and s satisfying conditions (a) and (b). Then, we obtain that $(p_1 - t_1)g_1 + (p_2 - t_2)g_2 + \cdots + (p_m - t_m)g_m + (r - s) = 0$. This implies that $h = r - s \in \mathcal{I}$.

Suppose $h \neq 0$, then $\text{in}(h) \in \langle \text{in}(\mathcal{I}) \rangle$, and so

$$\text{in}(h) = \sum_{i=1}^{m} \text{in}(g_i) m_i.$$

Therefore, $\text{in}(h) \in \bigcup_{i=1}^{m} \text{supp}(\text{in}(g_i)m_i)$, and hence, $\text{in}(h) \in \text{supp}(\text{in}(g_i)m_i)$ for some i. This implies that $\text{in}(g_i)|\text{in}(h)$ for some i. Since $h = r - s \neq 0$, $\text{in}(h) \in \text{supp}(r) \cup \text{supp}(s)$. So, there is a term of r or s which is divisible by $\text{in}(g_i)$. This contradicts the definition of the remainder. Therefore, $h = 0$, i.e., $r = s$.

2. $\Rightarrow$. Let $f \in \mathcal{I}$. Suppose that

$$f = p_1 g_1 + p_2 g_2 + \cdots + p_m g_m + r, \qquad (8.18)$$

where $p_i \in R$, is the the standard expression of f on dividing f by the set G. Hence, $r \in \mathcal{I}$. If $r \neq 0$, then $in(r) \in \langle in(\mathcal{I}) \rangle = \langle in(G) \rangle$. This implies that $in(r)$ is divisible by some $in(g_i)$, which contradicts the definition of the remainder. So, the remainder $r = 0$.

$\Leftarrow$. Let the standard expression of a polynomial $f \in R$ on dividing f by the set G has the remainder $r = 0$, i.e., has the form (8.18). Since $G = \{g_1, g_2, \ldots, g_m\}$ is a system of generators of $\mathcal{I}$, by Theorem 8.61, we obtain that $f \in \mathcal{I}$. $\qquad \square$

✳ Definition 8.63.

A polynomial $r \in R = K[x_1, x_2, \ldots, x_n]$, which is the remainder on dividing a polynomial $f \in R$ by a Gröbner basis G, is called the **reduced form** of f with respect to G and a monomial order $\leq$.

From Theorem 8.62(2), it follows that the division of a polynomial by a Gröbner basis is always unique, in the sense that the reduced form of every polynomial $f \in R$ is defined uniquely.

The next theorem can be considered as an equivalent definition of a Gröbner basis of an ideal $\mathcal{I}$.

✳ Theorem 8.64.

Let $\mathcal{I}$ be an ideal of the ring $R = K[x_1, x_2, \ldots, x_n]$. A finite subset $G = \{g_1, g_2, \ldots, g_k \mid g_i \in R\} \subset \mathcal{I}$ is a Gröbner basis of $\mathcal{I}$ if and only if the remainder on dividing each polynomial $f \in \mathcal{I}$ by G is equal to zero.

Proof.

$\Rightarrow$. This follows from Theorem 8.62(2).

$\Leftarrow$. Suppose that $f \in \mathcal{I}$ and the remainder on dividing f by a finite subset $G = \{g_1, g_2, \ldots, g_k \mid g_i \in R\} \subset \mathcal{I}$ is equal to zero. In this case, the standard expression of f on dividing by G has the form (8.18). Therefore, $in(f) = \max_{1 \leq i \leq n} [in(p_i g_i)]$, i.e., there exists an index i such that $in(f) = in(p_i g_i)$. Hence, $in(f) = in(p_i) in(g_i) \in \langle in(G) \rangle$.

So, $\langle in(\mathcal{I}) \rangle \subseteq \langle in(G) \rangle \subseteq \langle in(\mathcal{I}) \rangle$, which means that $\langle in(\mathcal{I}) \rangle = \langle in(G) \rangle$, i.e., G is a Gröbner basis of $\mathcal{I}$. $\qquad \square$

Note, that the first definition of a Gröbner basis, introduced by Bruno Buchberger in his doctor thesis, was given in the form of Theorem 8.64.

There are other equivalent definitions of a Gröbner basis. For example, the following theorem can be also considered as an equivalent definition of a Gröbner basis.

✳ Theorem 8.65.

Let $\mathcal{I}$ be an ideal of the ring $R = K[x_1, x_2, \ldots, x_n]$. A finite subset $G = \{g_1, g_2, \ldots, g_k \mid g_i \in R\} \subset \mathcal{I}$ is a Gröbner basis of $\mathcal{I}$ if and only if for each polynomial $f \in \mathcal{I}$ there exists a polynomial $g_i \in G$ such that $in(g_i)$ divides $in(f)$.

Proof.

⇒. Let $G = \{g_1, g_2, \ldots, g_k \mid g_i \in R\} \subset \mathcal{I}$ be a Gröbner basis of $\mathcal{I}$ and $f \in \mathcal{I}$. From Theorem 8.62(2), it follows that the remainder on dividing f by G is equal to zero, i.e., the standard expression of f has the form (8.18). Therefore, $\text{in}(f) = \max_{1 \le i \le n} [\text{in}(p_i g_i)]$, i.e., there exists an index i such that $\text{in}(f) = \text{in}(p_i g_i) = \text{in}(p_i)\text{in}(g_i)$, which means that $\text{in}(g_i)$ divides $\text{in}(f)$.

⇐. Let $f \in \mathcal{I}$ and $\text{in}(g_i)|\text{in}(f)$. Then, $\text{in}(f) \in \langle \text{in}(G)\rangle$. Hence, $\langle \text{in}(\mathcal{I})\rangle \subseteq \langle \text{in}(G)\rangle \subseteq \langle \text{in}(\mathcal{I})\rangle$, which means that $\langle \text{in}(\mathcal{I})\rangle = \langle \text{in}(G)\rangle$, i.e., G is a Gröbner basis of $\mathcal{I}$. □

8.7. *S*-polynomials

In the previous section, it was shown that, for each ideal $\mathcal{I}$ of $K[x_1, x_2, \ldots, x_n]$ with respect to a monomial order $\le$, there always exists a Gröbner basis of $\mathcal{I}$. However, this leaves the important practical question: "How can we verify that a given set $G \subset \mathcal{I}$ is a Gröbner basis of $\mathcal{I}$?" The answer on this question is given by Buchberger's criterion, which was proved by Bruno Buchberger in his doctor thesis. This criterion is given in terms of so-called *S*-polynomials that will be considered in this section.

Let $f, g \in K[x_1, x_2, \ldots, x_n]$, and $\text{in}(f) = u^\alpha$, $\text{in}(g) = u^\beta$, where $\alpha = (\alpha_1, \alpha_2, \ldots, \alpha_n)$, $\beta = (\beta_1, \beta_2, \ldots, \beta_n)$. We define

$$\text{lcm}(\text{in}(f), \text{in}(g)) = u^\gamma, \tag{8.19}$$

where $\gamma = (\gamma_1, \gamma_2, \ldots, \gamma_n)$ and $\gamma_i = \max(\alpha_i, \beta_i)$, and call it the **least common multiple** of monomials $\text{in}(f)$ and $\text{in}(g)$.

✳ Example 8.66.

Find the least common multiple of monomials $\text{in}(f)$ and $\text{in}(g)$ if $f = 4x^2y^2z + xz^2$ and $g = -3xyz^3 + 2xz$ with respect to $\le_{\text{lex}}$ and $z <_{\text{lex}} y <_{\text{lex}} x$.

We have $\text{in}(f) = x^2y^2z$, $\text{in}(g) = xyz^3$. Therefore, $\text{lcm}(\text{in}(f), \text{in}(g)) = x^2y^2z^3$.

✳ Definition 8.67.

Let $f, g \in K[x_1, x_2, \ldots, x_n]$ and $M = \text{lcm}(\text{in}(f), \text{in}(g))$, $m_1 = \dfrac{M}{\text{lt}(f)}$, $m_2 = \dfrac{M}{\text{lt}(g)}$. The polynomial

$$S[f, g] = m_1 f - m_2 g \tag{8.20}$$

is called the **S-polynomial** of f and g.

✳ Example 8.68.

Determine the *S*-polynomial of f and g if $f(x, y) = xy - 2y$ and $g(x, y) = y^2 - x^2$ with respect to the lexicographic order $\le_{\text{lex}}$ and $y <_{\text{lex}} x$.

Here, we have: $\text{in}(f) = \text{lt}(f) = xy$, $\text{lt}(g) = -x^2$, $\text{in}(g) = x^2$. Then, $M = \text{lcm}(\text{in}(f), \text{in}(g)) = \text{lcm}(xy, x^2) = x^2 y$

Therefore, $S[f,g] = \dfrac{M}{xy} f(x,y) - \dfrac{M}{-x^2} g(x,y) = xf(x,y) + yg(x,y) = x(xy - 2y) + y(2y^2 - x^2) = -2xy + 2y^3$ and $\text{in}(S) = xy$.

✳ Example 8.69.

Determine the S-polynomial of f and g if $f(x,y) = x^3 y^2 - x^2 y^3 + x$ and $g(x,y) = 3x^4 y + y^2$ with respect to the lexicographic order $\leq_{\text{lex}}$ and $y <_{\text{lex}} x$.

Here, we have $\text{lt}(f) = \text{in}(f) = x^3 y^2$; and
$\text{lt}(g) = 3x^4 y$; $\text{in}(g) = x^4 y$.
$M = \text{lcm}(x^3 y^2, x^4 y) = x^4 y^2$
$m_1 = \dfrac{M}{\text{lt}(f)} = x$, $m_2 = \dfrac{M}{\text{lt}(g)} = \dfrac{1}{3} y$

Therefore, $S[f,g] = m_1 f(x,y) - m_2 g(x,y) = xf(x,y) - \dfrac{1}{3} yg(x,y) = -x^3 y^3 + x^2 - \dfrac{1}{3} y^3$ and $\text{in}(S) = -x^3 y^3$.

✳ Example 8.70.

Determine the S-polynomial of f and g if $f(x,y,z) = x + 2y - z$ and $g(x,y,z) = -3x - y + 4z$ with respect to the monomial order $\leq$ and $z < y < x$.

We have: $\text{lt}(f) = \text{in}(f) = x$ and $\text{lt}(g) = -3x$, $\text{in}(g) = x$. Then,

$M = x$; $m_1 = 1$, $m_2 = -\dfrac{1}{3}$.

Therefore, $S[f,g] = f + \dfrac{1}{3} g = \dfrac{5}{3} y + \dfrac{1}{3} z$.

✳ Example 8.71.

Determine $S[f,g]$, where $f(x) = 2x^3 - 7x^2 + 9x - 4$ and $g(x) = 2x^2 - 5x + 4$.

Here, we have: $\text{lt}(f) = 2x^3$, $\text{in}(f) = x^3$ and $\text{in}(g) = 2x^2$, $\text{in}(g) = x^2$. Then,

$M = x^3$; $m_1 = \dfrac{1}{2}$, $m_2 = \dfrac{1}{2} x$

Then, $S[f,g] = \dfrac{1}{2} f(x) - \dfrac{1}{2} xg(x) = -x^2 + \dfrac{5}{2} x - 2$.

It is easy to see that, in this case, $S[f,g] = \gcd(f,g)$.

Remind that if G is a set of polynomials of the ring $R = K[x_1, x_2, \ldots, x_n]$ then the ideal $\langle G \rangle$ generated by G consists of linear combinations of polynomials of G, i.e.,

$$\langle G \rangle = \{p_1 f_1 + p_2 f_2 + \cdots + p_s f_s \; : \; f_i \in G, \; p_i \in R\}.$$

So from the definition of S-polynomials, we obtain the following statement:

✱ Proposition 8.72.

Let G be a finite non-empty set of polynomials of the ring $R = K[x_1, x_2, \ldots, x_n]$ and let $\langle G \rangle$ be an ideal of R generated by G. If $f_1, f_2 \in \langle G \rangle$, then $S[f_1, f_2] \in \langle G \rangle$.

✳ Lemma 8.73.

Let $f, f_i \in K[x_1, x_2, \ldots, x_n]$, $\mathrm{in}(f_i) = u^\beta \in T_n$ for all $i = 1, \ldots, m$. Assume that $f = \sum\limits_{i=1}^{m} c_i f_i$, where $u^\beta \in T_n$, $c_i \in K$. If $\mathrm{in}(f) < u^\beta$, then f can be written in the form:

$$f = \sum_{j,k} b_{jk} S[f_j, f_k],$$

where $b_{jk} \in K$.

Proof. First, we show the proof of this lemma for $m = 2$. Let $f = c_1 f_1 + c_2 f_2$ and $\mathrm{in}(f_1) = \mathrm{in}(f_2) = u^\beta$. We assume, for simplicity, that $\mathrm{lc}(f_1) = \mathrm{lc}(f_2) = 1$. In this case, $\mathrm{lcm}(\mathrm{in}(f_1), \mathrm{in}(f_2)) = u^\beta$ and $S[f_1, f_2] = f_1 - f_2$. Since $\mathrm{in}(f) < u^\beta$, $c_1 + c_2 = 0$, by Lemma 8.48. Therefore, $f = c_1 f_1 + c_2 f_2 = c_1(f_1 - f_2) + (c_1 + c_2)f_2 = c_1 S[f_1, f_2]$.

Let $m > 2$. Then, assuming that $\mathrm{lc}(f_i) = 1$ for all i, and taking into account that $\sum\limits_{i=1}^{m} c_i = 0$, by Lemma 8.48, we obtain that $f = c_1 f_1 + c_2 f_2 + \cdots + c_m f_m = c_1(f_1 - f_2) + (c_1 + c_2)(f_2 - f_3) + \cdots + (c_1 + c_2 + \cdots c_{m-1})(f_{m-1} - f_m) + (c_1 + c_2 + \cdots c_{m-1} + c_m)f_m = c_1 S[f_1, f_2] + (c_1 + c_2)S[f_2, f_3] + \cdots + (c_1 + c_2 + \cdots + c_{m-1})S[f_{m-1}, f_m]$.

If $\mathrm{lc}(f_i) = a_i \in K$, then $S[f_i, f_j] = \dfrac{1}{a_i} f_i - \dfrac{1}{a_j} f_j$ and $f = d_1 g_1 + d_2 g_2 + \cdots + d_m g_m$, where $d_i = c_i a_i \in K$ and $g_i = \dfrac{1}{a_i} f_i$. Then, $\mathrm{lc}(g_i) = 1$ and we have the previous case. $\square$

We are now ready to give the famous criterion that shows whether a finite non-empty subset G of an ideal $\mathcal{I}$ of $K[x_1, x_2, \ldots, x_n]$ is a Gröbner basis of $\mathcal{I}$. This criterion gives the answer in terms of S-polynomials of G.

✳ Theorem 8.74 (Buchberger's criterion).

Let $\mathcal{I}$ be an ideal of $K[x_1, x_2, \ldots, x_n]$ with basis $G = \{g_1, g_2, \ldots, g_m\}$. Then, G is a Gröbner basis of $\mathcal{I}$ if and only if the remainder on dividing every S-polynomial $S[g_i, g_j]$ by G is equal to zero, for all i, j with $i \neq j$.

Proof.

$\Rightarrow$. Let $G = \{g_1, g_2, \ldots, g_m\}$ be a Gröbner basis of an ideal $\mathcal{I} = \langle G \rangle$. Then, by Proposition 8.72, $S[g_i, g_j] \in \mathcal{I}$ for all polynomials $g_i, g_j \in G$. Hence, by Theorem 8.64, the remainder on dividing $S[g_i, g_j]$ by the set G is equal to zero.

$\Leftarrow$. Let $\mathcal{I} = \langle G \rangle$, where $G = \{g_1, g_2, \ldots, g_m\}$. If $f \in \mathcal{I}$, then f can be written in the form $f = p_1 g_1 + p_2 g_2 + \cdots + p_m g_m$. It is obvious that there are many ways to write f in such a form. We chose the expression for which $v = \max\limits_{1 \leq i \leq n} [\mathrm{in}(p_i g_i)]$ is the least.

By Lemma 8.48, $\text{in}(f) \le v$. If we have an equality, then there is an index i such that $\text{in}(f) = \text{in}(p_i g_i) = \text{in}(p_i)\text{in}(g_i)$, i.e., $\text{in}(f)$ is divisible by $\text{in}(g_i)$ for some i, and so, by Theorem 8.65, G is a Gröbner basis of $\mathcal{I}$.

Suppose that we have a strictly inequality $\text{in}(f) < v$. This means that there is a non-empty set $H = \{i \ : \ g_i \in G \ : \ \text{in}(p_i g_i) = v\}$ with $\#(H) > 1$. We consider a polynomial $h = \sum_{i \in H} t_i$, where $t_i = \text{lt}(p_i)g_i$. Then, $\text{in}(t_i) = v$ for all $i \in H$ and $\text{in}(h) < v$, since $\text{in}(f) < v$. Therefore, we can apply Lemma 8.73, and so there exist $b_{jk} \in K$ such that h can be written in the form:

$$h = \sum_{\substack{i,j \in H, \\ i \ne j}} b_{ij} S[t_i, t_j]. \tag{8.21}$$

Since $\text{in}(p_i g_i) = \text{in}(p_j g_j) = v$, $M_{ij} = \text{lcm}[\text{in}(p_i g_i), \text{in}(p_j g_j)] = v$ for all $i, j \in H$. Therefore,

$$S[t_i, t_j] = \frac{v}{\text{lt}(t_i)} t_i - \frac{v}{\text{lt}(t_j)} t_j = \frac{v}{\text{lt}(t_i)} \text{lt}(p_i)g_i - \frac{v}{\text{lt}(t_j)} \text{lt}(p_j)g_j =$$

$$= \frac{v}{\text{lt}(g_i)} g_i - \frac{v}{\text{lt}(g_j)} g_j,$$

since $\text{lt}(t_i) = \text{lt}(p_i)\text{lt}(g_i)$. Let $\text{lcm}[\text{in}(g_i), \text{in}(g_j)] = w_{ij} \in R$, then

$$S[g_i, g_j] = \frac{w_{ij}}{\text{lt}(g_i)} g_i - \frac{w_{ij}}{\text{lt}(g_j)} g_j.$$

Therefore,

$$S[t_i, t_j] = \frac{v}{w_{ij}} S[g_i, g_j],$$

for all $i, j \in H$. Since, by the assumption of the theorem, the remainder of an S-polynomial $S[g_i, g_j]$ on dividing by the set G is equal to zero, for all i, j, this is also true for $S[t_i, t_j]$ for all $i, j \in H$. Therefore, from the division with remainder we obtain that each $S[t_i, t_j]$ can be written in the form

$$S[t_i, t_j] = \sum_{k=1}^{m} h_{ijk} g_k \tag{8.22}$$

and so, by Lemma 8.48, we have:

$$\max_{1 \le k \le m} [\text{in}(h_{ijk} g_k)] = \text{in}(S[t_i, t_j]) < \max[\text{in}(p_i g_i), \text{in}(p_j g_j)] = v.$$

Substituting the expression (8.22) for $S[t_i, t_j]$ into (8.22) for h, and then, substituting h into f, we obtain the expression $f = \sum_{k=1}^{m} p'_i g_k$ with $\max_{1 \le k \le m} [\text{in}(p'_k g_k)] < v$. This contradicts the choice of the expression of f.

$\square$

8.8. Buchberger's Algorithm

In this section, we describe the original version of Buchberger's algorithm which is simple to use. It is not optimal, because an obtained Gröbner basis may not to be minimal, i.e., it can contain some number of unnecessary polynomials.

Buchberger's algorithm answers the question of what way, for a given finite set F of polynomials, can we find a finite set of polynomials G which is a Gröbner basis for the ideal $\langle \text{id}(F) \rangle$, i.e., $\langle \text{id}(F) \rangle = \langle \text{id}(G) \rangle$.

The implementation of Buchberger's algorithm is based on Theorem 8.73. At the beginning of the algorithm, we have a set of generators $F = \{f_1, f_2, \ldots, f_n\}$ of an ideal $\mathcal{I} = \langle f_1, f_2, \ldots, f_n \rangle$. First, we assume that $G = F$. On each step of the algorithm, we find a new basis of $\mathcal{I}$ beginning from G and computing an S-polynomial $S[f, g]$, for all pairs of polynomials $f, g \in G$ such that $f \neq g$, and then its remainder on dividing by the set H. If the reduced form h of some S-polynomial $S[f, g]$ is not equal to zero, then we add the polynomial h to G and obtain a new set $G := G \cup \{h\}$. In this case, we repeat our process with the new set G. The algorithm terminates when reduced forms of all S-polynomials of G with regard to the set G are equal to zero. Then, the obtained set G is a Gröbner basis of the ideal $\mathcal{I}$.

In what follows, we give the scheme of the original Buchberger's algorithm, which is given by Buchberger in his Ph.D. thesis.

Buchberger's algorithm

INPUT: $\mathcal{I} = \langle f_1, f_2, \ldots, f_k \rangle$; $F = \{f_1, f_2, \ldots, f_k\}$
OUTPUT: $G \subset \mathcal{I}$ is a Gröbner basis of $\mathcal{I}$
INITIALIZATION: $G := F$
$P := \{\{f_i, f_j\} \ : \ f_i, f_j \in G, \ i \neq j\}$
WHILE $P \neq \emptyset$ DO
    FOR each pair of polynomials $\{f, g\} \in P$ DO:
        $P := P \backslash \{f, g\}$
        Compute an S-polynomial $S[f, g]$.
        Compute the remainder h of $S[f, g]$ on dividing by the set G.
        IF $h \neq 0$ THEN
        $M := M \cup \{\{f, h\} \ : \forall f \in G\}$
        $G := G \cup \{h\}$, i.e., add h to the set G
        END IF
    END FOR
END DO
RETURN G

Since, on each step of this algorithm, the number of polynomials in the set G may decrease, we have the natural question: "Does Buchberger's algorithm always terminate?"

The answer to this question gives the following theorem.

✳ Theorem 8.75.

Buchberger's algorithm terminates in a finite number of steps and the obtained set G is a Gröbner basis of the ideal $\mathcal{I} = \langle F \rangle$ for a finite given set of polynomials F.

Proof.

On each step of Buchberger's algorithm, we compute the reduced form h of an S-polynomial. If the reduced form $h = 0$, then we set $H = G$, otherwise we get a new set $H = G \cup \{h\}$. In the last case, since $\mathrm{in}(h)$ is not divisible by any initial monomials of elements of G, we obtain $\langle \mathrm{in}(G) \rangle \subsetneq \langle \mathrm{in}(H) \rangle$.

Assume that Buchberger's algorithm does not terminate. Then, it generates a strictly increasing sequence of sets

$$H_1 \subset H_2 \subset \cdots \subset H_n \subset \cdots$$

and a strictly increasing sequence of ideals

$$\mathcal{I}_1 \subset \mathcal{I}_2 \subset \cdots \subset \mathcal{I}_n \subseteq \cdots$$

where $\mathcal{I}_n = \langle \mathrm{in}(H_n) \rangle$. However, this contradicts the Hilbert Basis theorem.
□

We show the implementation of Buchberger's algorithm on the following example.

✳ Example 8.76.

Consider the lexicographic order $\leq_{\mathrm{lex}}$ and $y <_{\mathrm{lex}} x$.

Find a Gröbner basis of the ideal $\mathcal{I} = \langle f_1, f_2 \rangle$, where $f_1 = xy - 2y$ and $f_2 = 2y^2 - x^2$.

Here, $F = \{f_1, f_2\}$, $\mathrm{lt}(f_1) = \mathrm{in}(f_1) = xy$, $\mathrm{lt}(f_2) = -x^2$ and $\mathrm{in}(f_2) = x^2$. First, we set $G = F$ and $H = G$.

1. Compute the S-polynomial $S[f_1, f_2]$:

   $M = \mathrm{lcm}(\mathrm{in}(f_1), \mathrm{in}(f_2)) = \mathrm{lcm}(xy, x^2) = x^2 y.$

   Then $S[f_1, f_2] = \dfrac{M}{xy} f_1(x, y) - \dfrac{M}{-x^2} f_2(x, y) = x f_1(x, y) + y f_2(x, y) =$
   $x(xy - 2y) + y(2y^2 - x^2) = 2y^3 - 2xy$ and $\mathrm{lt}(S) = -2xy$, $\mathrm{in}(S) = xy$.

2. Compute the remainder on dividing $S[f_1, f_2]$ by the set H:

   $\mathrm{in}(f_1) | \mathrm{in}(S)$

   $m_1 = \mathrm{lt}(S) / \mathrm{lt}(f_1) = -2$

   $h = S + 2f_1 = 2y^3 - 2xy + 2(xy - 2y) = 2y^3 - 4y \neq 0$ is the remainder on dividing of $S[f_1, f_2]$ by the set H.

   $\mathrm{lt}(h) = 2y^3$, $\mathrm{in}(h) = y^3$.

3. Since $h \neq 0$, we set $f_3 = h = 2y^3 - 4y$.

4. $G := G \cup \{f_3\} = \{f_1(x,y), f_2(x,y), f_3(x,y)\} = \{xy - 2y; -x^2 + 2y^2; 2y^3 - 4y\}$.

5. Since $H \neq G$, we repeat our process, setting $H := G = \{f_1(x,y), f_2(x,y), f_3(x,y)\}$.

6. Compute the S-polynomial $S[f_1, f_3]$:

$$M = \mathrm{lcm}(\mathrm{in}(f_1), \mathrm{in}(f_3)) = \mathrm{lcm}(xy, y^3) = xy^3$$

$$S[f_1, f_3] = \frac{M}{xy} f_1(x,y) - \frac{M}{2y^3} f_3(x,y) = y^2 f_1(x,y) - \frac{1}{2} x f_3(x,y) = 2xy - 2y^3.$$

Since $S[f_1, f_3] = 2xy - 2y^3 = 2(xy - 2y) - (2y^3 - 4y) = 2f_1(x,y) - f_3(x,y)$, its reduced form $h = 0$ upon division by H.

7. Compute the S-polynomial $S[f_2, f_3]$:

$$M = \mathrm{lcm}(\mathrm{in}(f_2), \mathrm{in}(f_3)) = \mathrm{lcm}(x^2, y^3) = x^2 y^3$$

$$S[f_2, f_3] = \frac{M}{-x^2} f_2(x,y) - \frac{M}{2y^3} f_3(x,y) = -y^3 f_2(x,y) - \frac{1}{2} x^2 f_3(x,y) = 2x^2 y - 2y^5.$$

Since $S[f_2, f_3] = -2y(-x^2 + 2y^2) - y^2(2y^3 - 4y) = -2y f_2(x,y) - y^2 f_3(x,y)$, its reduced form $h = 0$ upon division by H.

Therefore, $H = G$ and so $G = \{xy - 2y; -x^2 + 2y^2; 2y^3 - 4y\}$ is a Gröbner basis of the ideal $\mathcal{I}$.

8.9. Minimal and Reduced Gröbner Basis

Note, that a Gröbner basis is not determined uniquely. Moreover, in constructing a Gröbner basis, we do not require its minimality. If G is a Gröbner basis of an ideal $\mathcal{I}$, then it is easy to show that each finite set H such that $G \subset H \subset \mathcal{I}$ is also a Gröbner basis. For correctness of such non-minimality and to obtain some minimal uniquely determined Gröbner basis, we will use the method of reduction of a Gröbner basis. First of all, there are redundant polynomials in a Gröbner basis that can be eliminated.

The main question is: "How to remove redundant polynomials in a Gröbner basis?"

Let $G = \{g_1, g_2, \ldots, g_m\}$ be a Gröbner basis of an ideal $\mathcal{I}$. Since $\langle \mathrm{in}(\mathcal{I}) \rangle = \langle \mathrm{in}(g_1), \mathrm{in}(g_2), \ldots, \mathrm{in}(g_m) \rangle$, a polynomial g_i can be eliminated from the set G if there exists an index j such that $\mathrm{in}(g_i)$ is divisible by $\mathrm{in}(g_j)$. Then, we obtain a new basis $G \setminus \{g_i\}$.

> ✳ **Definition 8.77.**
> A Gröbner basis $G = \{g_1, g_2, \ldots, g_m\}$ of an ideal I is called **minimal** if $\text{in}(g_i)$ does not divide $\text{in}(g_j)$ for all $i, j = 1, 2, \ldots, m$ with $i \neq j$.

✳ **Example 8.78.**

Suppose that $G = \{xy - 2y; -x^2 + 2y^2; 2y^3 - 4y\}$ is a Gröbner basis of an ideal $\mathcal{I}$.

Here, we have $\text{in}(xy - 2y) = xy$, $\text{in}(-x^2 + 2y^2) = x^2$ and $\text{in}(2y^3 - 4y) = y^3$

Hence, G is a minimal Gröbner basis.

Example 8.79.

Suppose that $G = \{g_1, g_2, g_3, g_4, g_5\} = \{x^3 - 2xy; x^2y - 2y^2 + x; -x^2; -2xy; -2y^2 + x\}$ is a Gröbner basis with respect to the graded lexicographic order $\leq_{\text{grlex}}$.

Then, in this case, $\text{in}(g_1) = x^3$; $\text{in}(g_2) = x^2y$; $\text{in}(g_3) = x^2$; $\text{in}(g_4) = xy$; $\text{in}(g_5) = y^2$.

Since $\text{in}(g_1) = x\,\text{in}(g_3)$, the polynomial g_1 can be removed from the set G.

Since $\text{in}(g_2) = x\,\text{in}(g_4)$, the polynomial g_2 can be removed from the set G.

Therefore, $H = \{-x^2; -2xy; 2y^2 + x\}$ is a minimal Gröbner basis.

Note, that minimal Gröbner bases are not unique, i.e., for a given ideal $\mathcal{I}$, there are a number of different minimal Gröbner bases. However, they have a nice property:

> ✱ **Proposition 8.80.**
> Let $\mathcal{I}$ be an ideal of $K[x_1, x_2, \ldots, x_n]$ and with specified a monomial order $\leq$. Then, all minimal Gröbner bases have the same number of elements.

Proof. Suppose that $G = \{g_1, g_2, \ldots, g_m\}$ and $H = \{h_1, h_2, \ldots, h_k\}$ are two minimal Gröbner bases of an ideal $\mathcal{I}$. Without loss of generality, we can assume that $m \geq k$.

Since G and H are Gröbner bases of $\mathcal{I}$, $\langle\text{in}(\mathcal{I})\rangle = \langle\text{in}(G)\rangle = \langle\text{in}(H)\rangle$. Therefore, $\text{in}(g_1)$ divides some $\text{in}(h_i)$. We can assume that $i = 1$, i.e., $\text{in}(g_1)$ divides $\text{in}(h_1)$. Analogously, $\text{in}(h_1)$ divides some $\text{in}(g_j)$, therefore, $\text{in}(g_1)$ divides $\text{in}(g_j)$. Since G is a minimal Gröbner basis, this is possible if only $j = 1$ and $\text{in}(g_1) = \text{in}(h_1)$.

Continuing this process, we obtain that $\text{in}(g_i) = \text{in}(h_i)$ for $i = 1, \ldots, m$. If $m > k$, then we obtain that $\text{in}(g_{k+1})$ divides some $\text{in}(h_j) = \text{in}(g_j)$, $j \neq k+1$. Therefore, $\text{in}(g_{k+1}) = \text{in}(g_j)$ for $j \neq k+1$, that contradicts the minimality of G. Hence, $m = k$. $\qquad\square$

> ✳ **Definition 8.81.**
> A Gröbner basis $G = \{g_1, g_2, \ldots, g_k\}$ of an ideal $\mathcal{I}$ is called **reduced** if it satisfies the following conditions:
> 1) $\text{lc}(g_i) = 1$ for all $g_i \in G$;

2) no term $u \in \text{supp}(g_i)$ is divisible by $\text{in}(g_j)$ for all $i, j = 1, 2, \ldots, m$ with $i \neq j$.

In order to construct a reduced Gröbner basis of an ideal $\mathcal{I}$, we can use the following reduction algorithm:

Reduction algorithm

1. Construct a Gröbner basis $G = \{g_1, g_2, \ldots, m\}$ of $\mathcal{I}$.
2. Remove all $g_i \in G$ for which $\text{in}(g_i)$ is divisible by some $\text{in}(g_j)$ for $g_j \in G \backslash \{g_i\}$. So that, we obtain a set $G' = \{g_1', \ldots, g_s'\} \subseteq G$ which is a minimal Gröbner basis of $\mathcal{I}$.
3. Each $g_i' \in G'$ scales by $\dfrac{1}{\text{lc}(g_i')}$ making it monic, i.e., $h_i = \dfrac{g_i'}{\text{lc}(g_i')}$ with $\text{lc}(h_i) = 1$. Let $H = \{h_1, h_2, \ldots, h_s\}$
4. We change step by step each polynomial h of H by its reduced form on dividing h by $H \backslash \{h\}$ using the following scheme:
   1) r_1 is the remainder of h_1 on dividing by the set $H_1 = H \backslash \{h_1\}$.
   2) r_2 is the remainder of h_2 on dividing by the set $H_2 = (H_1 \backslash \{h_2\}) \cup \{r_1\}$.
   3) r_3 is the remainder of h_3 on dividing by the set $H_3 = (H_2 \backslash \{h_3\}) \cup \{r_2\}$.

$$\cdots \cdots \cdots \cdots \cdots \cdots \cdots \cdots \cdots \cdots \cdots \cdots \cdots$$

4) r_s is the remainder of h_s on division by the set $H_s = (H_{s-1} \backslash \{h_s\}) \cup \{r_{s-1}\}$.

The obtained set $R = H_s = \{r_1, r_2, \ldots, r_s\}$ is a reduced Gröbner basis of $\mathcal{I}$.

✳ Theorem 8.82.
Each ideal $\mathcal{I} \subset K[x_1, x_2, \ldots, x_n]$ has the unique reduced Gröbner basis with respect to the specified monomial ordering.

Proof.
Suppose that $G = \{g_1, g_2, \ldots, g_m\}$ and $H = \{h_1, h_2, \ldots, h_m\}$ are two reduced Gröbner bases of $\mathcal{I}$. We can assume that $\text{in}(g_i) = \text{in}(h_i)$ for all $i = 1, \ldots, m$.

Consider $g_i - h_i$. Since $G \subset \mathcal{I}$ and $H \subset \mathcal{I}$, $g_i - h_i \in \mathcal{I}$ for all i. If $g_i - h_i = 0$, then we are done. Suppose that $g_i - h_i \neq 0$ for some i. Since $g_i - h_i \in \mathcal{I}$ and G is a Gröbner basis of $\mathcal{I}$, the remainder of $g_i - h_i$ on dividing by G is zero, and so $\text{in}(g_i - h_i) = \max\limits_{1 \leq k \leq m} [\text{in}(p_k g_k)]$, by Theorem 8.54. Therefore, there is an index j such that $\text{in}(g_i - h_i)$ is divisible by $\text{in}(g_j)$. Since $\text{lc}(g_i) = \text{lc}(h_i) = 1$, $\text{lt}(g_i) = \text{lt}(h_i)$. Therefore, there exists a term of $g_i - h_i$ that belongs to $\text{supp}(g_i) \cup \text{supp}(h_i)$ and which is divisible by $\text{in}(g_j)$. However, this contradicts the definition of the reduced Gröbner basis. $\qquad \square$

✳ Example 8.83.
$H = \{x^2; xy; y^2 + 1/2x\}$ is the reduced Gröbner basis.

✳ **Example 8.84.**

Let $\mathcal{I} = \langle y, x + y, x^2 + xy + y^2 \rangle$ be an ideal in $K[x, y]$. Then, a Gröbner basis of $\mathcal{I}$ with respect to the graded lexicographic order $\leq_{\text{grlex}}$ and $y <_{\text{grlex}} x$ is $G = \{x, y, x + y, y^2, x^2 + xy + y^2\}$.

Here, $\text{in}(x) = x$, $\text{in}(y) = y$, $\text{in}(x + y) = x$, $\text{in}(y^2) = y^2$, $\text{in}(x^2 + xy + y^2) = x^2$. Since $\text{in}(x)|\text{in}(x + y)$, $\text{in}(y)|\text{in}(y^2)$, and $\text{in}(x)|\text{in}(x^2 + xy + y^2)$, we can eliminate from G the polynomials $x + y$, y^2 and $x^2 + xy + y^2$. So, $H = \{x, y\}$ is a minimal Gröbner basis of $\mathcal{I}$. It is also a reduced Gröbner basis of $\mathcal{I}$.

8.10. Applications of Gröbner Bases

A number of different problems from many fields of mathematics can be reduced to the problem of finding a Gröbner basis. The method of Gröbner bases gives a uniform approach for solving systems of polynomial equations and linear diophantine equations with polynomial coefficients, and it is used in many different fields of science, such as commutative and non-commutative algebra, invariant theory, coding theory, statistics, symbolic calculation and many others.

In algebra, Gröbner bases are often used to solve systems of polynomial equations.

Let F be a finite subset in $K[x_1, x_2, \ldots, x_n]$. An **affine variety** (or **affine algebraic set**) of F is a set of all common zeros of its elements:

$$V_K\{F\} = \{y \in K^n \ : \ f(y) = 0 \ \text{ for all } f \in F\}. \tag{8.23}$$

If $F = \{f_1, f_2, \ldots, f_m\}$ and $\mathcal{I} = \langle f_1, f_2, \ldots, f_m \rangle$ is an ideal of the ring $K[x_1, x_2, \ldots, x_n]$ generated by all elements of F, then

$$V(\mathcal{I}) = \{y \in K^n \ : \ f(y) = 0 \ \text{ for all } f \in \mathcal{I}\}. \tag{8.24}$$

It is easy to see that

$$V_K(\mathcal{I}) = V_K(f_1, f_2, \ldots, f_m) = V_K(F) \tag{8.25}$$

Hence, the set F itself is not so important, but the ideal $\mathcal{I}$ is generated by this set. In particular, if $\mathcal{I} = \langle G \rangle$, where G is a Gröbner basis of $\mathcal{I}$, we can change generators $f_1, f_2, \ldots, f_m$ of the ideal $\mathcal{I}$ by generators which form a Gröbner basis G and we obtain that:

$$V_K(\mathcal{I}) = V_K(\langle G \rangle) = V_K(G) \tag{8.26}$$

One of the important questions is "How to verify whether an affine variety $V_K(F)$ is an empty set?" The answer to this question gives the following corollary of Hilbert Zeros Theorem, which is also called the Weak Hilbert Zeros Theorem.

> ✻ **Theorem 8.85.**
> If K is an algebraically closed field (e.g., $K = \mathbb{C}$) and $\mathcal{I}$ is an ideal of the ring $K[x_1, x_2, \ldots, x_n]$, then an affine variety $V_K(\mathcal{I})$ is an empty set if and only if $1 \in \mathcal{I}$, i.e., $\mathcal{I} = K[x_1, x_2, \ldots, x_n]$.

From this theorem, taking into account the equality (8.26), we get the following corollary:

> ✻ **Theorem 8.86.**
> If K is an algebraically closed field and G is a Gröbner basis of an ideal $\mathcal{I}$ of the ring $K[x_1, x_2, \ldots, x_n]$, then an affine variety $V_K(\mathcal{I})$ is an empty set if and only if $1 \in \langle G \rangle$.

Note that if $F = \{f_1, f_2, \ldots, f_m\}$, then $V_K(F)$ is a set of solutions of the system of equations:

$$
\begin{cases}
f_1(x_1, x_2, \ldots, x_n) & = & 0 \\
f_2(x_1, x_2, \ldots, x_n) & = & 0 \\
\cdots\cdots\cdots\cdots\cdots\cdots & \cdots & \cdots \\
f_m(x_1, x_2, \ldots, x_n) & = & 0
\end{cases}
\tag{8.27}
$$

Another important question is: "How to verify that a given system of polynomial equations has a solution? And if it has a solution, then how many solutions does it have?" The answer to these questions can be found using the following theorem, which is a corollary from Theorem 8.85.

> ✻ **Theorem 8.87.**
> Let K be an algebraically closed field, $f_1, f_2, \ldots, f_m \in K[x_1, x_2, \ldots, x_n]$. A system of polynomial equations (8.27) has a solution in K^n if and only if the ideal $\mathcal{I} = \langle f_1, f_2, \ldots, f_m \rangle$ in $K[x_1, x_2, \ldots, x_n]$ is proper, i.e., $1 \notin \mathcal{I}$.

✻ **Example 8.88.**

Find the points of intersection between two ellipses defined by the equations:

$$2x^2 - 6x + y^2 - 2y + 4 = 0$$

$$x^2 - 3x + 2y^2 - 4y + 2 = 0$$

This problem is equivalent to solving the system of equations:

$$
\begin{cases}
2x^2 - 6x + y^2 - 2y + 4 & = & 0 \\
x^2 - 3x + 2y^2 - 4y + 2 & = & 0
\end{cases}
$$

To solve this system, we consider the ideal $\mathcal{I} = \langle f, g \rangle$, where $f = 2x^2 - 6x + y^2 - 2y + 4$ and $g = x^2 - 3x + 2y^2 - 4y + 2$. We find a Gröbner basis of this ideal with the lexicographic order $\leq_{\text{lex}}$ and $y \leq_{\text{lex}} x$. Here, $\text{in}(f) = x^2$ and $\text{in}(g) = x^2$. Using Buchberger's algorithm, we obtain that $S[f, g] = f - 2g = -3y^2 + 6y$. Therefore, we add the polynomial $h = y^2 - 2y$

to the list of generators of the ideal $\mathcal{I}$. Using Buchberger's algorithm again to the set $G_1 = \{f, g, h\}$, we obtain:

$$S[f, h] = y^2 f - 2x^2 h = 4x^2 y - 6xy^2 + y^4 - 2y^3 + 4y^2 = 4y(x^2 - 3x + 2y^2 - 4y + 2) +$$

$$+ (-6x + y^2 - 8y + 4)(y^2 - 2y) = 4yg + (-6x + y^2 - 8y + 4)h$$

and

$$S[g, h] = y^2 g - x^2 h = 2x^2 y - 3xy^2 + 2y^4 - 4y^3 + 2y^2 = 2y(x^2 - 3x + 2y^2 - 4y + 2) +$$

$$+ (-3x + 2y^2 - 4y + 2)(y^2 - 2y) = 2yf + (-3x + 2y^2 - 4y + 2)h.$$

Hence, $G_1 = \{2x^2 - 6x + y^2 - 2y + 4, x^2 - 3x + 2y^2 - 4y + 2, y^2 - 2y\}$ is a Gröbner basis of the ideal $\mathcal{I}$. Since $in(f) = in(g) = x^2$ and $x^2 - 3x + 2y^2 - 4y + 2 = 2(y^2 - 2y) + (x^2 - 3x + 2)$, $G = \{x^2 - 3x + 2, y^2 - 2y\}$ is the reduced Gröbner basis of the ideal $\mathcal{I}$. Therefore, the equivalent system of equations has the following form:

$$\begin{cases} x^2 - 3x + 2 &= 0 \\ y^2 - 2y &= 0 \end{cases}$$

which has four solutions: $u_1 = (1, 0)$, $u_2 = (1, 2)$, $u_3 = (2, 0)$, $u_4 = (2, 2)$.

✳ Example 8.89.
Solve the system of equations:

$$\begin{cases} f(x) = x^3 + 2x^2 - 5x + 2 &= 0 \\ g(x) = x^2 + 3x - 4 &= 0 \end{cases}$$

Here, $in(f) = x^3$ and $in(g) = x^2$. So, $lcm(f, g) = x^3$.
Using Buchberger's algorithm, we obtain that
$S[f, g] = f(x) - xg(x) = -x^2 - 5x + 6$, and $in(S) = x^2$ is divisible by $in(g) = x^2$. Therefore,
$S[f, g] = -g(x) - 2x + 2$ and $h(x) = -2(x - 1) \neq 0$ is the remainder of f on dividing by $g(s)$. So, $G = \{f(x) = x^3 + 2x^2 - 5x + 2, g(x) = x^2 + 3x - 4, h(x) = -2(x - 1)\}$ is a Gröbner basis of the ideal $\mathcal{I} = \langle f, g \rangle$. It is easy to see that $f(x)$ and $g(x)$ are divisible by $h(x)$, and so the reduced Gröbner basis is $G = \{x - 1\}$. Thus, $\mathcal{I} = \langle f, g \rangle = \langle x - 1 \rangle$ and the system of equations $f = g = 0$ is equivalent to the equation $x - 1 = 0$ that have the unique solution $x = 1$.

Applying the Euclidean algorithm in order to find the greatest common divisor of polynomials f, g, it is easy to show that $lcd(f, g) = x - 1$, i.e., $G = \{lcd(f, g)\} = \{x - 1\}$.

✳ Example 8.90.
Solve the system of linear equations:

$$\begin{cases} 3x - 6y - 2z &= 0 \\ 2x - 4y + 4w &= 0 \\ x - 2y - z - w &= 0 \end{cases}$$

with corresponding matrix:

$$A = \begin{pmatrix} 3 & -6 & -2 & 0 \\ 2 & -4 & 0 & 4 \\ 1 & -2 & -1 & -1 \end{pmatrix}$$

Here, we have the ideal $\mathcal{I} = \langle 3x - 6y - 2z; 2x - 4y + 4w; x - 2y - z - w \rangle$ with the monomial ordering: $w < z < y < x$.

Then, $G = \{3x - 6y - 2z; 2x - 4y + 4w; x - 2y - z - w; -4z - 12w\}$ is a Gröbner basis of the ideal $\mathcal{I}$, and $G_1 = \{x - 2y - z - w; z + 3w\}$ is a minimal Gröbner basis which corresponds row echelon form of the matrix A:

$$B = \begin{pmatrix} 1 & -2 & -1 & -1 \\ 0 & 0 & 1 & 3 \\ 0 & 0 & 0 & 0 \end{pmatrix}.$$

The set $G_2 = \{x - 2y + 2w; z + 3w\}$ is a reduced Gröbner basis which corresponds to the matrix in the reduced normal form of the matrix A:

$$C = \begin{pmatrix} 1 & -2 & 0 & 2 \\ 0 & 0 & 1 & 3 \\ 0 & 0 & 0 & 0 \end{pmatrix}$$

These examples show that Burchberger's algorithm is some kind of generalization of both the Euclidean algorithm for computing the greatest common divisors of polynomials in one variable and the Gaussian elimination algorithm for solving systems of linear equations.

The realization of algorithms for finding Gröbner bases can be found in the most mathematical systems of programming such as MATHEMATICS, MAPLE, MAGMA and others.

8.11. Exercises

1. Express the symmetric polynomial P in terms of the elementary symmetric polynomials if:

   1) $P(x, y, z) = x^4 + y^4 + z^4$

   2) $P(x, y, z) = xy^2 + xz^2 + yx^2 + yz^2 + zx^2 + zy^2$

   3) $P(x, y, z) = x^2y^2 + x^2z^2 + y^2z^2$

2. Verify whether or not $G = \{g_1, g_2\}$, where $g_1 = x^3 + 2x^2 - 5x + 2$, $g_2 = x^2 + 3x - 4$ is a Gröbner basis.

3. Verify whether or not $G = \{g_1, g_2, g_3, g_4\}$, where $g_1 = y^2 + z^2$, $g_2 = x^2y + yz$, $g_3 = z^3 + xy$, $g_4 = x^2z^2 - y^2z$ is a Gröbner basis.

4. Using Buchberger's algorithm, find a minimal Gröbner basis for the set $F = \{x^2y - 1, xy^2 - x\}$ with the graded lexicographic order $\leq_{\text{grlex}}$ and $y <_{\text{grlex}} x$.

5. Using Buchberger's algorithm, find a Gröbner basis for the ideal generated by polynomials $f_1 = x^3 - 3xy$, $f_2 = x^2y - 2y^2 + x$ with the graded lexicographic order $\leq_{\text{grlex}}$ and $z <_{\text{grlex}} y <_{\text{grlex}} x$.

6. Using Buchberger's algorithm, find the reduced Gröbner basis for the ideal $\mathcal{I} = \langle x^2 + y^2 + z^2 - 1, x^2 + y^2 + z^2 - 2x, 2x - 3y - z \rangle$ with the graded lexicographic order $\leq_{\text{grlex}}$ and $z <_{\text{grlex}} y <_{\text{grlex}} x$.

7. Using Buchberger's algorithm, find a minimal Gröbner basis for the ideal $\mathcal{I} = \langle x^2 + y^2 + z^2 - 1, x^2 - y + z^2, x - z \rangle$ with the graded lexicographic order $\leq_{\text{grlex}}$ and $z <_{\text{grlex}} y <_{\text{grlex}} x$.

8. Find a Gröbner basis for the ideal $\mathcal{I} = \langle x^2 + y, x^4 + 2x^2y + y^2 + 3 \rangle$ with the graded lexicographic order $\leq_{\text{grlex}}$ and $z <_{\text{grlex}} y <_{\text{grlex}} x$.

9. Verify whether or not the polynomial $x^3 + x + 1$ belongs to the ideal $\mathcal{I} = \langle xz, xy - z, yz - x \rangle$.

10. Verify whether or not the polynomial

    (a) $h = -4x^2y^2z^2 + y^6 + 3z^5$

    (b) $f = xy - 5z^2 + x$

    belongs to the ideal $\mathcal{I} = \langle xz - y^2, x^3 - z^2 \rangle$ with the lexicographic order $\leq_{\text{grlex}}$ and $z <_{\text{grlex}} y <_{\text{grlex}} x$.

11. Verify whether or not the polynomial

    (a) $h = xy^3 + y^5 - z^2 - z^3$

    (b) $f = -x^4 + xy + x^3yz - xz^2$

    belongs to the ideal $\mathcal{I} = \langle x^2y - z, -x^3 + y \rangle$ with the graded lexicographic order $\leq_{\text{grlex}}$ and $z <_{\text{grlex}} y <_{\text{grlex}} x$.

12. Solve the system of linear equations:

$$\begin{cases} 2x + 2y + 2z & = & 0 \\ -2x + 5y + 2z & = & 1 \\ 8x + y + 4z & = & -1 \end{cases}$$

using a) the Gaussian elimination method; b) the method of Gröbner bases.

13. Solve the system of linear equations:

$$\begin{cases} -2y + 3z & = & 1 \\ 3x + 6y - 3z & = & -2 \\ 6x + 6y + 3z & = & 5 \end{cases}$$

using a) the Gaussian elimination method; b) the method of Gröbner bases.

14. Find $\gcd(f, g)$, where $f(x) = 3x^4 + x^3 + 2x^2 + 1$ and $g(x) = x^2 + 4x + 2$. Solve the system of equations:

$$\begin{cases} 3x^4 + x^3 + 2x^2 + 1 & = & 0 \\ x^2 + 4x + 2 & = & 0 \end{cases}$$

15. Solve the system of polynomial equations:

$$\begin{cases} x^2 + y^2 + z^2 & = & 1 \\ x^2 + y^2 + z^2 - 2x & = & 0 \\ 2x - 3y - z & = & 0 \end{cases}$$

References

[1] Adams, W.W. and P. Loustaunau. 1994. An introduction to Gröbner Bases, Amer. Math. Society, Providence.

[2] Becker, T. and V. Weispfenning. 1993. Gröbner Bases: A Computational Approach to Commutative Algebra, Springer-Verlag.

[3] Blum-Smith, B. and S. Coskey. 2017. The fundamental theorem on symmetric polynomials: History's Whiff of Galois theory. First Whiff of Galois Theory. The College Mathematics Journal 48(1): 18–29.

[4] Buchberger, B. 1965. Ein Algorithmus zum Auffinden der Basiselemente des Restklassenringes nach einen nulldimensionalen Polynomideal. Ph.D. Thesis, Inst. University of Innsbruck, Innsbruck, Austria.

[5] Buchberger, B. 1979. A criterion for detecting unnecessary reductions in the construction of Gröbner Bases. Lecture Notes in Computer Science, Vol. 72, Springer-Verlag, 3–21.

[6] Buchberger, B. 1983. A note on the complexity of constructing Gröbner bases. Lecture Notesin Comp. Sci. Vol. 162, Springer-Verlag, Berlin-New York, 137–145.

[7] Buchberger, B. 1985. Gröbner Bases: An algorithmic method in polynomial ideal theory. pp. 184–232. *In*: Bose, N.K. [ed.]. 1985. Recent Trends in Multidimensional System Theory, Reidel Publishing Company, Dordrecht.

[8] Buchberger, B. 1998. Introduction to Gröbner Bases. pp. 3-31. *In*: Buchberger, B. and F. Winker [eds.]. 1998. Gröbner Bases and Applications, Vol. 251 of London Mathematical Society Series Cambridge University Press.

[9] Buchberger, B. 2001. Gröbner Bases and systems theory. Multidimensional Systems and Sygnal Processing 12(3-4): 223–251.

[10] Buchberger, B. 2006. An algorithm for finding the basis elements of the residue class ring of a zero dimensional polynomial ideal. J. Symbolic Comp. 41(3-4): 475–511.

[11] Cox, D., J. Little and D. O'Shea. 1992. Ideals, Varieties, and Algorithms: An Introduction to Computational Algebraic Geometry and Commutative Algebra, Springer-Verlag.

[12] Daoub, H.E.S. 2012. The fundamental theorem on symmetric polynomials. The Teaching of Mathematics 15(1): 55–59.

[13] Dummit, D.S. and R.M. Foote. 2004. Abstract Algebra (3rd Ed.), John Wiley & Sons.

[14] Eisenbud, D. 1995. Commutative Algebra with a view toward Algebraic Geometry, Vol. 150 of GTM, Springer.

[15] Ene, V. and J. Herzof. 2012. Gröbner Bases in Commutative Algebra, AMS.

[16] Gianni, P., B. Trager and G. Zacharias. 1988. Gröbner bases and primary decomposition of polynomial ideals. J. Symb. Comp. 6: 149–167.

[17] Herzog, J. and T. Hibi. 2010. Monomial Ideals, Graduate Text in Mathematics, 260, Springer.

[18] Iyad, A. Ajwa, Zhuojun, Liu and Paul S. Wang. 1995. Gröbner Bases Algorithm, ICM Technical Reports.
http://symbolicnet.mcs.kent.edu/icm/reports/index1995.html.

[19] Joyner, D., R. Kreminski and J. Turisco. 2004. Applied Abstract Algebra, The Johns Hopkins University Press.

[20] Judson, Th.W. 1994. Abstract Algebra. Theory and Applications, PWS Publishing Company, Michigan.

[21] Lazard, D. 1983. Gröbner Bases, Gaussian Elimination and Resolution of Systems of Algebraic Equations. pp.146–157. *In:* Proc. Eurocal 83: 162.

[22] Lazard, D. 1985. Ideal bases and primary decomposition: Case of two variables. J. Symb. Comp. 1: 261–270.

[23] Lazard, D. 1993. Systems of algebraic equations (algorithms and complexity). *In:* Eisenbud, D. and L. Robbiano [eds.] 1993. Computational Algebraic Geometry and Commutative Algebra, Cambridge University Press, London.

[24] Sturmfels, B. 2005. What is a Gröbner Basis? Notices of the AMS 52(10): 2–3.

Chapter 9

Finite Fields and their Applications

"The most painful thing about mathematics is how far away you are from being able to use it after you have learned it."

James Newman

Finite fields, which are also called **Galois fields**, are very important algebraic structures. First, they were considered by Evariste Galois (1811-1832), who was one of the most outstanding mathematicians in history.

Galois fields play a significant role in various fields of mathematics: Number theory, group theory, projective geometry and others. They also have many interesting practical applications, in particular, in coding theory and cryptography. Galois fields are the basis for different cryptographic systems and correcting codes which are used in various applications of data transmission.

In this chapter, we consider some main properties of finite fields and their applications in coding theory and cryptographic algorithms, such as Diffie-Hellman's algorithm and the ElGamal algorithm.

9.1. Properties of Finite Fields

Some finite fields and their properties were considered in previous chapters. Examples of finite fields are:

- $\mathbb{Z}_p$, where p is a prime.

- A field containing q elements, where $q = p^n$, $n > 1$, p is a prime: F_2, F_4, F_8, ..., F_3, F_9, ..., F_5, F_{25},...,.

In this section, we show that there are no other finite fields up to isomorphism.

✻ Lemma 9.1.

In each field K of characteristic p, the following equality holds

$$(a + b)^p = a^p + b^p$$

for all $a, b \in K$.

Proof.

Let K be a field of characteristic p and $a, b \in K$. Then,

$$(a + b)^p = a^p + \sum_{i=1}^{p-1} \binom{p}{i} a^i b^{p-i} + b^p \tag{9.1}$$

Since the number $\binom{p}{i}$, for $i = 1, 2, \ldots, p-1$, is divided by p, by Theorem 6.6, we obtain that, in the field of characteristic p, the following equality holds:

$$\binom{p}{i} a^i b^{p-i} = 0 \quad \text{for} \quad i = 1, 2, \ldots, p - 1.$$

Hence, from (9.1), we obtain that $(a + b)^p = a^p + b^p$, as required. $\qquad \square$

✻ Lemma 9.2.

In each field K of characteristic p, the following equality holds

$$(a + b)^{p^n} = a^{p^n} + b^{p^n} \tag{9.2}$$

for all $a, b \in K$ and each natural number n.

Proof.

We perform the proof by mathematical induction on n. For $n = 0$, the equality (9.2) is trivial. If $n = 1$, the equality (9.2) coincides with (9.1). Suppose that the equality (9.2) is correct for each k, where $1 \le k \le n$. Then,

$$(a + b)^{p^{n+1}} = ((a + b)^p)^{p^n} = (a^p + b^p)^{p^n} = (a^p)^{p^n} + (b^p)^{p^n} = a^{p^{n+1}} + b^{p^{n+1}}.$$

This means that (9.2) is correct also for $n + 1$. $\qquad \square$

✻ Theorem 9.3.

If a field F_q has q elements and char $F_q = p$, then $q = p^n$.

Proof.

If $|F_q| = q$ and char $F_q = p$, then, by Theorem 6.7, the field F_q contains a subfield K isomorphic to the field $\mathbb{Z}_p$. Since the field F_q is an extension of the field K and F_q is a finite field, F_q can be considered as a vector space over $K \cong \mathbb{Z}_p$ with finite dimension $[F_q : K] = n$. Let $\alpha_1, \alpha_2, \ldots \alpha_n$ be a basis of the field F_q over K. Then,

$$F_q = \{c_1 \alpha_1 + c_2 \alpha_2 + \cdots + c_n \alpha_n \ : \ c_i \in K\}.$$

Since $|K| = p$, each element $c_i \in K$ has p different values. Therefore, $q = p^n$.
□

✳ Definition 9.4.
A finite field F_q, where $q = p^n$, is called the **Galois field** of order p^n and is denoted by $GF(p^n)$ or $GF(q)$.

> **✳ Theorem 9.5.**
> If F_q is a finite field, where $q = p^n$, then each element of F_q satisfies the equation $x^q - x = 0$ and F_q is a set of all roots of this equation. Inversely, for each power $q = p^n$ the splitting field of the polynomial $x^q - x$ over a field F_q is a field having q elements.

Proof.
Let $G = F_q^* = F_q \backslash \{0\}$ be the multiplicative group of the field F_q, then $|G| = q - 1$. If $\alpha \in G$, then, by Lagrange's theorem, $\alpha^{q-1} = 1$, so $\alpha^q = \alpha$ for all $\alpha \in F_q$. Therefore, each element of the field F_q is a root of the polynomial $f(x) = x^q - x$. Since $|F_q| = q$ and the number of roots of the polynomial $f(x)$ is $\leq q$, the polynomial $f(x)$ has exactly q different roots in F_q. This means that F_q is a splitting field of $f(x)$.

Inversely, let K be a splitting field of the polynomial $f(x) = x^q - x \in F_q[x]$, where $q = p^n$. Since $\text{char}\, F_q = p$ and $q = p^n$,

$$f'(x) = qx^{q-1} - 1 = -1 \neq 0.$$

By Corollary 5.51, this means that the polynomial $f(x)$ has no multiple roots in the splitting field, i.e., $f(x)$ has exactly q different roots. We show that all q roots of $f(x)$ form a splitting field of $f(x)$.

Let $a, b \in S$, where S is the set of all roots of $f(x)$. Hence, $a^q = a$ and $b^q = b$. Then, taking Lemma 9.2 into account, we obtain that:

$$(a + b)^q = a^q + b^q = a + b$$

$$(ab)^q = a^q b^q = ab$$

$$(a^{-1})^q = a^{-q} = (a^q)^{-1} = a^{-1},$$

hence, $a + b, ab, a^{-1} \in S$. Therefore, S is a field and $S \subseteq K$. Since each element of the field F_q is a root of the polynomial $g(x) = x^p - x$ and $g(x)|f(x)$, $F_p \subset S$. Therefore, we have $F_p \subset S \subseteq K$, that shows that S is a splitting field of the polynomial $f(x)$. Hence, $S = K$ and $|K| = q$. □

> **✳ Corollary 9.6.**
> For each prime p and each positive integer n there is one up to isomorphism the Galois field $F_q = GF(p^n)$, where $q = p^n$.

✳ Example 9.7.

Construct a field containing 4 elements.

Since $4 = 2^2$, it suffers to choose an irreducible polynomial $f(x) \in \mathbb{Z}_2[x]$ of degree 2. Let $f(x) = x^2 + x + 1 \in \mathbb{Z}_2[x]$. Since $f(0) = f(1) = 1$, $f(x)$ has no roots in $\mathbb{Z}_2$. This means that $f(x)$ is irreducible in $\mathbb{Z}_2$. Therefore,

$$GF(4) = \mathbb{Z}_2/\langle x^2 + x + 1\rangle = \mathbb{Z}(\alpha) = \{0, 1, \alpha, \alpha + 1 \ : \ \alpha^2 + \alpha + 1 = 0\}.$$

The tables of addition and multiplication in $GF(4)$ have the following forms:

+	0	1	α	$\alpha + 1$
0	0	1	α	$\alpha + 1$
1	1	0	$\alpha + 1$	α
α	α	$\alpha + 1$	0	1
$\alpha + 1$	$\alpha + 1$	α	1	0

$\cdot$	0	1	α	$\alpha + 1$
0	0	0	0	0
1	0	1	α	$\alpha + 1$
α	0	α	$\alpha + 1$	1
$\alpha + 1$	0	$\alpha + 1$	1	α

Note that $GF(4) \neq \mathbb{Z}_4$. Recall that $\mathbb{Z}_4 = \{0, 1, 2, 3\}$ is the ring of remainders modulo 4 with zero divisors and the table of multiplication in $\mathbb{Z}_4$ has the following form:

$\cdot$	0	1	2	3
0	0	0	0	0
1	0	1	2	3
2	0	2	0	2
3	0	3	2	1

✳ **Theorem 9.8.**

Each subfield of the Galois field $F_q = GF(p^n)$ has p^m elements, where $m|n$. Inversely, if $m|n$ for $m > 0$, then there is a uniquely determined subfield of $GF(p^n)$ isomorphic to the field $GF(p^m)$.

Proof.

Let K be a subfield of a field F_q. Then, K is an extension of a field E which is isomorphic to the field $\mathbb{Z}_p$, i.e., we have a sequence of fields: $E \subseteq K \subseteq F_q$. By Theorem 6.13:

$$[F_q : E] = [F_q : K][K : E],$$

hence, we obtain that $m|n$.

Inversely, if $m|n$ for $m > 0$ then $(p^m - 1)|(p^n - 1)$. Hence, it follows that the polynomial $x^{p^m - 1} - 1$ divides the polynomial $x^{p^n - 1} - 1$. Therefore, each root of the polynomial $x^{p^m} - x$ is a root of the polynomial $x^{p^n} - x$. So, the field F_q contains as a subfield the splitting field of the polynomial $x^{p^m} - x$ that isomorphic to the field $GF(p^m)$. □

9.2. Multiplicative Group of a Finite Field

In this section, we study the structure of multiplicative group F_q^* of a finite field F_q.

Let $F_q = GF(p^n)$, where $q = p^n$ and p is a prime. We show that there is an element $g \in F_q$ such that each non-zero element of F_q is a power of g, i.e., $F_q = \{0, 1, g, g^2, \ldots, g^{q-2}\}$ and $g^{q-1} = 1$.

✳ **Theorem 9.9.**

1) The multiplicative group F_q^* of a finite field F_q is cyclic of order $q - 1$.
2) If g is a generator of the multiplicative group F_q^* then g^k is also a generator of this group if and only if $\gcd(k, q - 1) = 1$.
3) There are exactly $\varphi(q - 1)$ distinct generators of the group F_q^*.

Proof.

Let $d | (q - 1)$. We denote by $G(d) \subseteq F_q^*$ the set of all elements F_q^* whose order is equal to d, and $\psi(d) = |G(d)|$. Then, an element $g \in G(d)$ and all its powers $1, g, g^2, \ldots, g^{d-1}$ are roots of the polynomial $f(x) = x^d - 1$, which has at most d roots in F_q. Moreover, each element of order d is a root of this polynomial.

By Theorem 2.27, $\operatorname{ord}(g^k) = d$ if and only if $\gcd(d, k) = 1$. Hence, it follows that $\psi(d) = \varphi(d)$ if $\psi(d) > 0$. We show that $\psi(d) > 0$ for each $d | (q - 1)$. Suppose that $\psi(d) = 0$ for some $d | (q - 1)$. Then, taking Theorem 7.8 into account, we have:

$$q - 1 = \sum_{d|(q-1)} \psi(d) < \sum_{d|(q-1)} \varphi(d) = q - 1.$$

This contradiction shows that $\psi(d) = \varphi(d)$ for each $d | (q - 1)$. In particular, $\psi(q - 1) = \varphi(q - 1)$, i.e., there are exactly $\varphi(q - 1)$ elements of the group F_q^* of order $q - 1$. Therefore, by Theorem 3.43, the group F_q^* is cyclic.

The parts 2 and 3 of this theorem follow from Theorem 3.45. □

In particular, if $q = p$ is a prime, from Theorem 9.9, we obtain the following corollary:

✳ **Corollary 9.10.**

1. If p is a prime, then $G = \mathbb{Z}_p^*$ is a cyclic group of order $p - 1$.
2. If g is a generator of the multiplicative group $\mathbb{Z}_p^*$, then g^k is a generator of the group $\mathbb{Z}_p^*$ if and only if $\gcd(k, q - 1) = 1$.
3. There are exactly $\varphi(p - 1)$ distinct generators of the group $\mathbb{Z}_p^*$.

✳ **Example 9.11.**

Consider the group $G = \mathbb{Z}_{13}^*$ which is cyclic by Theorem 9.9 and $|G| = 12$. The order of the element $g = 2 \in G$ is equal to $|G| = 12$. Hence, by Theorem 3.43, it follows that $g = 2$ is a generator of the group G, i.e.,

$$G = \{1, 2, 2^2, \ldots, 2^{12} (\operatorname{mod} 13) \ : \ 2^{13} \equiv 1 \, (\operatorname{mod} 13)\}.$$

Since $\varphi(12) = 4$ there are 4 distinct generators of the group G, namely: $2 = 2^1$, $2^5 \equiv 6 \,(\mathrm{mod}\,13)$, $2^7 \equiv 11 \,(\mathrm{mod}\,13)$, $2^{11} \equiv 7 \,(\mathrm{mod}\,13)$.

> * **Definition 9.12.**
> An element g of a finite field $F_q = GF(p^n)$ is called a **generator** (or a **primitive element**) of the field F_q if g is a generator of the multiplicative group F_q^*, i.e., $m = q - 1$ is the minimal positive integer such that $g^m = 1$.

From Theorem 9.5, it follows the construction algorithm of a finite field F_q, where $q = p^n$. To this aim, we should to construct a splitting field of the polynomial $f(x) = x^q - x \in F_p[x]$.

* Example 9.13.
Construct the field F_9.

Since $9 = 3^2$, by Theorem 9.4, each element of F_9 is a root of the polynomial $x^9 - x$ and F_9 is the set of all these roots. The field F_9 is a splitting field of the polynomial $x^9 - x$ over the field $F_3 = \mathbb{Z}_3$.

$$x^9 - x = x(x^8 - 1) = x(x^2 - 1)(x^2 + 1)(x^2 + x - 1)(x^2 - x - 1) =$$

$$= x(x + 1)(x + 2)(x^2 + 1)(x^2 + x + 2)(x^2 + 2x + 2).$$

Since $\varphi(9 - 1) = \varphi(2^3) = 4$, there are 4 distinct generators of the field F_9.

The elements $0, 1, 2$ are the roots of the polynomials x, $x + 2$ and $x + 1$, respectively. If α is a root of the polynomial $x^2 + 1$, then its order is equal to 4 but not 8. Therefore, α is not a generator of F_9. Therefore, α can be root of either the polynomial $x^2 + 2x + 2$ or $x^2 + 2x + 2$. Suppose that α is a root of the polynomial $x^2 + 2x + 2$. Then,

$$\alpha^1 = \alpha, \alpha^2 = \alpha + 1, \alpha^3 = 2\alpha + 1, \alpha^4 = 2, \alpha^5 = 2\alpha, \alpha^6 = 2\alpha + 2, \alpha^7 = \alpha + 2, \alpha^8 = 1.$$

Therefore, α is a generator of F_9 and so:

$$F_9 = \{0, \alpha^1 = \alpha, \alpha^2, \alpha^3, \alpha^4, \alpha^5, \alpha^6, \alpha^7, \alpha^8 = 1\}$$

The polynomial $x^2 + 2x + 2$ has 2 roots: α and $\alpha^3 = 2\alpha + 1$. Analogously, we can show that each root of the polynomial $x^2 + x + 2$ over F_3 is a generator of F_9.

9.3. Primitive Roots and Indexes. Discrete Logarithm Problem

In this section, we present one of the most difficult computational problems: The problem of discrete logarithm. First, we introduce the notions of primitive roots and indexes and prove some their basic properties.

Let $n \in \mathbb{Z}^+$ and $\mathbb{Z}_n^* = \{a \in \mathbb{Z}_n \; : \; \gcd(a, n) = 1\}$ be the multiplicative group of units modulo n. As was shown earlier, $|\mathbb{Z}_n^*| = \varphi(n)$.

※ **Definition 9.14.**
    The **order modulo** n of an element $a \in \mathbb{Z}_n^*$ is the smallest positive integer k such that
$$a^k \equiv 1 \,(\mathrm{mod}\, n)$$
and is denoted by $\mathrm{ord}_n(a)$ or $\mathrm{ord}(a, n)$.

✳ **Example 9.15.**
    Let $n = 11$, then:
    $\mathrm{ord}_{11}(2) = \mathrm{ord}_{11}(6) = \mathrm{ord}_{11}(7) = \mathrm{ord}_{11}(8) = 10$.
    $\mathrm{ord}_{11}(3) = \mathrm{ord}_{11}(4) = \mathrm{ord}_{11}(5) = \mathrm{ord}_{11}(9) = 5$.
    $\mathrm{ord}_{11}(10) = 2$.

Taking into account Theorem 2.43 and the results of section 2.2, we easily obtain the following theorem, which gives basic properties of the orders of elements of $\mathbb{Z}_n^*$.

✳ **Theorem 9.16.**
Let $n, m \in \mathbb{Z}^+$ and $a \in \mathbb{Z}_n^*$ with $\mathrm{ord}_n(a) = k$. Then,
1) If $a^m \equiv 1 \,(\mathrm{mod}\, n)$, then $k|m$.
2) $k|\varphi(n)$. In particular, if $n = p$ is a prime, then $\mathrm{ord}_p(a)|(p-1)$.
3) If $r, s \in \mathbb{Z}$, then $a^r \equiv a^s \,(\mathrm{mod}\, n)$ if and only if $r \equiv s \,(\mathrm{mod}\, k)$.
4) Elements $1, a, a^2, \ldots, a^{k-1}$ are distinct integers modulo n.
5) $\mathrm{ord}_n(a^m) = k/\gcd(k, m)$.
6) $\mathrm{ord}_n(a^m) = k$ if and only if $\gcd(k, m) = 1$.

From this theorem, it follows that $\mathrm{ord}_n(a) \leq \varphi(n)$ for all $a \in \mathbb{Z}_n^*$. Hence, we have a natural question: can the following equality hold $\mathrm{ord}_n(a) = \varphi(n)$? It turns out that "Yes". There are such integers a, n for which $\mathrm{ord}_n(a) = \varphi(n)$, and these numbers have wide applications.

※ **Definition 9.17.**
    Let $n \in \mathbb{Z}^+$. An element g of the multiplicative group $\mathbb{Z}_n^*$ is called a **primitive root modulo** n if and only if $\mathrm{ord}_n(g) = \varphi(n)$.

In a particular case, when $n = p$ is a prime, $\mathbb{Z}_p^*$ is a cyclic group. An element $g \in \mathbb{Z}_p^*$ is a primitive root modulo p if $\mathrm{ord}_p(g) = \varphi(p) = p - 1$. In this case, the element g is also a generator of the cyclic group $\mathbb{Z}_p^*$. Inversely, each generator of the cyclic group $\mathbb{Z}_p^*$ is a primitive root modulo p. Hence, we have the following theorem:

> **❋ Theorem 9.18.**
> Let p be a prime. Then, an element $g \in \mathbb{Z}_p^*$ is a primitive root modulo p if and only if g is a generator of this group, i.e.,
> $$\mathbb{Z}_p^* = \langle g \rangle = \{1, g, g^2, \ldots, g^{p-1} \ : \ g^p = 1\}$$

❋ Example 9.19.

Verify whether or not the integer 7 is a primitive root modulo 45.

We have that $\gcd(7, 45) = 1$ and $\varphi(n) = \varphi(45) = \varphi(3^2 \cdot 5) = \varphi(3^2)\varphi(5) = 24$.

Construct the following table:

k	0	1	2	3	4	5	6	7	8	9	10	11	12
7^k	1	7	4	28	16	22	19	43	31	37	34	13	1

From this table, we see that $\text{ord}_{45}(7) = 12$. Since $12 < 24$, the number 7 is not a primitive root modulo 45.

Note that, by Theorem 9.16(2), it suffers to verify not all integers but only the divisors of $\varphi(45) = 24$, i.e., integers 1, 2, 3, 4, 6 and 12.

❋ Example 9.20.

Verify whether or not the integer 2 is a primitive root modulo 13.

We have $\gcd(2, 13) = 1$ and $\varphi(n) = \varphi(13) = 12$. We will find the order of 2 modulo 13 among the divisors of 12: $\{1, 2, 3, 4, 6, 12\}$.

Since $2^2 \equiv 4 \,(\text{mod}\,13)$, $2^3 \equiv 8 \,(\text{mod}\,13)$, $2^4 \equiv 3 \,(\text{mod}\,13)$, $2^6 \equiv 12 \,(\text{mod}\,13)$, $2^{12} \equiv 1 \,(\text{mod}\,13)$, we obtain that $\text{ord}_{13}(2) = 12 = \varphi(13)$, i.e., 2 is a primitive root modulo 13.

❋ Example 9.21.

Verify whether or not the integer 7 is a primitive root modulo 46.

We have that $\gcd(7, 46) = 1$ and
$\varphi(n) = \varphi(46) = \varphi(2)\varphi(23) = 22$.

We will find the order of 7 modulo 46 among divisors of 22: $\{1, 2, 11, 22\}$.

Since $7^2 \equiv 3 \not\equiv 1 \,(\text{mod}\,46)$, $7^{11} \equiv 15 \not\equiv 1 \,(\text{mod}\,46)$ and $7^{22} \equiv 1 \,(\text{mod}\,46)$, we obtain that $\text{ord}_{46}(7) = 22 = \varphi(46)$, i.e., 7 is a primitive root modulo 46.

> **❋ Theorem 9.22.**
> If a is a primitive root modulo n, then
> $$a^x \equiv a^y \,(\text{mod}\,n) \Longleftrightarrow x \equiv y \,(\text{mod}\,\varphi(n)).$$

Proof.

1. Let $a^x \equiv a^y \,(\text{mod}\,n)$. Since a is a primitive root $(\text{mod}\,n)$, $\text{ord}_n(a) = \varphi(n)$. Then, from Theorem 9.16(3), it follows that $x \equiv y \,(\text{mod}\,\varphi(n))$.

2. Suppose that $x \equiv y \,(\mathrm{mod}\,\varphi(n))$. Then, $x = y + \varphi(n)s$ for some $s \in \mathbb{Z}$. Since, by Euler's Theorem (Theorem 7.9), $a^{\varphi(n)} \equiv 1 \,(\mathrm{mod}\,n)$, we get that

$$a^x \equiv a^{y+\varphi(n)s} \equiv a^y (a^{\varphi(n)})^s \equiv a^y \,(\mathrm{mod}\,n).$$

□

✳ Theorem 9.23.
If there is a primitive root modulo n, then there exist exactly $\varphi(\varphi(n))$ distinct primitive roots modulo n. In particular, if p is a prime, then there exist exactly $\varphi(p-1)$ distinct primitive roots modulo p.

Proof.
From Theorem 3.43, it follows that, if there is a primitive root modulo n, then the group $\mathbb{Z}_n^*$ is cyclic of order $\varphi(n)$. Therefore, by theorem 3.45(2), there are exactly $\varphi(\varphi(n))$ distinct primitive roots modulo n. In particular, if p is a prime, then the group $\mathbb{Z}_p^*$ is cyclic and it has generators which are primitive roots modulo p and their number is equal to $\varphi(p-1)$. □

✳ Example 9.24.
Let $p = 5$, then $p - 1 = 4$ and $\mathbb{Z}_5^* = \{1, 2, 3, 4\}$.
Since $\varphi(4) = 2$, we have two distinct primitive roots modulo 5:
1) $g = 2$ is a primitive root modulo 5, since:
$g^1 = 2$, $g^2 = 4$, $g^3 = 3$, $g^4 = 1$;
2) $h = 3$ is also a primitive root modulo 5, since:
$h^1 = 3$, $h^2 = 4$, $h^3 = 2$, $h^4 = 1$.

As it turns out, for most composite numbers, there are no primitive roots, since the group $\mathbb{Z}_n^*$ is not always cyclic. From the theory of groups, it is known that the group $\mathbb{Z}_n^*$ is cyclic if and only if $n = 2, 4, p^k, 2p^k$, where $p > 2$ is a prime. Hence, we obtain the following theorem that shows for which numbers n there exist primitive roots.

✳ Theorem 9.25.
Primitive roots modulo n exist if and only if $n = 2,\ 4,\ p^k,\ 2p^k$, where $p > 2$ is a prime and k is a positive integer.

✳ Example 9.26.
Find the number of primitive roots modulo $n = 15,\ 54,\ 64$.

1. If $n = 15 = 3 \cdot 5$, then there are no primitive roots modulo 15.

2. If $n = 54 = 2 \cdot 3^3$, then there exist primitive roots modulo 54 and their number is equal to $\varphi(\varphi(54)) = \varphi(18) = 6$.

3. If $n = 64 = 2^6$, then there are no primitive roots modulo 64.

So far, there is no simple general formula for computing primitive roots and there is no effective (in the polynomial time) algorithm for finding primitive roots.

✳ **Definition 9.27.**

Let $n \in \mathbb{Z}^+$ and $a \in \mathbb{Z}_n^*$ be a primitive root modulo n. If $b \in \mathbb{Z}_n^*$, then the smallest positive integer k such that $b \equiv a^k \,(\mathrm{mod}\,n)$ is called the **index** (or the **discrete logarithm**) of b modulo n to the base a and write $k = \mathrm{ind}_a b$ (or $k = \log_a b$), i.e.,

$$b \equiv a^{\mathrm{ind}_a b} \,(\mathrm{mod}\,n) \tag{9.3}$$

or

$$b \equiv a^{\log_a b} \,(\mathrm{mod}\,n) \tag{9.4}$$

✶ **Example 9.28.**

Find $\mathrm{ind}_2 11$ modulo 13.

We have that $\gcd(2, 13) = \gcd(11, 13) = 1$.

From Example 9.20, it follows that 2 is a primitive root modulo 13.

To find $\mathrm{ind}_2 11$, it is necessary to solve the congruence: $2^k \equiv 11 \,(\mathrm{mod}\,13)$.

Since $2^1 \equiv 2 \,(\mathrm{mod}\,13)$, $2^2 \equiv 4 \,(\mathrm{mod}\,13)$, $2^3 \equiv 8 \,(\mathrm{mod}\,13)$, $2^4 \equiv 3 \,(\mathrm{mod}\,13)$, $2^5 \equiv 6 \,(\mathrm{mod}\,13)$, $2^6 \equiv 12 \,(\mathrm{mod}\,13)$, $2^7 \equiv 11 \,(\mathrm{mod}\,13)$, we obtain that $\mathrm{ind}_2 11 = 7$ modulo 13.

✹ **Theorem 9.29.**

Let a be a primitive root modulo n, and $b \in \mathbb{Z}_n^*$. Then,

$$a^k \equiv b \,(\mathrm{mod}\,n) \iff k \equiv \mathrm{ind}_a b \,(\mathrm{mod}\,\varphi(n)) \tag{9.5}$$

Proof.

By definition, $b \equiv a^{\mathrm{ind}_a b} \equiv a^k \,(\mathrm{mod}\,n)$. Hence, from Theorem 9.22, it follows that $k \equiv \mathrm{ind}_a b \,(\mathrm{mod}\,\varphi(n))$. □

The next theorem shows that the basic properties of the index are very similar to properties of the ordinary logarithm (therefore, it is also called the discrete logarithm).

✹ **Theorem 9.30.**

Let a be a primitive root modulo n. Then, there hold the following properties:

1) $\mathrm{ind}_a 1 \equiv 0 \,(\mathrm{mod}\,\varphi(n))$.

2) $\mathrm{ind}_a (xy) \equiv \mathrm{ind}_a x + \mathrm{ind}_a y \,(\mathrm{mod}\,\varphi(n))$ for all $x, y \in \mathbb{Z}_n^*$.

3) $\mathrm{ind}_a (x^k) \equiv k \cdot \mathrm{ind}_a x \,(\mathrm{mod}\,\varphi(n))$ for all $x \in \mathbb{Z}_n^*$ and $k \in \mathbb{Z}^+$.

Proof.

1. From Theorem 9.29, it follows that $a^0 \equiv 1 \pmod{n}$ if and only if $\text{ind}_a 1 \equiv 0 \pmod{\varphi(n)}$.

2. If $c = \text{ind}_a x$ and $d = \text{ind}_a y$, then $x = a^c$ and $y = a^d$. Therefore, $xy = a^c a^d = a^{c+d}$. Hence, by Theorem 9.27, we have:

$$c + d = \text{ind}_a x + \text{ind}_a y \equiv \text{ind}_a(xy) \pmod{n}.$$

3. This follows directly from the previous part of the theorem. □

The notion of the discrete logarithm can be generalized for an arbitrary cyclic group.

✳ **Definition 9.31.**
Let G be a finite multiplicative cyclic group of order $n \geq 2$ with generator $g \in G$ and let $b \in G$. The **discrete logarithm** of b to the base g, denoted by $\log_g b$, is an integer $1 \leq x \leq n - 1$ such that

$$g^x = b \tag{9.6}$$

in the group G.

✳ **Example 9.32.**
Find $\log_2 7$ in the group $G = \mathbb{Z}_{19}^*$, i.e., find the integer $1 \leq x \leq 18$ such that $2^x \equiv 7 \pmod{19}$.

The element $g = 2$ is a generator of G. To find x, we compute step by step:
$2^2 \equiv 4 \pmod{19}$, $2^3 \equiv 8 \pmod{19}$, $2^4 \equiv 16 \pmod{19}$, $2^5 \equiv 13 \pmod{19}$, $2^6 \equiv 7 \pmod{19}$.
Hence, $x = \log_2 7 = 6$.

✳ **Example 9.33.**
Let $p = 13$. Then, $p - 1 = 12$ and $\varphi(12) = 4$. Therefore, we have 4 primitive roots modulo 13: $g = 2, 6, 7, 11$.

For the generator $g = 2 \in \mathbb{Z}_{13}^*$, we give the values $\log_g b$ for different elements $b \in \mathbb{Z}_{13}^*$:

b	1	2	3	4	5	6	7	8	9	10	11	12
$\log_g b$	0	1	4	2	9	5	11	3	8	10	7	6

In particular, from this table we have:

- $3 \cdot 7 \equiv 8 \pmod{13}$ and $\log_2(3 \cdot 7) \equiv \log_2(8) \equiv 3 \pmod{12}$, $\log_2(3) + \log_2(7) \equiv 4 + 11 \equiv 3 \pmod{12}$, so $\log_2(3 \cdot 7) \equiv \log_2(3) + \log_2(7) \pmod{12}$.

- $3^3 \equiv 1 \,(\mathrm{mod}\,13)$ and $\log_2(3^3) \equiv \log_2(1) \equiv 0 \,(\mathrm{mod}\,12)$, $3\log_2(3) \equiv 3 \cdot 4 \equiv 0 \,(\mathrm{mod}\,12)$, so $\log_2(3^3) \equiv 3\log_2(3) \,(\mathrm{mod}\,12)$.

✳ Example 9.34.

Let G be the multiplicative group of the finite field F_9, i.e., $G = F_9^*$. Suppose that α is a root of the irreducible polynomial $x^2 + 2x + 2 \in F_3[x]$. Then, by Example 9.13, α is a generator of G. Since $\alpha^2 + 2\alpha + 2 = 0$, we obtain that $\alpha^2 = \alpha + 1$. Therefore, $\alpha^3 = \alpha^2 + \alpha = 2\alpha + 1$ and $\alpha^4 = 2\alpha^2 + \alpha = 2(\alpha + 1) + \alpha = 2$. Hence, the discrete logarithm of the element $2 \in G$ to the base α is the integer 4.

Note that the notion of the discrete logarithm can also be generated for an arbitrary finite multiplicative group.

✳ Definition 9.35.

Let G be a finite multiplicative group and let $b \in G$. The **discrete logarithm** of b to the base $g \in G$, denoted $\log_g b$, is an integer x such that

$$g^x = b \qquad (9.6)$$

in the group G, if such an integer exists.

The **discrete logarithm problem** is the problem of finding the integer $x = \log_g b$ such that $g^x = b$ for given elements b and g of a finite group G.

Finding the discrete logarithm depends on the selected group. For some groups, this problem is not difficult. However, for most groups, the problem turns out to be surprisingly difficult. In general, the discrete logarithm does not always exist, and if it even exists it is not always unique. It is believed that the discrete logarithm problem is as hard as the problem of factoring integers. The discrete logarithm problem is very important and has significant applications in cryptography. The security of many cryptographic algorithms, such as the ElGamal cryptosystem and the Diffie-Hellman scheme considered in the next section, relies on the presumed hardness of the discrete logarithm problem.

9.4. Diffie-Hellman Scheme. ElGamal Cryptosystem

The main defect of cryptographic algorithms with public key (such as algorithm RSA) is their slow action compared to classical symmetric algorithms. Therefore, cryptographic algorithms with public key are often used to establish the common private key used for symmetric algorithms. The first algorithm to establish the common private key was the algorithm proposed by W. Diffie and M.E. Hellmam in their crucial paper in 1976 [11].

W. Diffie and M.E. Hellman showed that the proposed algorithm allows the common private key (named by the Diffie-Hellman key-exchange scheme) to be established using insecure communication channel. The security of this protocol relies on the **assumption of Diffie-Hellman**, which assumes that finding g^{ab}, where g is an element of a given group and $a, b \in \mathbb{Z}$, provided that only g^a and g^b are known, is a very difficult computing problem and will not be resolved in feasible time.

Unfortunately, their suggestion is now rather a hypothesis, which, so far, is not proved. In other words, the equivalence between breaking the Diffie-Hellman key-exchange scheme and calculating the discrete logarithms was not proved.

Algorithm Task

Two users (Alice and Bob) want to use symmetric cryptographic algorithm. First, they want to establish their common key, which is an element of a multiplicative group F_q^* of a finite field F_q, for encrypting a message. Moreover, the procedure of key exchange must be implemented using insecure communication channel.

I. Diffie-Hellman Key Exchange (for the group $G = \mathbb{Z}_p^*$, where p is a prime)

1. Alice and Bob chose the group $G = \mathbb{Z}_p^*$ of order $|G| = p - 1$ and a primitive root g of G.
2. Alice chooses a random integer $a \in \{1, 2, \ldots, p - 2\}$.
3. Alice computes $A = g^a$ and sends A to Bob using insecure communication channel.
4. Bob chooses a random integer $b \in \{1, 2, \ldots, p - 2\}$.
5. Bob computes $B = g^b$ and sends B to Alice using insecure communication channel.
6. Alice computes $B^a = g^{ab}$.
7. Bob computes $A^b = g^{ab}$.
8. Their common private key is $K = g^{ab}$.

Note, that in this key-exchange scheme a prime p, a primitive root g of the group $G = \mathbb{Z}_p^*$ and powers $A = g^a$, $B = g^b$ are given for the public communication, while Alice and Bob keep in secret the integers a, b and the common key K.

✳ Example 9.36.

Let $p = 17$. We choose the element $g = 3 \in \mathbb{Z}_{17}^*$ as a primitive root modulo 17.

Alice chooses the integer $a = 7$ and calculates:

$$A \equiv g^a \equiv 3^7 \equiv 11 \, (\mathrm{mod} \, 17).$$

Bob chooses the integer $b = 4$ and calculates:

$$B \equiv g^b \equiv 3^4 \equiv 13 \, (\mathrm{mod} \, 17).$$

Therefore, the agreed common key is:

$$K \equiv A^b \equiv 11^4 \equiv 4 \,(\mathrm{mod}\,17)$$

$$K \equiv B^a \equiv 13^7 \equiv 4 \,(\mathrm{mod}\,17)$$

The Diffie-Hellman key-exchange scheme described above can be generalized for arbitrary finite cyclic group. Below, we give the variant of this scheme for multiplicative cyclic group F_q^* of a finite field F_q.

Diffie-Hellman key-exchange scheme (for the group $G = F_q^*$, where $q = p^k$ and p is a prime)

1. Alice and Bob chose the group $G = F_q^*$ of order $|G| = q - 1$, where $q = p^k$ and p is a prime, and a primitive root g of the group G.
2. Alice chooses a random integer $a \in \{1, 2, \ldots, q - 2\}$.
3. Alice computes $A = g^a \in F_q^*$ and sends A to Bob.
4. Bob chooses a random integer $b \in \{1, 2, \ldots, q - 2\}$.
5. Bob computes $B = g^b \in F_q^*$ and sends B to Alice.
6. Alice computes $B^a = g^{ab} \in F_q^*$.
7. Bob computes $A^b = g^{ab} \in F_q^*$.
8. Their common key is $K = g^{ab} \in F_q^*$.

III. ElGamal cryptosystem

The Diffie-Hellman cryptosystem was elaborated for security of the key-exchange procedure. This protocol allows the key-exchange to be secured even if there is a person who observes the process of exchanging key, while it can not be used for encryption and decryption of a message. There are another public-key cryptographic algorithms for secure information transfer that rely on the difficulty of computing discrete logarithms and which can be used to encrypt and decrypt a message. One of these algorithms is the ElGamal cryptosystem, proposed in 1985 by Egyptian Taher ElGamal [14]. This algorithm can be considered as some improved version of the Diffie-Hellman key exchange algorithm.

The ElGamal algorithm can be defined for each finite cyclic group. Below, we give the ElGamal algorithm for the multiplicative cyclic group F_q^* of a finite field F_q.

I. Key Generation

1. Alice chooses a finite field F_q, where $q = p^k$ and p is a prime, and a generator g of the cyclic group $G = F_q^*$.

2. Alice randomly chooses an integer $2 \le a \le q - 1$ and calculates $A = g^a \in F_q^*$.

   Then, (q, g, A) is Alice's public key and a is her private key.

II. Encryption of the message

1. Bob randomly chooses a number $b \in \{2, \ldots, q-1\}$ and computes $B = g^b \in F_q^*$.

2. For encryption of the message M, Bob calculates the encryption text:

$$C = A^b M = g^{ab} M \in F_q^*.$$

and sends a pair of elements (B, C) to Alice.

III. Decryption of the message

For decryption of the message, Alice calculates: $k = q - 1 - a$ and $B^k C = M$.

We show the correctness of decryption of the message by the ElGamal cryptosystem. Since, by Lagrange's theorem, $g^{q-1} = 1$, we have:

$$B^k C = g^{bk} C = g^{b(q-1-a)} A^b M = g^{b(q-1)} g^{-ab} A^b M =$$

$$= (g^{(q-1)})^b (g^a)^{-b} A^b M = A^{-b} A^b M = M.$$

Note that the ElGamal cryptosystem is a non-determinate algorithm, since the content of the encryption message depends not only on the original text but also on randomly chosen numbers, thanks to which one original text can be encrypted in many different ways. It is important to note that in the ElGamal cryptosystem the obtained data to transfer are sometimes twice as long as the original text.

✳ Example 9.37.

Alice chooses the group $G = F_p^*$ of order $|G| = p - 1$, where $p = 23$ and a generator $g = 7$ of the group G.

Alice chooses a random integer $a = 6$ and calculates:

$$A \equiv g^a \equiv 7^6 \equiv 4 \,(\mathrm{mod}\,23).$$

Hence, $(23, 7, 4)$ is Alice's public key, and $a = 6$ is her private key.

To encrypt the message $M = 7$, Bob chooses a random integer $b = 3$ and calculates:

$$B \equiv g^b \equiv 7^3 \equiv 21 \,(\mathrm{mod}\,23).$$

Furthermore, Bob computes the encryption message:

$$C = A^b M \equiv 4^3 \cdot 7 \equiv 11 \,(\mathrm{mod}\,23)$$

and sends the pair of numbers $(B, C) = (21, 11)$ to Alice.

To decrypt the obtained message, Alice use her private key $a = 6$ and calculates: $k = p - 1 - a = 23 - 1 - 6 = 16$. Then, she computes:

$$M = B^k C \equiv 21^{16} \cdot 11 \equiv 7 \,(\mathrm{mod}\,23).$$

9.5. Error Detecting and Error Correcting Codes

In this section, we present the application of finite fields for construction of codes. Coding theory is concerned with transmission of information. In practice, the transmission channels are often both noisy and insecure, therefore, they can change the message and, as a result, we receive a distorted message. Error-correcting codes, or simply codes, are used to protect digitally encoded information and to increase its reliability during transmission through noisy and unreliable channels. Codes help us to detect errors in distorted messages and correct them.

We give some examples of the most important and popular codes, such as polynomial codes, cyclic codes and BCH codes.

9.5.1. Basic notions of algebraic coding theory

First of all, we give some necessary notions which are used in algebraic coding theory. Coding theory is originated from C. Shannon's classic paper [39] on information theory, published in 1948.

Each message is transmitted as a string of some symbols from a chosen alphabet M. In the most common case, these symbols are 0 and 1, i.e., the elements of the field F_2. However, the symbols which are the elements of other sets are also used. Since modern communications are digital, only finite sets are considered.

> ✳ **Definition 9.38.**
> A subset C of a free monoid M^* over a chosen alphabet M is called a **code** if each equality $a_1 a_2 \ldots a_n = b_1 b_2 \ldots b_m$, for all $n, m \in \mathbb{Z}^+$ and all $a_i, b_j \in C$, implies that $n = m$ and $a_i = b_i$, for all $i = 1, 2, \ldots, n$.
> Elements of the set C are called **codewords** (or **codevectors**).

The next step in coding is to split the message, which is the string of symbols, into blocks of fixed length m. Hence, each codeword can be considered to have a fixed length m.

> ✳ **Definition 9.39.**
> A code C is called a **block code** if all its codewords have the same number m of symbols chosen from the alphabet M. This number m is called the **length** of the code C. We will consider only block codes and call them codes.

In all our considerations in this chapter, we will consider as a chosen alphabet a finite field F, which is called the **coding alphabet**. So, we assume that a coding alphabet F is the Galois field $GF(q) = F_q$ containing q elements and each codeword is a finite string (or a vector) of elements of F. Therefore, m-code over F is a subset $C \subset F^m$.

> ✳ **Definition 9.40.**
>   A m-code $C \subset F^m$ over the field $F = \mathbb{Z}_2 = \{0, 1\}$ is called a **binary code**, and so each binary codeword is just a finite string of length m of 0's and 1's.
>   If C is a block code over a field $GF(q)$ and $|C| = 1$ or $|C| = q^n$, then C is called a **trivial code**.
>   A code $C \subset F^m$ is called a **linear code** over a field F if C is a vector subspace over F. The dimension of a vector subspace C over F is called the **dimension** of the linear code C and denoted by $\dim(C)$.

Informally, we can say that encoding is a record of a message as a codeword sequence, while decoding is the conversion of a codeword sequence into a message. The crucial idea of coding scheme is to increase the length of transmitting message by adding to each codeword a few symbols which help to correct errors. The encoding procedure suggests that, together with an original message, we send some extra information which does not change the content of the message but helps us to find and, if it is possible, to correct errors.

✳ **Examples 9.41.**
  The simplest example of a code for detecting errors is the **parity check code**.
  Let the field $F = \mathbb{Z}_2 = \{0, 1\}$ be the coding alphabet. Each codeword of length n is given as a vector $(\alpha_1, \alpha_2, \ldots, \alpha_n)$, where $\alpha_i \in \mathbb{Z}_2$. If $\alpha_{n+1} \equiv \alpha_1 + \alpha_2 + \cdots + \alpha_n \pmod{2}$ then,

$$\alpha_1 + \alpha_2 + \cdots + \alpha_n + \alpha_{n+1} \equiv 0 \pmod{2}. \tag{9.8}$$

Therefore, we can increase each codeword $(\alpha_1, \alpha_2, \ldots, \alpha_n)$ by adding one extra symbol α_{n+1} which is the sum modulo 2 of all other symbols of this codeword. So, we have the encoding function $\xi : F^n \to F^{n+1}$ given by

$$\xi(\alpha_1, \alpha_2, \ldots, \alpha_n) = (\alpha_1, \alpha_2, \ldots, \alpha_n, \alpha_{n+1}), \tag{9.9}$$

where $\alpha_{n+1} \equiv \alpha_1 + \alpha_2 + \cdots + \alpha_n \pmod{2}$.
  The verification of an error is based on checking equality (9.8). So, the decoding function $\eta : F^{n+1} \to F^n \cup \{\text{error}\}$ is given by

$$\eta(\beta_1, \ldots, \beta_n, \beta_{n+1}) = \begin{cases} (\beta_1, \ldots, \beta_n), & \text{if } \beta_1 + \cdots + \beta_{n+1} \equiv 0 \pmod{2}, \\ \text{error}, & \text{if } \beta_1 + \cdots + \beta_{n+1} \equiv 1 \pmod{2}. \end{cases} \tag{9.10}$$

The parity check code works if we have a single error or when the number of errors is odd. It shows us when an odd number of errors occurs, but it does not enable us to correct errors.

  In general, if $F = GF(q)$ is the Galois field, then the parity check code of length $n + 1$ is given by the encoding function of the form $\xi : F^n \to F^{n+1}$, where

$$\xi(x_1, x_2, \ldots, x_n) = (x_1, x_2, \ldots, x_n, -\sum_{i=1}^{n} x_i)$$

for any vector $(x_1, x_2, \ldots, x_n) \in F^n$. So, at the end of each sending codeword, we add an element opposite to the sum of all elements of this codeword. Then, to verify an error, we check the equality: $x_1 + x_2 + \ldots + x_n + x_{n+1} = 0$ in the field F.

✳ Example 9.42.

Find the binary parity check code of length $n = 4$.

The codewords of the length $n = 3$ are:

(0 0 0), (0 0 1), (0 1 1), (0 1 0), (1 0 0), (1 0 1), (1 1 0), (1 1 1).

Note, that when we consider a vector as a message, we will omit commas and write, for example, (0 0 1) instead of (0, 0, 1).

So, the binary parity check code of length $n = 4$ consists of codewords:

(0 0 0 0), (0 0 1 1), (0 1 1 0), (0 1 0 1), (1 0 0 1), (1 0 1 0), (1 1 0 0), (1 1 1 1).

Since the transmission channel often has noise, the obtaining message can have errors. If $\mathbf{u} = (x_1, x_2, \ldots, x_k) \in F^k$ is a sending vector and $\mathbf{v} = (y_1, y_2, \ldots, y_k) \in F^k$ is a receiving vector, then the number of errors is equal to the number of coordinates in which $\mathbf{u}$ and $\mathbf{v}$ differ. The following notion is a very useful characteristic of the error-correction capability of codes.

✳ Definition 9.43.

The **Hamming distance** $d(\mathbf{u}, \mathbf{v})$ between two elements $\mathbf{u}, \mathbf{v} \in F^n$ is the number of coordinates in which vectors $\mathbf{u} = (u_1, u_2, \ldots, u_n)$ and $\mathbf{v} = (v_1, v_2, \ldots, v_n)$ differ, that is:

$$d(\mathbf{u}, \mathbf{v}) = \#\{i \ : \ u_i \neq v_i\}. \tag{9.11}$$

✳ Theorem 9.44.

The Hamming distances has the properties:

1) $d(\mathbf{u}, \mathbf{v}) \geq 0$ and $d(\mathbf{u}, \mathbf{v}) = 0$ if and only if $\mathbf{u} = \mathbf{v}$;

2) $d(\mathbf{u}, \mathbf{v}) = d(\mathbf{v}, \mathbf{u})$ - symmetry

3) $d(\mathbf{u}, \mathbf{v}) \leq d(\mathbf{u}, \mathbf{w}) + d(\mathbf{w}, \mathbf{v})$ - triangle inequality

for all vectors $\mathbf{u}, \mathbf{v}, \mathbf{w} \in F^n$

Proof.

The first two properties follow directly from the definition of the Hamming distance. So, it only remains to prove the third property. Since $d(\mathbf{u}, \mathbf{v})$ is equal to the minimal number of changing coordinates to obtain a vector $\mathbf{w}$ from a vector $\mathbf{u}$, and $d(\mathbf{w}, \mathbf{v})$ is equal to the minimal number of changing coordinates to obtain a vector $\mathbf{v}$ from a vector $\mathbf{w}$, $d(\mathbf{u}, \mathbf{v}) + d(\mathbf{w}, \mathbf{v})$ is equal to the minimal number of changing coordinates to obtain a vector $\mathbf{v}$ from a vector $\mathbf{u}$. Therefore, $d(\mathbf{u}, \mathbf{v}) \leq d(\mathbf{u}, \mathbf{w}) + d(\mathbf{w}, \mathbf{v})$. $\qquad\square$

From Theorem 9.44, it follows that the function $d : F^n \times F^n \to \mathbb{N}$ defined by (9.11) satisfies all axioms of metrics and, therefore, it is called the **Hamming metrics** and the pair (F^n, d) forms a metric space which is called a **Hamming space**.

> ✳ **Definition 9.45.**
> The **Hamming weight** of a vector $\mathbf{u} \in F^n$ in the Hamming metrics is $d(\mathbf{u}, \mathbf{0})$, i.e., the number of non-zero coordinates in the vector $\mathbf{u}$ and it is denoted by $wt(\mathbf{u})$.

Note, that the Hamming distance between two vectors $\mathbf{u}, \mathbf{v} \in F^n$ for a linear code is the weight of the difference of these vectors $\mathbf{u} - \mathbf{v}$:

$$d(\mathbf{u}, \mathbf{v}) = d(\mathbf{u} - \mathbf{v}, \mathbf{0}) = wt(\mathbf{u} - \mathbf{v}). \qquad (9.12)$$

✳ **Examples 9.46.**

1. Find the Hamming distance between vectors

   $\mathbf{u} = (1\ 1\ 0\ 0\ 0\ 0\ 1\ 1\ 0)$ and $\mathbf{v} = (1\ 0\ 0\ 1\ 1\ 0\ 0\ 0\ 1\ 1)$.

   Since the number of distinct coordinates of vectors $\mathbf{u}$ and $\mathbf{v}$ is equal to 4, $d(\mathbf{u}, \mathbf{v}) = 4$.

2. Find the weight of the vector $\mathbf{u} = (1\ 0\ 0\ 1\ 1\ 0\ 0\ 1\ 1\ 0)$ in the Hamming metric.

   Since the vector $\mathbf{u}$ has 5 non-zero coordinates, the weight of the vector $\mathbf{u}$ is equal to $wt(\mathbf{u}) = d(\mathbf{u}, \mathbf{0}) = 5$.

> ✳ **Definition 9.47.**
> The **minimum distance** of a code C is the distance between the closest codewords in the Hamming metrics, that is:
>
> $$d_{\min}(C) = \min\{d(\mathbf{u}, \mathbf{v}) \ : \mathbf{u}, \mathbf{v} \in C, \ \mathbf{u} \neq \mathbf{v}\}. \qquad (9.13)$$

Note, that if C is a linear code, then, from (9.11)-(9.13), it follows that

$$d_{\min}(C) = \min\{d(\mathbf{u}, \mathbf{0}) \ : \mathbf{u} \in C, \mathbf{u} \neq \mathbf{0}\} = wt_{\min}(C). \qquad (9.14)$$

✳ Example 9.48.

Let $C = \{\mathbf{u}_1 = (0\ 0\ 1\ 0\ 0),\ \mathbf{u}_2 = (1\ 0\ 0\ 1\ 0),\ \mathbf{u}_3 = (1\ 1\ 0\ 0\ 1),\ \mathbf{u}_4 = (0\ 0\ 1\ 1\ 1)\}$. Then,

$$d(\mathbf{u}_1, \mathbf{u}_2) = 3,\ \ d(\mathbf{u}_1, \mathbf{u}_3) = 4,\ \ d(\mathbf{u}_1, \mathbf{u}_4) = 2,\ \ d(\mathbf{u}_2, \mathbf{u}_3) = 3,\ \ d(\mathbf{u}_2, \mathbf{u}_4) = 3,\ \ d(\mathbf{u}_3, \mathbf{u}_4) = 4.$$

So, $d_{\min}(C) = 2$.

The concept of minimum distance is an important parameter of a code because it determines the error-correction capability of a code that can be seen from the following theorems.

✳ Theorem 9.49.

A code $C \subset F^n$ detects up to t errors if and only if $d_{\min}(C) \geq t + 1$.

Proof.

$\Rightarrow$. Suppose that a code $C \subset F^n$ detects up to t errors and $d_{\min}(C) \leq t$. Then, there are, at most, two codewords $\mathbf{u}, \mathbf{v} \in C$ such that $d(\mathbf{u}, \mathbf{v}) \leq t$. So, the codeword $\mathbf{u}$ can be converted into $\mathbf{v}$ with no more that t errors, and this will not be detected.

$\Leftarrow$. Let $\mathbf{u} \in C$ be a sending codeword and $\mathbf{v} \in F^n$ a receiving vector. Assume that s errors occur in transmission. Then, $d(\mathbf{u}, \mathbf{v}) = s$. Since $d_{\min}(C) > t$, each vector with $0 < s \leq t$ cannot be a codeword. Therefore, in this case, we can notice that an error has occurred, and so C detects up to t errors. □

✳ Theorem 9.50.

A code $C \subset F^n$ corrects up to t errors if and only if $d_{\min}(C) \geq 2t + 1$. (These codes are often called t-**error-correcting codes.**

Proof.

$\Rightarrow$. Suppose that a code $C \subset F^n$ has $d_{\min}(C) < 2t + 1$. Then, there are wordcodes $\mathbf{u}, \mathbf{v} \in C$ such that $d(\mathbf{u}, \mathbf{v}) = k \leq 2t$. Suppose that during the transmission the codeword $\mathbf{u}$ is received with t errors. Consider the chain of words which can arise in this transmission

$$\mathbf{u} = \mathbf{u}_1 \to \mathbf{u}_2 \to \cdots \to \mathbf{u}_t \to \mathbf{u}_{t+1} \to \cdots \to \mathbf{u}_k = \mathbf{v}$$

where each transition from a vector $\mathbf{u}_i$ to a vector $\mathbf{u}_{i+1}$ changes only one coordinate of the previous vector $\mathbf{u}_i$. Since the received vector has t errors, we can consider that the received vector through this transmission is the vector $\mathbf{u}_t$ having t errors. In this case, since $k \leq 2t$ and $d(\mathbf{u}_t, \mathbf{v}) \leq t$, we can consider $\mathbf{v}$ as the encoded vector for $\mathbf{u}_t$, which is not equal to $\mathbf{u}$.

$\Leftarrow$. Let $C \subset F^n$ be a code with minimum distance $> 2t$. Suppose that $\mathbf{u} \in C$ is a sent codeword, and an obtained vector $\mathbf{v} \in F^n$ has, at most, $e \leq t$ errors. We will show that $\mathbf{u}$ is the closest codeword of C to $\mathbf{v}$. Suppose that this is not the case and there is a codeword $\mathbf{w} \neq \mathbf{u}$ that is closer to $\mathbf{v}$ than $\mathbf{u}$, i.e.,

$$d(\mathbf{v}, \mathbf{w}) \leq d(\mathbf{v}, \mathbf{u}) \leq e.$$

Then, by the triangle inequality, we have:

$$2t + 1 \leq d(\mathbf{u}, \mathbf{w}) \leq d(\mathbf{u}, \mathbf{v}) + d(\mathbf{v}, \mathbf{w}) \leq e + e = 2e \leq 2t.$$

This contradiction shows that the codeword $\mathbf{w} \neq \mathbf{u}$ does not exist, i.e., $\mathbf{u}$ is the closest codeword to the vector $\mathbf{v}$ in the code C. □

9.5.2. Linear codes

In this chapter, we will consider the most important family of codes that are linear codes for which we can apply the tools of linear algebra over a field.

Each linear error-correcting code can be described as a pair of two functions: Encoding function and decoding function.

> ✳ **Definition 9.51.**
> The **encoding function** (or the **coding algorithm**) is an injection:
>
> $$\xi : F^k \to F^n. \tag{9.15}$$
>
> In this case, the code C is the image of the mapping ξ. The original information is a vector $\mathbf{u} = (x_1, x_2, \ldots, x_k) \in F^k$, while a vector $\xi(\mathbf{u}) = (y_1, y_2, \ldots, y_n) \in F^n$ is a codeword.
>
> The **decoding function** (or the **decoding algorithm**) is a mapping:
>
> $$\eta : F^n \to F^k \cup \{\text{error}\}. \tag{9.16}$$
>
> such that $\eta(\xi(\mathbf{u})) = \mathbf{u}$ for all $\mathbf{u} \in F^k$.

> ✳ **Definition 9.52.**
> A linear code C, which is a k-dimensional vector subspace of an n-dimensional vector space over a field F, is called an (n, k)-**code** over F. If the minimal distance of an (n, k)-code C over a field F is equal to d, then C is called a linear (n, k, d)-**code**.

Assume that C is a linear (n, k)-code over a finite field F. In this case, in order to describe C, we only need to know basis elements of C. Let $(\mathbf{f}_1, \mathbf{f}_2, \ldots, \mathbf{f}_k)$ be a basis of C as a k-dimensional vector space over F. Then, we can write them in the standard basis $(\mathbf{e}_1, \mathbf{e}_2, \ldots, \mathbf{e}_n)$ of the vector space F^n as follows:

$$
\begin{aligned}
\mathbf{f}_1 &= (f_{11}, \ f_{12}, \ \ldots, \ f_{1n}) \\
\mathbf{f}_2 &= (f_{21}, \ f_{22}, \ \ldots, \ f_{2n}) \\
\ldots \quad \ldots & \quad \ldots \quad \ldots \quad \ldots \quad \ldots \\
\mathbf{f}_k &= (f_{k1}, \ f_{k2}, \ \ldots, \ f_{kn})
\end{aligned}
$$

where $f_{ij} \in F$. Therefore, any codeword $\mathbf{v} \in C$ can be written in the form:

$$\mathbf{v} = \sum_{i=1}^{k} u_i \mathbf{f}_i = \mathbf{uG} \tag{9.17}$$

where $\mathbf{u} = (u_1, u_2, \ldots, u_k) \in F^k$ and

$$\mathbf{G} = \begin{pmatrix} f_{11} & f_{12} & \cdots & f_{1n} \\ f_{21} & f_{22} & \cdots & f_{2n} \\ \vdots & \vdots & \ddots & \vdots \\ f_{k1} & f_{k2} & \cdots & f_{kn} \end{pmatrix} \in M_{k \times n}(F) \tag{9.18}$$

The matrix $\mathbf{G}$ is called a **generator matrix** of the code C. It is clear that the rank of $\mathbf{G}$ is equal to k. Since the rows of a generator matrix form a basis of the code C, this matrix uniquely defines the encoding function $\xi : F^k \to F^n$ given by $\xi(\mathbf{u}) = \mathbf{uG}$ for each $\mathbf{u} = (u_1, u_2, \ldots, u_k) \in F^k$. Since any codeword has the form $\mathbf{v} = \xi(\mathbf{u})$ for some $\mathbf{u} \in F^k$, we obtain that

$$C = \{\mathbf{uG} \ : \ \mathbf{u} \in F^k, \mathbf{G} \in M_{k \times n}(F)\}. \tag{9.19}$$

If $\mathbf{u} = (u_1, u_2, \ldots, u_k)$ is an original message then $\mathbf{v} = \mathbf{uG}$ is its codeword.

✳ **Examples 9.53.**

1. The matrix $\mathbf{G} \in M_{k \times n}(\mathbb{Z}_2)$ of the form $\begin{pmatrix} 1 & 0 & 0 & 0 \\ 1 & 1 & 1 & 1 \end{pmatrix}$ generates the codewords:

$$\mathbf{u}_1 \mathbf{G} = (0\ 0)\mathbf{G} = (0\ 0\ 0\ 0)$$
$$\mathbf{u}_2 \mathbf{G} = (0\ 1)\mathbf{G} = (1\ 1\ 1\ 1)$$
$$\mathbf{u}_3 \mathbf{G} = (1\ 0)\mathbf{G} = (1\ 0\ 0\ 0)$$
$$\mathbf{u}_4 \mathbf{G} = (1\ 1)\mathbf{G} = (0\ 1\ 1\ 1)$$

So, the binary code is given as follows:

$$C_1 = \{(0\ 0\ 0\ 0),\ (0\ 1\ 1\ 1),\ (1\ 0\ 0\ 0), (1\ 1\ 1\ 1)\}.$$

2. The matrix $\mathbf{G} \in M_{k \times n}(\mathbb{Z}_2)$ of the form $\begin{pmatrix} 0 & 1 & 1 & 1 \\ 1 & 0 & 0 & 0 \end{pmatrix}$ generates the codewords:

$$\mathbf{u}_1 \mathbf{G} = (0\ 0)\mathbf{G} = (0\ 0\ 0\ 0)$$
$$\mathbf{u}_2 \mathbf{G} = (0\ 1)\mathbf{G} = (1\ 0\ 0\ 0)$$
$$\mathbf{u}_3 \mathbf{G} = (1\ 0)\mathbf{G} = (0\ 1\ 1\ 1)$$
$$\mathbf{u}_4 \mathbf{G} = (1\ 1)\mathbf{G} = (1\ 1\ 1\ 1)$$

So, we have the binary code as follows:

$$C_2 = \{(0\ 0\ 0\ 0),\ (0\ 1\ 1\ 1),\ (1\ 0\ 0\ 0),\ (1\ 1\ 1\ 1)\}.$$

It is easy to see that $C_2 = C_1$.

These examples show that different matrices can generate the same code. As we know, from the course of linear algebra, the form of a matrix depends on the choosing a basis in the vector subspace C.

✴ Examples 9.54.

Let C be the parity check $(n, n+1)$-code considered in Example 9.42. Let

$$
\begin{aligned}
\mathbf{e}_1 &= (1\ 0\ \ldots\ 0\ 0), \\
\mathbf{e}_2 &= (0\ 1\ \ldots\ 0\ 0), \\
&\ldots \quad \ldots \\
\mathbf{e}_n &= (0\ 0\ \ldots\ 0\ 1)
\end{aligned}
$$

be the standard basis of the vector space F^n and let ξ be an encoding function given by (9.15). Then,

$$
\begin{aligned}
\xi(\mathbf{e}_1) &= (1\ 0\ \ldots\ 0\ 1), \\
\xi(\mathbf{e}_2) &= (0\ 1\ \ldots\ 0\ 1), \\
&\ldots \quad \ldots \\
\xi(\mathbf{e}_n) &= (0\ 0\ \ldots\ 1\ 1),
\end{aligned}
$$

So the generator matrix of this code has the form:

$$
G = \begin{pmatrix}
1 & 0 & \cdots & 0 & 1 \\
0 & 1 & \cdots & 0 & 1 \\
\vdots & \vdots & \ddots & \vdots & \vdots \\
0 & 0 & \cdots & 1 & 1
\end{pmatrix}
$$

Another important type of matrices associated with a code C is its parity check matrix.

If C is a linear (n, k)-code over a field F, then the set of all vectors which are orthogonal to every vector in a linear code $C \subset F^n$ is also a subspace of F^n and, thus, another linear code called the **dual code** of C and denoted by $C^\perp$. So that

$$C^\perp = \{\mathbf{u} \in F^n\ :\ \mathbf{u} \cdot \mathbf{v} = \mathbf{0} \text{ for all } \mathbf{v} \in C\}.$$

If C is an (n, k)-code, then it can be shown that $C^\perp$ is an $(n, n - k)$-code.

✴ Definition 9.55.

A generator matrix for $C^\perp$ is called a **parity check matrix** (or the **checking matrix**) of the code C.

So, a matrix $\mathbf{H} \in M_{(n-k) \times k}(F)$ is a parity check matrix of an (n, k)-code C over a field F if and only if

$$C = \{\mathbf{v} \in F^k\ :\ \mathbf{v}\mathbf{H}^T = \mathbf{0}\}. \tag{9.20}$$

It is easy to see that the following identities hold

$$\mathbf{G}\mathbf{H}^T = \mathbf{H}\mathbf{G}^T = \mathbf{0}. \tag{9.21}$$

Generator and parity check matrices are not defined uniquely and their forms depend on the choice of bases in the spaces F^k and F^n. Since a generator matrix $\mathbf{G} \in M_{k \times n}(F)$ has the rank k, from linear algebra, it is known that each such matrix can be reduced to its row reduced echelon form by elementary row operations:

$$\mathbf{G} = [\, \mathbf{I}_k \mid \mathbf{P} \,] \tag{9.22}$$

where $\mathbf{I}_k \in M_{k \times k}(F)$ is the identity matrix and $\mathbf{P}$ is an $k \times (n - k)$-matrix. If a generator matrix G has the form (9.22), we say that G is in **standard form**.

A linear code which admits a generator matrix in standard form is called a **systematic code**. Two linear codes C_1 and C_2 of length n over a field F are called **equivalent** if $|C_1| = |C_2|$ and they can be transformed into each other using only position permutations, i.e., there is a permutation $\sigma \in S_n$ and automorphisms $\pi_1, \ldots, \pi_n$ of F such that, if $(c_1, c_2, \ldots, c_n) \in C_1$, then $\sigma(\pi_1(c_1), \ldots, \pi_n(c_n)) \in C_2$. It is easy to show that each linear (n, k)-code is equivalent to some systematic code with a generator matrix in standard form. Each systematic code with a generator matrix in standard form (9.22) has a parity check matrix of the following form:

$$\mathbf{H}_1 = [\, -\mathbf{P}^T \mid \mathbf{I}_{n-k} \,] \tag{9.23}$$

where $\mathbf{I}_{n-k} \in M_{(n-k) \times (n-k)}(F)$ is the identity matrix. So, each linear (n, k)-code has a parity check matrix of dimension $(n - k) \times n$.

✳ Example 9.56.

Let

$$\mathbf{G} = \begin{pmatrix} 1 & 0 & 0 & 0 & 1 \\ 0 & 1 & 0 & 1 & 0 \\ 0 & 0 & 1 & 1 & 1 \end{pmatrix} = [\, \mathbf{I}_3 \mid \mathbf{P} \,]$$

be a generator matrix of a binary (5,3)-code C over the field $F = \mathbb{Z}_2$. Then, a parity check matrix of the code C has the form:

$$\mathbf{H} = \begin{pmatrix} 0 & 1 & 1 & 1 & 0 \\ 1 & 0 & 1 & 0 & 1 \end{pmatrix} = [\, -\mathbf{P}^T \mid \mathbf{I}_2 \,].$$

From (9.20), it follows that a word $c = (c_1, c_2, c_3, c_4, c_5) \in C$ if and only if it is a solution of a system of linear equations:

$$\begin{cases} c_2 + c_3 + c_4 = 0 \\ c_1 + c_3 + c_5 = 0 \end{cases}$$

The following bound on the parameters of a linear code is important.

❋ Theorem 9.57.
Every linear (n, k, d)-code $C \subset F^n$ satisfies:

$$d \le n - k + 1. \tag{9.24}$$

Proof. Since every linear (n, k, d)-code C is equivalent to some systematic code C_1 with matrix $\mathbf{G}_1$ in standard form (9.22), each row vector $\mathbf{u}$ has the Hamming weight $wt(\mathbf{u}) \le n - k + 1$. Hence, $d = d_{\min}(C) \le n - k + 1$, because C for a linear code $d = wt_{\min}(C)$, by (9.14). □

9.5.3. Polynomial codes

In this subsection, we consider one particular class of linear codes, namely, which are polynomial codes. In this consideration, we will identify a codeword of length n with a polynomial of degree at most $n - 1$.

Recall that, by $F_n[x]$, we denote the set of all polynomials of degree $< n$ over a field F. If $f(x), g(x) \in F_n[x]$, then $f(x) + g(x) \in F_n[x]$ and $\alpha f(x) \in F_n[x]$ for an arbitrary element $\alpha \in F$. Therefore, $F_n[x]$ is a vector space over the field F and $\dim_F F_n[x] = n$.

Let $u(x) \in F_n[x]$ be a polynomial of degree $n - 1$:

$$u(x) = a_0 + a_1 x + a_2 x^2 + \cdots + a_{n-1} x^{n-1}, \tag{9.25}$$

where $a_i \in F$. The polynomial $u(x)$ corresponds to a vector $\mathbf{u} = (a_0, a_1, a_2, \ldots, a_{n-1}) \in F^n$ and, inversely, each vector $\mathbf{u} = (a_0, a_1, a_2, \ldots, a_{n-1}) \in F^n$ corresponds to the polynomial (9.22).

Consider the mapping:

$$\varphi : F^n \to F_n[x] \tag{9.26}$$

with $\varphi(\mathbf{u}) = \varphi(a_0, a_1, a_2, \ldots, a_{n-1}) = u(x) = a_0 + a_1 x + a_2 x^2 + \cdots + a_{n-1} x^{n-1}$. It is easy to show that φ is a linear map which is an isomorphism of vector spaces. Therefore, in our consideration, the vector space F^n can be replaced by the vector space $F_n[x]$.

❋ Example 9.58.
Let $F = \mathbb{Z}_2$ and $n = 6$. The vector $(1, 0, 1, 0, 0, 1) \in F^6$ corresponds to the polynomial

$$1 + 0 \cdot x + 1 \cdot x^2 + 0 \cdot x^3 + 0 \cdot x^4 + 1 \cdot x^5 = 1 + x^2 + x^5 \in F_6[x].$$

The polynomial $x + x^2 + x^4 \in F_6[x]$ corresponds to the vector: $(0, 1, 1, 0, 1, 0) \in F^6$.

Let $g(x) \in F[x]$ be a polynomial of degree $n - k$ and consider a mapping:

$$\xi : F_k[x] \to F_n[x] \tag{9.27}$$

given by:
$$\xi(f(x)) = f(x)g(x) \tag{9.28}$$

for an arbitrary polynomial $f(x) \in F_k[x]$.

If $\xi(f(x)) = 0$, then $f(x)g(x) = 0$. Hence, $f(x) = 0$, since $\deg[f(x)g(x)] < n$ and $F[x]$ is an integral domain. So, ξ is an injection. Therefore, the mapping ξ defines a linear (n, k)-code that is equal to $\mathrm{Im}(\xi)$.

> ✳ **Definition 9.59.**
>
> A linear (n, k)-code over a field F whose codewords are exactly those polynomials of degree $\leq n - 1$ which are divided by $g(x)$ of degree $n - k$, i.e.,
> $$C = \{f(x)g(x) \; : \; f(x) \in F_k[x]\} \subset F_n[x] \tag{9.29}$$
> is called a **polynomial** (n, k)**-code** and the polynomial $g(x)$ is called the **generator polynomial** of this code.

Since a polynomial (n, k)-code is generated by the generator polynomial $g(x)$ of degree $n - k$ and $\dim(C) = k$, we obtain that:

$$\dim(C) = n - \deg(g(x)) \tag{9.30}$$

✱ **Examples 9.60.**

1. The polynomial $g(x) = 1 + x$ generates a polynomial $(3,2)$-code over the field $F = \mathbb{Z}_2$. The ring of polynomials $F_2[x]$ consists of 4 elements:

$$0, 1, x, 1 + x$$

Since $0 \cdot (1 + x) = 0$; $\quad 1 \cdot (1 + x) = 1 + x$;

$x \cdot (1 + x) = x + x^2$; $\quad (1 + x) \cdot (1 + x) = 1 + x^2$;

the linear $(3,2)$-code over the field $\mathbb{Z}_2$ with the generator polynomial $g(x) = 1 + x$ contains 4 elements:

$$0, \; 1 + x, \; x + x^2, \; 1 + x^2,$$

which correspond to 4 codewords:

$$(0\ 0\ 0), \; (1\ 1\ \ 0), \; (0\ 1\ 1), \; (1\ 0\ 1).$$

2. The polynomial $g(x) = x + x^2$ generates a polynomial $(5,3)$-code over the field $F = \mathbb{Z}_2$. The ring of polynomials $F_3[x]$ consists of 8 elements:

$$0, 1, x, 1 + x, x^2, 1 + x^2, 1 + x + x^2, x + x^2$$

Since

$$0 \cdot (x + x^2) = 0; \quad 1 \cdot (x + x^2) = x + x^2;$$
$$x \cdot (x + x^2) = x^2 + x^3; \quad (1 + x) \cdot (x + x^2) = x + x^3;$$
$$x^2 \cdot (x + x^2) = x^3 + x^4; \quad (1 + x^2) \cdot (x + x^2) = x + x^2 + x^3 + x^4;$$
$$(1 + x + x^2) \cdot (x + x^2) = x + x^4; \quad (x + x^2) \cdot (x + x^2) = x^2 + x^4,$$

the linear (5,3)-code over the field $\mathbb{Z}_2$ with the generator polynomial $g(x) = x + x^2$ contains 8 elements:

$$0, \ x + x^2, \ x^2 + x^3, \ x + x^3, \ x^3 + x^4, \ x + x^2 + x^3 + x^4, \ x + x^4, \ x^2 + x^4,$$

which correspond to 8 codewords:

$$(0\ 0\ 0\ 0\ 0), \ (0\ 1\ 1\ 0\ 0), \ (0\ 0\ 1\ 1\ 0), \ (0\ 1\ 0\ 1\ 0),$$

$$(0\ 0\ 0\ 1\ 1), \ (0\ 1\ 1\ 1\ 1), \ (0\ 1\ 0\ 0\ 1), \ (0\ 0\ 1\ 0\ 1).$$

✳ **Theorem 9.61.**
Let C_g be a polynomial (n, k)-code over a field F with the generator polynomial $g(x) \in F[x]$ of degree $n - k$. If $h(x) \in F[x]$ is a polynomial of degree n and $g(x)$ divides $h(x)$, then C_g is a principal ideal in a factor ring $F[x]/\langle h(x) \rangle$ generated by the polynomial $g(x)$.

Proof.
Let $h(x) \in F[x]$ be a polynomial of degree n and $g(x)|h(x)$, i.e., $h(x) = g(x)u(x)$. By definition, an arbitrary element $c(x) \in C_g$ has the form: $c(x) = f(x)g(x)$, where $f(x) \in F_k[x]$. Let $s(x) \in F[x]$, then there are polynomials $q(x), r(x) \in F[x]$ such that $c(x)s(x) = q(x)h(x) + r(x)$ and either $r(x) = 0$ or $0 \leq \deg(r(x)) < n$. Therefore, $r(x) \in F_n[x]$. Since $g(x)|h(x)$ and $g(x)|c(x)$, we also have that $g(x)|r(x)$. In this way, $r(x) \in C_g$. So, $c(x)s(x) = q(x)h(x) + r(x) \equiv r(x) \, (\mathrm{mod}\, J)$, where $J = \langle h(x) \rangle$, and $r(x) \in C_g$. This means that C_g is an ideal of the ring $F[x]/\langle h(x) \rangle$ and it is generated by $g(x)$, i.e., C_g is a principal ideal. □

✳ **Theorem 9.62.**
Let $g(x) \in F[x]$ be a polynomial of degree $n - k$ and let $h(x) \in F[x]$ be a polynomial of degree n. If a vector subspace $C \subset F_n[x]$ is a principal ideal of a factor ring $F[x]/\langle h(x) \rangle$ generated by the polynomial $g(x)$, then $g(x)|h(x)$ and C is a polynomial (n, k)-code with the generator polynomial $g(x)$.

Proof.
Since $\deg(g(x)) = n - k$ and $\deg(h(x)) = n$, there are polynomials $q(x), r(x) \in F[x]$ such that

$$h(x) = q(x)g(x) + r(x) \tag{9.31}$$

and either $r(x) = 0$ or

$$0 \leq \deg(r(x)) < \deg(g(x)) = n - k. \tag{9.32}$$

Let $J = \langle h(x) \rangle$. Then, from (9.31), we obtain that $r(x) \equiv -g(x)q(x) \pmod{J}$. By assumption, C is a principal ideal of the ring $F[x]/J$ generated by the polynomial $g(x)$, therefore,

$$\deg(r(x)) \geq \deg(g(x)) = n - k \tag{9.33}$$

Comparing with (9.32), we obtain that $r(x) = 0$. So, $h(x) = g(x)q(x)$, which proves that $g(x)|h(x)$.

Let $c(x) \in C \subset F_n[x]$. Then, $c(x) = g(x)u(x) \in F[x]/J$, because C is a principle ideal of $F[x]/J$ generated by $g(x)$. Therefore, $\deg(c(x)) = \deg(g(x)u(x)) = \deg(g(x)) + \deg(u(x)) = (n - k) + \deg(u(x))$. Since $\deg(c(x)) \leq n - 1$, we obtain that $\deg(u(x)) \leq k - 1$, i.e., $u(x) \in F_k[x]$. Therefore, $c(x) = g(x)u(x)$, where $u(x) \in F_k[x]$, which proves that C is a polynomial (n, k)-code with the generator polynomial $g(x)$. □

To construct a polynomial code with the generator polynomial $g(x)$, we can use the following algorithm.

Algorithm of construction of a polynomial code

Let C be a polynomial (n, k)-code generated by a polynomial $g(x)$ with $\deg g(x) = n - k$.

Any original word of length k can be written as a polynomial $f(x)$ with $\deg f(x) = k - 1$.

1. The polynomial $f(x)$ is multiplied by a monomial x^{n-k}.

2. The polynomial $x^{n-k}f(x)$ is divided with reminder by the generator polynomial $g(x)$: $x^{n-k}f(x) = g(x)q(x) + r(x)$, where either $r(x) = 0$ or $0 \leq \deg r(x) < \deg g(x) = n - k$.

3. The polynomial $v(x) = g(x)q(x) = x^{n-k}f(x) - r(x)$ is a polynomial of degree $n - 1$. It is a codeword of C, which corresponds to the original word $f(x)$.

Since $0 \leq \deg r(x) < n-k$, only the first k coefficients of the highest powers of the polynomial $v(x)$ represent the **information digits** of the original message. The last $n - k$ coefficients of the lowest powers of $v(x)$ are called **control digits** or **checking bits**.

✳ Example 9.63.
Let $F = \mathbb{Z}_2$. The polynomial $g(x) = 1 + x^2 + x^3 + x^4 \in \mathbb{Z}_2[x]$ can generate a polynomial (7,3)-code. In this case, $n = 7$, $k = 3$, $n - k = 4$.

Suppose we have the original message: (1 0 1), which corresponds to the polynomial $f(x) = 1 + x^2$. We compute:

$$x^4 f(x) = x^4 + x^6 = (1 + x^2 + x^3 + x^4)(1 + x + x^2) + (1 + x),$$

where $r(x) = 1 + x$ is the remainder. Then,

$$v(x) = x^4 f(x) - r(x) = 1 + x + x^4 + x^6$$

is the polynomial corresponding to the codeword: (1 1 0 0 1 0 1).

The first 4 digits of this codeword are 1 1 0 0, which are the control digits, and the last 3 digits 1 0 1 the information digits, which form the original message.

✳ **Example 9.64.**

Encode the message given by the word (1 1 0) in the polynomial (6,3)-code with generator polynomial $g(x) = 1 + x + x^3 \in \mathbb{Z}_2[x]$.

In this case, $k = 3$, $n - k = 3$, $n = 3$. The number of all possible messages is equal to $2^k = 2^3 = 8$. The message (1 1 0) corresponds to the polynomial $f(x) = 1 + x$. We compute

$$x^3 f(x) = x^3 + x^4 = (1 + x + x^3)(1 + x) + (1 + x^2),$$

where $r(x) = 1 + x^2$ is the remainder. Then, the polynomial

$$v(x) = x^3 f(x) - r(x) = 1 + x^2 + x^3 + x^4$$

is the polynomial corresponding to the codeword: (1 0 1 1 1 0).

The first 3 digits of the codeword are 1 0 1 and they are the control digits, and the last 3 digits of this codeword are 1 1 0 and they are information symbols, so they form the original message.

To detect errors in an encoded message in a polynomial code, we can use a simple method. The obtained codeword must be divided by the generator polynomial $g(x)$. If the remainder on division of a polynomial corresponding to the obtained codeword by $g(x)$ is not equal to zero, then the message is obtained with errors.

To correct errors in an encoded message, we can use some other algorithms for correcting errors, one of which is the algorithm of BCH codes, which will be considered below.

✳ **Example 9.65.**

Let $F = \mathbb{Z}_2$, $n = 5$, $k = 3$ and $g(x) = 1 + x + x^2$ is the generator polynomial of the polynomial (5, 3)-code. Verify whether or not the codewords

$$(1\ 1\ 0\ 1\ 1), \ (1\ 0\ 0\ 1\ 0), \ (1\ 1\ 1\ 1\ 1)$$

contain errors.

These codewords correspond to polynomials: $1 + x + x^3 + x^4$, $1 + x^3$, $1 + x + x^2 + x^3 + x^4$.

We compute the remainders on dividing these polynomials by the generator polynomial $g(x)$.

1. $1 + x + x^3 + x^4 = (1 + x^2)(1 + x + x^2)$. Since the remainder is equal to 0, the codeword (1 1 0 1 1) most likely contains no errors.

2. $1 + x^3 = (1 + x)(1 + x + x^2)$. Since the remainder is equal to 0, the codeword (1 0 0 1 0) most likely contains no errors.

3. $1 + x + x^2 + x^3 + x^4 = x^2(1 + x + x^2) + (1 + x)$. Since the remainder $r(x) = 1 + x \neq 0$, the codeword (1 1 1 1 1) contains errors.

9.5.4. Cyclic codes

In this subsection, we consider one of the most important and interesting examples of linear codes, namely, cyclic codes. They can also be considered as a particular example of polynomial codes.

✳ **Definition 9.66.**

A linear code $C \subset F^n$ is called **cyclic** if for every codeword $\mathbf{u} = (a_0, a_1, a_2, \ldots, a_{n-1}) \in C$ the vector $\mathbf{u}' = (a_{n-1}, a_0, a_1, a_2, \ldots, a_{n-2})$ obtained by a cyclic shift from the vector $\mathbf{u}$ is also a codeword of C.

✴ **Example 9.67.**

1. The binary linear code $C = \{(0\ 0\ 0), (1\ 0\ 1), (0\ 1\ 1), (1\ 1\ 0)\} \in \mathbb{Z}_2^3$ is a cyclic code of length 3.

2. The code with the generator matrix $\mathbf{G} = \begin{pmatrix} 1 & 0 & 0 & 1 & 0 & 0 \\ 0 & 1 & 0 & 0 & 1 & 1 \\ 0 & 0 & 1 & 0 & 0 & 1 \end{pmatrix}$ is a cyclic code of length 6. Really, we have:

$$(0\ 0\ 0)\mathbf{G} = (0\ 0\ 0\ 0\ 0\ 0); \quad (0\ 0\ 1)\mathbf{G} = (0\ 0\ 1\ 0\ 0\ 1);$$

$$(0\ 1\ 0)\mathbf{G} = (0\ 1\ 0\ 0\ 1\ 0); \quad (0\ 1\ 1)\mathbf{G} = (0\ 1\ 1\ 0\ 1\ 1);$$

$$(1\ 0\ 0)\mathbf{G} = (1\ 0\ 0\ 1\ 0\ 0); \quad (1\ 0\ 1)\mathbf{G} = (1\ 0\ 1\ 1\ 0\ 1);$$

$$(1\ 1\ 0)\mathbf{G} = (1\ 1\ 0\ 1\ 1\ 0); \quad (1\ 1\ 1)\mathbf{G} = (1\ 1\ 1\ 1\ 1\ 1).$$

3. The binary linear code $C = \{(0\ 0\ 0\ 0), (1\ 0\ 0\ 1), (0\ 1\ 1\ 0), (1\ 1\ 1\ 1)\}$ is not cyclic because the word (1 1 0 0) does not belong to C.

Linear cyclic codes are a particular case of polynomial codes. Each codeword $\mathbf{u} = (a_0, a_1, a_2, \ldots, a_{n-1})$ can be represented by the polynomial $u(x) = a_0 + a_1 x + a_2 x^2 + \cdots + a_{n-1} x^{n-1}$. Then, a cyclic shift can be represented as result of multiplication: $xu(x) = a_0 x + a_1 x^2 + a_2 x^3 + \cdots + a_{n-1} x^n = a_0 t + a_1 x^2 + a_2 x^3 + \cdots + a_{n-1} = a_{n-1} + a_0 x + a_1 x^2 + a_2 x^3 + \cdots + a_{n-1}$ if we consider that $x^n = 1$. In this case, a codeword of a cyclic code C corresponds to a polynomial of the factor ring $F[x]/\langle x^n - 1 \rangle$ and a cyclic code can be interpreted as a subset in $F[x]/\langle x^n - 1 \rangle$ by the mapping $\mathbf{u} \to u(x)$. Since the structure of the ring $F[x]/\langle x^n - 1 \rangle$ is well known, we obtain an extra tool to study cyclic codes.

✳ **Theorem 9.68.**
A linear subspace $C \subset F_n[x]$ over a field F is a cyclic code if and only if it is an ideal of the ring $F[x]/\langle x^n - 1 \rangle$.

Proof.
Let C be a linear cyclic code of length n. Then, each codeword of C can be represented as a polynomial $u(x) = a_0 + a_1 x + a_2 x^2 + \cdots + a_{n-1} x^{n-1} \in C$, where $x^n = 1$, which means that $u(x) \in F[x]/\langle x^n - 1 \rangle$. Since C is a cyclic code, $xu(x) = a_{n-1} + a_0 x + a_1 x^2 + \cdots + a_{n-2} x^{n-1}$ also belongs to C. It is obvious that $x^k u(x) \in C$ for any $k \in \mathbb{N}$. Since C is a linear code, an arbitrary linear combination of codewords $u(x), xu(x), \ldots, x^k u(x)$ also belongs to C, i.e., $u(x)f(x) \in C$ for any polynomial $f(x) \in F[x]$, which means that C is an ideal of $F[x]/\langle x^n - 1 \rangle$.

Inversely, let C be an ideal of the ring $F[x]/\langle x^n - 1 \rangle$. Let $u(x) = a_0 + a_1 x + a_2 x^2 + \cdots + a_{n-1} x^{n-1} \in C$. Since C is an ideal of $F[x]/\langle x^n - 1 \rangle$, the polynomial $xu(x) = a_{n-1} + a_0 x + a_1 x^2 + \cdots + a_{n-2} x^{n-1}$ also belongs to C, since $x^n = 1$. Therefore, C is a cyclic code, since C is also a linear subspace of $F_n[x]$. $\qquad\square$

✳ **Theorem 9.69.**
Each ideal of the ring $F[x]/\langle x^n - 1 \rangle$ is principal. For any non-zero ideal $I \subset F[x]/\langle x^n - 1 \rangle$, there exists a unique monic polynomial $g(x)$ such that it is the generator polynomial of I and a proper divisor of $x^n - 1$.

Proof.
Let I be a non-zero ideal of the ring $F[x]/\langle x^n - 1 \rangle$. Then, I contains a monic polynomial $g(x)$ of the least degree in I. Let $f(x) \in I$ and $f(x) \neq ag(x)$ for all $a \in F$. Then, by choosing $g(x)$, $\deg(f(x)) > \deg(g(x))$. Therefore, there exist polynomials $q(x), r(x) \in F[x]$ such that $f(x) = q(x)g(x) + r(x)$ and either $r(x) = 0$ or $0 \leq \deg(r(x)) < \deg(g(x))$. Since $f(x), g(x) \in I$, we have $r(x) \in I$ as well. Then, by choosing $g(x)$, $r(x) = 0$, i.e., $f(x) = q(x)g(x)$. So, I is a principal ideal. If I is generated by two monic polynomials $g_1(x)$ and $g_2(x)$, then $g_1(x)|g_2(x)$ and $g_2(x)|g_1(x)$. Since $g_1(x), g_2(x)$ are both monic polynomials, $g_1(x) = g_2(x)$.

Since $I = \langle g(x) \rangle$ is a non-zero ideal of $F[x]/\langle x^n - 1 \rangle$, $x^n - 1 \in I$. Therefore, $g(x) | x^n - 1$ and $g(x)$ is a proper divisor of $x^n - 1$, since $\deg(g(x)) < n$ by construction. □

From Theorems 9.68 and 9.69, it follows that there is a bijection between principal ideals of the ring $F[x]/\langle x^n - 1 \rangle$ and cyclic codes of length n. Hence, there exists a bijection between cyclic codes of length n and proper divisors of the polynomial $x^n - 1$.

Therefore, we have the following corollary.

✳ **Theorem 9.70.**

Each linear subspace $C \subset F^n$ is a cyclic code of length n over a field F if and only if it is a principle ideal of the ring $F[x]/\langle x^n - 1 \rangle$ of the form $C = \langle g(x) \rangle$, where $g(x)$ is a monic polynomial which divides the polynomial $x^n - 1$.

✳ **Example 9.71.**

The binary cyclic code $C = \{(0\ 0\ 0), (1\ 1\ 0),\ (1\ 0\ 1),\ (0\ 1\ 1)\}$ to length 3 from Example 9.67(1) corresponds to the set of polynomials: $I = \{0, 1 + x, 1 + x^2, x + x^2\}$ with $x^3 = 1$. I is a principal ideal $I = \langle 1 + x \rangle$ of the factor ring $R = \mathbb{Z}_2[x]/\langle x^3 - 1 \rangle$. Really, the ring R contains 8 elements: $0, 1, x, x^2, 1+x, 1+x^2, x+x^2, 1+x+x^2$ and $1 \cdot (1+x) = 1+x$, $x(1+x) = x + x^2$, $x^2(1+x) = x^2 + 1$, $(1+x)(1+x) = 1+x^2$, $(1+x^2)(1+x) = x+x^2$, $(x + x^2)(1+x) = 1+x$, $(1+x+x^2)(1+x) = 0$.

The generator polynomial $1 + x$ of the ideal I is a proper divisor of $x^3 - 1$, since $x^3 - 1 = (1+x)(1+x+x^2)$ in $\mathbb{Z}_2[x]$.

✳ **Example 9.72.**

Find a binary cyclic code of length 7.

Each binary cyclic code of length 7 corresponds to a principle ideal in $\mathbb{Z}_2[x]/\langle x^7 - 1 \rangle$ generated by a polynomial $g(x) \in F_6[x]$ which divides the polynomial $x^7 - 1$. Since

$$x^7 - 1 = (1+x)(1+x+x^3)(1+x^2+x^3)$$

is the factorization of $x^7 - 1$ into a product of irreducible polynomials of $\mathbb{Z}_2[x]$, the generator polynomial of a cyclic code of length 7 can be chosen as one of polynomials which is a divisor of $x^7 - 1$. Suppose we choose the generator polynomial $g(x) = (1+x)(1+x+x^3) = 1 + x^2 + x^3 + x^4$. Since $F_3[x]$ has 8 elements:

$$0,\ 1,\ x,\ 1+x,\ x^2,\ 1+x^2,\ 1+x+x^2,\ x+x^2,$$

we obtain the following 8 polynomials belonging to the cyclic code of length 7:

$0 \cdot g(x) = 0$
$1 \cdot g(x) = 1 + x^2 + x^3 + x^4$
$x \cdot g(x) = x + x^3 + x^4 + x^5$
$(1 + x) \cdot g(x) = 1 + x + x^2 + x^5$

$$x^2 \cdot g(x) = x^2 + x^3 + x^5 + x^6$$
$$(1 + x^2) \cdot g(x) = 1 + x^3 + x^5 + x^6$$
$$(x + x^2) \cdot g(x) = x + x^2 + x^3 + x^6$$
$$(1 + x + x^2) \cdot g(x) = 1 + x + x^4 + x^6$$

and corresponding codewords:

$$(0\ 0\ 0\ 0\ 0\ 0\ 0),\ (1\ 0\ 1\ 1\ 1\ 0\ 0),\ (0\ 1\ 0\ 1\ 1\ 1\ 0),\ (1\ 1\ 1\ 0\ 0\ 1\ 0),$$

$$(0\ 0\ 1\ 1\ 0\ 1\ 1),\ (1\ 0\ 0\ 1\ 0\ 1\ 1),\ (0\ 1\ 1\ 1\ 0\ 0\ 1),\ (1\ 1\ 0\ 0\ 1\ 0\ 1).$$

If C is a linear cyclic code of length n with the generator polynomial $g(x) = g_0 + g_1 x + g_2 x^2 + \cdots + g_{n-k-1} x^{n-k-1} + g_{n-k} x^{n-k}$, then $\dim(C) = n - k$ and the generator matrix $\mathbf{G} \in M_{(n-k) \times n}(F)$, which is obtained as a result of multiplication of the polynomial $g(x)$ by the basis elements $1, x, x^2, \ldots, x^{n-k}$ of the vector space $F_k[x]$, has the form:

$$\mathbf{G} = \begin{pmatrix} g_0 & g_1 & g_2 & \cdots & g_{n-k-1} & g_{n-k} & 0 & \cdots & 0 \\ 0 & g_0 & g_1 & \cdots & g_{n-k-2} & g_{n-k-1} & g_{n-k} & \cdots & 0 \\ \vdots & \vdots & \vdots & \ddots & \vdots & \vdots & \vdots & \ddots & \vdots \\ 0 & 0 & 0 & \cdots & g_0 & g_1 & g_2 & \cdots & g_{n-k} \end{pmatrix} \tag{9.34}$$

✳ **Definition 9.73.**

Let C be a linear cyclic code of length n with the generator polynomial $g(x)$. Then, a polynomial $h(x)$ such that

$$g(x)h(x) = x^n - 1 \tag{9.35}$$

is called the **parity check polynomial** of C.

It is easy to show that $c(x) \in C$, where C is a cyclic code of length n, if and only if

$$c(x)h(x) \equiv 0 \,(\mathrm{mod}\, x^n - 1).$$

The parity check matrix $\mathbf{H} \in M_{k \times n}(F)$ corresponding to the parity check polynomial $h(x) = h_0 + h_1 x + h_2 x^2 + \cdots + h_{k-1} x^{k-1} + h_k x^k$ has the form:

$$\mathbf{H} = \begin{pmatrix} h_k & h_{k-1} & \cdots & h_2 & h_1 & h_0 & 0 & \cdots & 0 & 0 \\ 0 & h_k & \cdots & h_3 & h_2 & h_1 & h_0 & \cdots & 0 & 0 \\ \vdots & \vdots & \ddots & \vdots & \vdots & \vdots & \vdots & \ddots & \vdots & \vdots \\ 0 & 0 & \cdots & h_k & h_{k-1} & h_{k-2} & h_{k-3} & \cdots & h_0 & 0 \\ 0 & 0 & \cdots & 0 & h_k & h_{k-1} & h_{k-2} & \cdots & h_1 & h_0 \end{pmatrix} \tag{9.36}$$

❋ Example 9.74.

Consider the binary cyclic code of length 7 generated by polynomial $g(x) = (1 + x)(1 + x + x^3) = 1 + x^2 + x^3 + x^4$. Then, the parity check polynomial $h(x) = 1 + x^2 + x^3$. In this case, $n = 7$, $n - k = 4$, $k = 3$. The corresponding generator and parity check matrices have the following forms:

$$
G = \begin{pmatrix} 1 & 0 & 1 & 1 & 1 & 0 & 0 \\ 0 & 1 & 0 & 1 & 1 & 1 & 0 \\ 0 & 0 & 1 & 0 & 1 & 1 & 1 \end{pmatrix}, \quad
H = \begin{pmatrix} 1 & 1 & 0 & 1 & 0 & 0 & 0 \\ 0 & 1 & 1 & 0 & 1 & 0 & 0 \\ 0 & 0 & 1 & 1 & 0 & 1 & 0 \\ 0 & 0 & 0 & 1 & 1 & 0 & 1 \end{pmatrix}
$$

9.5.5. BCH codes

The **BCH codes**, or **Bose-Chaudhuri-Hocquenghem codes**, form an important class of linear cyclic codes which enable us to correct multiple errors. They are particular examples of cyclic polynomials codes with generator polynomials over finite fields. This class of multiple-error correcting codes is a remarkable generalization of the cyclic Hamming codes that correct only one error. Binary BCH codes were invented by A. Hocquenghem in 1959 [21] and independently by R.C. Bose and D.K. Ray-Chaudhuri in 1960 [9]. They are very important because of their vastness and sufficiently simple scheme of decoding.

Let $F = GF(q)$ be the Galois field containing $q = p^m$ elements, where p is a prime. Let α be a primitive element of the field F. If $n = q - 1 = p^m - 1$, then $F = \{0, \alpha, \alpha^2, \ldots, \alpha^{n-1}, \alpha^n = 1\}$.

We choose a positive integer $d \leq p^m - 1$. For each positive integer $1 \leq i \leq d$, let $m_i(x) \in \mathbb{Z}_p[x]$ be the **minimal polynomial** of the element α^i, i.e., a non-zero monic polynomial of the least degree whose root is α^i. We define the polynomial:

$$
g(x) = \mathrm{lcm}(m_1(x), m_2(x), \ldots, m_{d-1}(x)) \tag{9.37}
$$

✳ Definition 9.75.

A polynomial code with the generator polynomial $g(x)$ given by (9.37) is called the **primitive narrow-sense BCH code** of length n and design distance d.

Since the polynomial $g(x) \in \mathbb{Z}_p[x]$ divides $x^n - 1$, the primitive narrow-sense BCH code is the cyclic polynomial code, by Theorem 9.70.

❋ Remark 9.76.

In the general case for construction of BCH codes, we choose an element $\alpha \in F$ with $\mathrm{ord}(\alpha) = k \leq n$ (recall that k is a minimal positive integer such that $\alpha^k = 1$) instead of primitive elements of a field F. In this case, codewords are of length k. Also, in order to construct the general BCH codes, the generator polynomial may have the form

$$g(x) = \text{lcm}(m_c(x), \ldots, m_{c+d-2}(x))$$

In this section, we will consider only the primitive narrow-sense BCH codes.

✳ **Theorem 9.77.**

A primitive narrow-sense BCH code of length $n = p^m - 1$ and design distance d with the generator polynomial (9.37) has minimum distance at least d.

Proof.

Let C be a primitive narrow-sense BCH code of length $n = p^m - 1$. Since C is a linear code, its minimum distance is equal to its minimum weight. Suppose that there is a codeword $u(x) = c_0 + c_1 x + \cdots c_{n-1} x^{n-1} \in C$ such that $wt(u(x)) = k < d$. Let α be a primitive element of the field F, $m_i(x)$ be the minimal polynomial of α^i, and the generator polynomial $g(x)$ is given in the form (9.37). Then, all elements α^i are roots of $u(x)$, i.e., $u(\alpha^i) = 0$, for $i = 1, 2, \ldots, d - 1$. Therefore, we have the system of linear equations:

$$u(\alpha) = u(\alpha^2) = \cdots = u(\alpha^{d-1}) = 0$$

Since $wt(u(x)) = k < d$, there are only $k < d$ indexes $i \in \{0, 1, 2, \ldots, n - 1\}$ for which $c_i \neq 0$. Therefore, we can consider only the set of indexes $S = \{i_1, i_2, \cdots, i_k\}$ such that $i_1 < i_2 < \ldots < i_k$ and $c_{i_j} \neq 0$ if and only if $i_j \in S$. Hence, we obtain the equivalent system of equations:

$$\begin{cases} c_{i_1}\alpha^{i_1} + c_{i_2}\alpha^{i_2} + \cdots + c_{i_k}\alpha^{i_k} & = 0 \\ c_{i_1}\alpha^{2i_1} + c_{i_2}\alpha^{2i_2} + \cdots + c_{i_k}\alpha^{2i_k} & = 0 \\ \cdots & \cdots \\ c_{i_1}\alpha^{(d-1)i_1} + c_{i_2}\alpha^{(d-1)i_2} + \cdots + c_{i_k}\alpha^{(d-1)i_k} & = 0 \end{cases}$$

Choosing the first k equations of this system, we obtain the following system:

$$\begin{cases} b_1 x_1 + b_2 x_2 + \cdots + b_k x_k & = 0 \\ b_1^2 x_1 + b_2^2 x_2 + \cdots + b_k^2 x_k & = 0 \\ \cdots & \cdots \\ b_1^k x_1 + b_2^k x_2 + \cdots + b_k^k x_k & = 0 \end{cases} \qquad (9.38)$$

where $b_j = \alpha^{i_j}$, $x_j = c_{i_j}$. Let

$$\mathbf{B} = \begin{pmatrix} b_1 & b_2 & \cdots & b_k \\ b_1^2 & b_2^2 & \cdots & b_k^2 \\ \cdots & \cdots & \cdots & \cdots \\ b_1^k & b_2^k & \cdots & b_k^k \end{pmatrix}.$$

Then,

$$\det(\mathbf{B}) = \prod_{j=1}^{k} b_j \det \begin{pmatrix} 1 & 1 & \cdots & 1 \\ b_1 & b_2 & \cdots & b_k \\ \cdots & \cdots & & \cdots \\ b_1^{k-1} & b_2^{k-1} & \cdots & b_k^{k-1} \end{pmatrix} = \prod_{j=1}^{k} b_j \prod_{1 \le j < s \le k} (b_s - b_j).$$

Since $b_j = \alpha^{i_j} \ne \alpha^{i_s} = b_s$ for $j \ne s$, $\det(\mathbf{B}) \ne 0$ and the system (9.38) has only the trivial solution: $x_i = c_{i_j} = 0$ for all i. This contradiction proves the theorem. $\qquad \square$

The next theorem shows importance and practice of BCH codes. Its proof follows from the construction of primitive narrow-sense BCH codes and Theorems 9.77 and 9.52.

✱ **Theorem 9.78.**
For arbitrary positive integers m and $t \le p^{m-1} - 1$, there is a primitive narrow-sense BCH code of length $n = p^m - 1$ which corrects up to t errors.

✱ **Example 9.79.**
Find a primitive narrow-sense BCH code of length $n = 3^2 - 1$ with designed distance 5 which corrects up to 2 errors.

In this case, $p = 3$, $m = 2$, $n = 8$ and $t = 2$.

Consider the irreducible polynomial $p(x) = x^2 + 2x + 2 \in \mathbb{Z}_3[x]$ and the field:

$$F = \mathbb{Z}_3[x]/\langle x^2 + 2x + 2 \rangle = \{c_0 + c_1\alpha + c_2\alpha^2 \ : \ \alpha^2 = 1 + \alpha, c_i \in \mathbb{Z}_3\}.$$

Then, $p(\alpha) = 0$ and $\alpha^8 = 1$. From Example 9.13, we have

$$\alpha^1 = \alpha, \alpha^2 = \alpha+1, \alpha^3 = 2\alpha+1, \alpha^4 = 2, \alpha^5 = 2\alpha, \alpha^6 = 2\alpha+2, \alpha^7 = \alpha+2, \alpha^8 = 1,$$

which shows that α is a primitive element of F. Since each element of F is a root of the polynomial

$$x^9 - x = x(x+1)(x+2)(x^2+1)(x^2+x+2)(x^2+2x+2)$$

we can easy verify that minimal polynomial of α^i have the forms:
$m_1(x) = m_3(x) = m_5(x) = m_7(x) = x^2+2x+2, \quad m_2(x) = m_6(x) = x^2+1,$
$m_4(x) = x + 1$.

Since $2 \cdot t + 1 = 5 = d$, $d - 1 = 4$. So, we must take 4 minimal polynomials to construct the generator polynomial for a primitive narrow-sense BCH code of length 8 with minimum distance at least 5 which corrects up to 2 errors:

$$g(x) = \text{lcm}(m_1(x), m_2(x), m_3(x), m_4(x)) =$$

$$= (x+1)(x^2+1)(x^2+2x+2) = x^5 + 2x^3 + 2x^2 + x + 2$$

✳ Remark 9.80.

Note, that if $p = 2$, then $m_i(x) \in \mathbb{Z}_2[x]$. Therefore, $[m_i(x)]^2 = m_i(x^2)$. So, α^{2i} is a root of a polynomial $m_i(x)$, which means that the minimal polynomials in (9.37) with even indexes can be omitted.

✳ Examples 9.81.

1. Find the binary primitive narrow-sense BCH code of length $n = 2^4 - 1$ which corrects up to 2 errors.

   In this case, $m = 4$, $q = 16$, $q - 1 = 15$ and $t = 2 < 7$. Consider the irreducible polynomial $p(x) = x^4 + x^3 + 1 \in \mathbb{Z}_2[x]$ and the field:

   $$F = \mathbb{Z}_2[x]/\langle x^4 + x^3 + 1 \rangle = \{c_0 + c_1\alpha + c_2\alpha^2 + c_3\alpha^3 : \alpha^4 = 1 + \alpha^3, c_i \in \mathbb{Z}_2\}.$$

   Then, $p(\alpha) = 0$ and $\alpha^{15} = 1$.

   Since $\alpha^4 = \alpha^3 + 1$, $\alpha^5 = \alpha^3 + \alpha + 1$, $\alpha^6 = \alpha^3 + \alpha^2 + \alpha + 1$, $\alpha^7 = \alpha^2 + \alpha + 1$, $\alpha^8 = \alpha^3 + \alpha^2 + \alpha$, $\alpha^9 = \alpha^2 + 1$, $\alpha^{10} = \alpha^3 + \alpha$, $\alpha^{11} = \alpha^3 + \alpha^2 + 1$, $\alpha^{12} = \alpha + 1$, $\alpha^{13} = \alpha^2 + \alpha$, $\alpha^{14} = \alpha^3 + \alpha^2$, $\alpha^{15} = 1$, we obtain that α is a primitive root of F. Any element of F is a root of $x^{15} - x$ which is decomposed into the product of irreducible polynomials:

   $$x^{16} - x = x(x+1)(x^2 + x + 1)(x^4 + x + 1)(x^4 + x^3 + 1)(x^4 + x^3 + x^2 + x + 1).$$

   Let $m_1(x)$, $m_3(x)$ be the minimal polynomials of elements α and α^3, i.e., $m_1(\alpha) = 0$, $m_3(\alpha^3) = 0$ and $m_1(x)$ and $m_3(x)$ are proper divisors of $x^{15} - 1$. From the definition of the field F, we immediately obtain that $m_1(x) = p(x) = x^4 + x + 1$. Since $(\alpha^3)^5 = \alpha^{15} = 1$, α^3 is a root of the polynomial $x^5 + 1 = (x+1)(x^4 + x^3 + x^2 + x + 1)$, where $x^4 + x^3 + x^2 + x + 1$ is an irreducible polynomial, $m_3(x) = x^4 + x^3 + x^2 + x + 1$ is the minimal polynomial of α^3. Since, by Remark 9.80, $m_4(x) = m_2(x) = m_1(x)$,

   $$g(x) = \text{lcm}(m_1(x), m_3(x)) = (x^4 + x^3 + 1)(x^4 + x^3 + x^2 + x + 1) =$$

   $$= x^8 + x^4 + x^2 + x + 1$$

   is the generator polynomial of a primitive BCH code with minimum distance at least 5, since, by construction and Theorem 9.77, $d - 1 = 4$, i.e., $d = 5$.

   This BCH code corrects up to 2 errors, which follows from Theorem 9.52, because $2t + 1 = d = 5$ implies $t = 2$.

2. Find the binary primitive narrow-sense BCH code of length $n = 2^4 - 1$ which corrects up to 3 errors.

   We can consider the same field F as in the previous example.

   For this example, $2 \cdot t + 1 = 6 + 1 = 7 = d$. Since $d - 1 = 6$, we must take 6 minimal polynomials. From Remark 9.80, it follows that we can omit all polynomials with even indexes. So, the generator polynomial of this BCH code with minimum distance 5 has the form

$$g(x) = \text{lcm}(m_1(x), m_3(x), m_5(x)),$$

where $m_5(\alpha^5) = 0$. The roots of $m_5(x)$ are α^5 and α^{10}. Hence, $m_5(x) = (x - \alpha^5)(x - \alpha^{10})$. Since $\alpha^5 = \alpha^3 + \alpha + 1$, $\alpha^{10} = \alpha^2 + \alpha^4$, implying that $\alpha^5 + \alpha^{10} = 1$, we obtain that $m_5(x) = x^2 + x + 1$. Therefore,

$$g(x) = (x^4 + x^3 + 1)(x^4 + x^3 + x^2 + x + 1)(x^2 + x + 1) =$$

$$= x^{10} + x^9 + x^8 + x^6 + x^5 + x^2 + 1$$

is the generator polynomial of a primitive BCH code with minimum distance at least 7, since, $m_6(x) = m_1(x)$ and so, by construction and Theorem 9.77, $d - 1 = 6$, i.e., $d = 7$.

This BCH code corrects up to 3 errors, which follows from Theorem 9.52, because $2t + 1 = d = 7$ implies $t = 3$.

9.6. Exercises

1. Find the order of the element:

    (a) 2 modulo 35
    (b) 3 modulo 55
    (c) 7 modulo 24
    (d) 4 modulo 21
    (e) 5 modulo 18
    (f) 9 modulo 10

2. Find an element of order 103 in the multiplicative group of remainder classes modulo 1237.

3. Verify whether or not the integer:

    (a) 2 is a primitive root modulo 29
    (b) 5 is a primitive root modulo 29
    (c) 11 is a primitive root modulo 31

4. Find all primitive roots modulo:

    (a) $n = 6$
    (b) $n = 14$

(c) $p = 11$

(d) $p = 19$

5. Find all primitive roots modulo 17 if it is known that 3 is a primitive root modulo 17.

6. Find all primitive roots modulo 19 if it is known that 2 is a primitive root modulo 19.

7. Find the index of the number:

   (a) 15 to base 6 modulo 109

   (b) 6 to base 3 modulo 7

8. Verify whether or not $2^{23} \equiv 2^6 \pmod{11}$.

9. Find the discrete logarithm of 13 to base 2 modulo 23.

10. In order to choose the key, Alice and Bob agree on the numbers $p = 37$ and $g = 2$. Alice sent to Bob the number 27 and Bob sent to Alice the number 17. Find their common private key.

11. In the ElGamal scheme, Alice chooses the prime integer $p = 31$ and the minimal primitive root modulo 31. As her private key, Alice chooses the integer 13. Bob sent to Alice a message in the form $(B, C) = (19, 8)$. Find Alice's public key, Bob's private key and the original message.

12. Alice obtains the message in the form $B = 30, C = 7$ using the ElGamal scheme. Her public key is the pair of numbers $p = 43, g = 3$ and her private key is $a = 3$. Find the original message.

13. Let $p = 53, g = 2, B = 30$ be Bob's public key, using the ElGamal scheme. Alice used them to generate the cipher $(24, 37)$. Find the corresponding original text.

14. Find the weight of each of the following vectors and the distance between the pairs of given vectors:

    (a) $(0\ 0\ 1\ 1)$, $(1\ 1\ 1\ 1)$

    (b) $(0\ 0\ 1\ 1)$, $(1\ 1\ 0\ 0)$

    (c) $(1\ 1\ 0\ 1\ 1\ 0\ 0\ 1)$, $(1\ 0\ 0\ 1\ 1\ 0\ 0\ 1)$

    (d) $(0\ 0\ 1\ 1\ 1\ 0\ 0\ 1)$, $(0\ 0\ 0\ 0\ 1\ 0\ 0\ 1)$

15. Find all linear codes of length 3 and the corresponding polynomials.

16. Find all binary cyclic codes of length 4.

17. Find the generator polynomials for binary cyclic codes of length 5.

18. Find the binary cyclic code of length 5 whose the generator polynomial is $g(x) = x^2 + x + 1$.

References

[1] Adleman, L.M. 1979. A subexponential algorithm for the discrete logarithm problem with application to cryptography. pp. 55–60. *In*: Proceedings of the 20th Annual IEEE Symposium on Foundations of Computer Science, IEEE Press.

[2] Adleman, L.M. and J. DeMarrais. 1993. A subexponential algorithm for the discrete logarithm over all finite fields. Math. Comp. 61: 1–15.

[3] Ash, R.B. 2007. Basic Abstract Algebra, Dover Publications.

[4] Asmuth, C.A. and J. Bloom. 1983. A modular approach to key safeguarding. IEEE Transactions on Information Theory IT 29(2): 208–210.

[5] Berlekamp, E.R. 1968. Algebraic Coding Theory, McGraw-Hill Company, New York.

[6] Blake, I., X. Gao, A. Menezes, R. Mullen, S. Vanstone and T. Yaghoobian. 1992. Applications of Finite Fields, Kluwer Acad. Publ.

[7] Blakley, G.R. 1979. Safeguarding cryptographic keys. pp. 313–317. *In*: Proc. AFIPS 1979 National Computer Conference, AFIPS.

[8] Birkhoff, G. and T.C. Bartee. 1970. Modern Applied Algebra, New York, McGraw-Hill.

[9] Bose, R.C. and D.K. Ray-Chaudhuri. 1960. On a class of error correcting binary group codes. Information and Control 3(1): 68–79.

[10] Coppersmith, D., A.D. Odlyzko and R. Schroeppel. 1986. Discrete logarithms in $GF(p)$. Algorithmica 1: 1–15.

[11] Diffie, W. and M.E. Hellman. 1976. New directions in cryptography. IEEE Trans. Information Theory 22(5): 644–654.

[12] Dummit, D.S. and R.M. Foote. 2004. Abstract Algebra (3rd Ed.), John Wiley & Sons.

[13] Dummit, E. 2016. Cryptography (part 3): Discrete Logarithms in Cryptography, Vol. 1.01.
https://math.la.asu.edu/http://www1.spms.ntu.edu.sg/~frederique/AA11.
pdfdummit/docs/cryptography_3_discrete_logarithms_in_cryptography.pdf

[14] ElGamal, T. 1985. A public key cryptosystem and a signature scheme based on discrete logarithms. IEEE Trans. Information Theory 31: 469–472.

[15] Garrett, P. 2001. Making, Breaking Codes: Introduction to Cryptography, Prentice-Hall.

[16] Gilbert, W.J. and W.K. Nicholson. 2004. Modern Algebra with Applications, John Wiley & Sons, New Jersey.

[17] Goldreich, O., D. Ron and M. Sudan. 2000. Chinese reaindering with errors. IEEE Transactions on Information Theory 46(4): 1330–1338.

[18] Gorenstein, D., W.W. Peterson and N. Zierler. 1960. Two-error correcting bose-chaudhuri codes are quasi-perfect. Information and Control 3(3): 291–294.

[19] Hamming, R.W. 1950. Error detecting and error correcting codes. The Bell System Technical Journal 26: 147–160.

[20] Hill, R. 1991. A First Course in Coding Theory, Oxford University Press.

[21] Hocquenghem, A. 1959. Codes correcteurs d'erreurs, Chiffres (in French), Paris 2: 147–156.

[22] HØholdt, T. and J. Justesen. 2017. A Course in Error-Correcting Codes (2nd Ed.), EMS Textbooks in Mathematics, EMS.

[23] Judson, Th.W. 1994. Abstract Algebra. Theory and Applications, PWS Publishing Company, Michigan.

[24] Koblitz, N. 1998. Algebraic Aspects of Cryptography, Springer-Verlag Berlin-Heidelberg.

[25] Lidl, R. and G. Pilz. 1997. Applied Abstract Algebra, (2nd Ed.), Springer-Verlag, New York.

[26] Lidl, R. and H. Niederreiter. 1994. Introduction to Finite Fields and their Applications, Cambridge, Cambridge University Press.

[27] Macwilliams, F.J. and N.J.A. Sloane. 1977. The Theory of Error-Correcting Codes, North-Holland, Amsterdam.

[28] Massey, J.L. 1983. Logarithms in finite cyclic groups-cryptographic issues. pp. 17–25. *In*: Proc. 4th Benelux Symposium on Information Theory, 1983.

[29] McCurley, K. 1990. The discrete logarithm problem, cryptology and computational number theory. pp. 49–74. *In*: Proc. Symp. Appl. Math., 42.

[30] McEliece, R.J. 1987. Finite Fields for Computer Scientists and Engineers, Kluwer.

[31] Menezes, A., I. Blake, S. Gao, R. Mullin, S. Vanstone and T. Yaghoobian. [eds.]. 1993. Applications of Finite Fields Kluwer Academic Publishers.

[32] Odlyzko, A.M. 1985. Discrete logarithms in finite fields and their cryptographic significance. Lecture Notes in Computer Science, Springer-Verlag 209: 224–314.

[33] Peterson, W.W. and E.J. Weldon, Jr. 1972. Error-Correcting Codes, The MIT Press, Cambridge.

[34] Pilitowska, A. 2008. Algebraiczne aspekty teorii kodów, Oficyna Wydawnicza Politechniki Warszawskiej, Warszawa.

[35] Pless, V. 1990. Introduction to Finite Fields and their Applications, Springer-Verlag.

[36] Pohlig, S. and M. Hellman. 1978. An improved algorithm for computing logarithms over $GF(p)$ and its cryptographic significance. IEEE Transactions on Information Theory 24: 106–110.

[37] Reed, I. and Chen, Xuemin. 1999. Error-Control Coding for Data Networks, Boston, MA: Kluwer Academic Publishers.

[38] Roman, S. 1997. Introduction to Coding and Information Theory, New York, Springer-Verlag.

[39] Shannon, C. 1948. Mathematical theory of communication, the bell system. Technical Journal 27(379-423): 623–656.

[40] Slinko, A. 2015. Algebra for Applications, Springer.

[41] Welsh, D. 1989. Codes and Cryptography, Oxford University Press.

[42] Yan, Song Y. 2000. Number Theory for Computing, Springer-Verlag, Berlin, Heidelberg.

Chapter 10

Finite Dimensional Algebras

"The art of doing mathematics consists in finding that special case which contains all the germs of generality."
David Hilbert

"We must not only obtain Wisdom: we must enjoy her."
Marcus Tullius Cicero

The theory of finite dimensional algebras is one of the most fundamental fields of modern algebra and has various applications in other branches of mathematics and theoretical physics.

Algebras are algebraic structures which connect structures of rings and vector spaces. The most important examples of algebras form quadratic matrices, polynomials and formal power series over fields. The first examples of non-commutative algebras were given in the works of W.R. Hamilton, A. Cayley and H. Grassman in 1843-1845. These examples are algebras of quaternions, octonions, biquaternion (i.e., quaternions over the field of complex numbers) and external algebras.

The study of hypercomplex numbers, i.e., finite dimensional algebras over the field of real or complex numbers, was initiated by B. Pierce (1870). First, the construction of Cayley-Dickson was used for extension of complex numbers. This construction shows how quaternions, octonions, sedeonions and other algebras are obtained in the natural way from the real numbers. The theorems of Hurwitz and Frobenius prove the limitation of the number of these algebras with natural properties, such as commutativity, associativity, alternativity and others. Significant results in the theory of hypercomplex numbers were presented in the works of F.E. Molin in 1897 and E.Cartan in 1898. Their results were generalized by J.M. Wedderburn in 1907 for finite dimensional associative algebras over arbitrary fields.

10.1. Quaternions and their Properties

✴ **Definition 10.1.**

Quaternions are the elements of a four-dimensional space over the field of real numbers of the form:

$$q = a_1 + a_2 i + a_3 j + a_4 k$$

where $1, i, j, k$ are the basis elements:

$$1 = (1, 0, 0, 0)$$

$$i = (0, 1, 0, 0)$$

$$j = (0, 0, 1, 0)$$

$$k = (0, 0, 0, 1)$$

for which multiplication is defined in such a way that:

$$i^2 = j^2 = k^2 = ijk = -1 \qquad (10.1)$$

Quaternions were first invented by Irish mathematician professor of Trinity College in Dublin, sir William Rowan Hamilton, after extensive studying in 1843. This happened on October 16th of that year when he was walking along the Royal Canal on his way to a meeting of the Irish Academy in Dublin and this was documented by him into the stone of the Brougham bridge, where Hamilton scratched the fundamental formula of quaternions (10.1). It is accepted that the set of quaternions is denoted by the letter $\mathbb{H}$ in honor of Hamilton:

$$\mathbb{H} = \{a_1 + a_2 i + a_3 j + a_4 k \; : \; a_r \in \mathbb{R}, r = 1, 2, 3, 4\} \qquad (10.2)$$

where $i^2 = j^2 = k^2 = -1, ij = -ji = k$.

Sir William Rowan Hamilton (1805-1865)

We can define the operations of addition and multiplication on the set of quaternions $\mathbb{H}$.

For arbitrary quaternions $q_1 = a_1 + a_2 i + a_3 j + a_4 k, q_2 = b_1 + b_2 i + b_3 j + b_4 k$, we define the operation of addition as follows:

$$q_1 + q_2 = (a_1 + b_1) + (a_2 + b_2)i + (a_3 + b_3)j + (a_4 + b_4)k \qquad (10.3)$$

A neutral element in the set $\mathbb{H}$, with respect to addition, is the element $0 + 0 \cdot i + 0 \cdot j + 0 \cdot k$, which is called the zero element and is denoted by $\mathbf{0}$, i.e.,

$q + 0 = 0 = q$. An element $-q = -a_1 - a_2i - a_3j - a_4k$ is an inverse element to an element $a_1 + a_2i + a_3j + a_4k$ with respect to addition, since $q + (-q) = \mathbf{0}$.

Since addition of quaternions is reduced to addition of real coordinates, we immediately obtain the next theorem:

✳ **Theorem 10.2.**

Quaternions form an Abelian additive group with respect to addition, i.e., for all elements $q, q_1, q_2, q_3 \in \mathbb{H}$ the following conditions hold:

1. $q_1 + q_2 = q_2 + q_1$
2. $(q_1 + q_2) + q_3 = q_1 + (q_2 + q_3)$
3. $q + \mathbf{0} = \mathbf{0} + q = q$
4. $q + (-q) = \mathbf{0}$

Multiplication of quaternions is based on multiplication of basis elements that can be written in the form of the following table:

$\cdot$	1	i	j	k
1	1	i	j	k
i	i	-1	k	$-j$
j	j	$-k$	-1	i
k	k	j	$-i$	-1

Table 10.1

From this table, it can be noted that $ij = -ji$, $ik = -kj$, $jk = -kj$, which means that multiplication is not commutative and that multiplication of basis elements is cyclic, i.e., $ij = k$, $ki = j$, $jk = i$, which is easy to remember using the following graph of multiplication, proposed by John Baez:

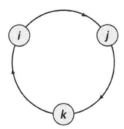

Figure 10.1: Graph of multiplication for quaternions.

Multiplication can be extended to all quaternions by means of Table 10.1. So, we obtain:

$$q_1q_2 = (a_1 + a_2i + a_3j + a_4k)(b_1 + b_2i + b_3j + b_4k) =$$

$$= (a_1b_1 - a_2b_2 - a_3b_3 - a_4b_4) + (a_1b_2 + a_2b_1 + a_3b_4 - a_4b_3)i +$$

$$+ (a_1b_3 - a_2b_4 + a_3b_1 + a_4b_2)j + (a_1b_4 + a_2b_3 - a_3b_2 + a_4b_1)k \qquad (10.4)$$

In the general case, $q_1q_2 \neq q_2q_1$, for example, $ij \neq ji$.

The neutral element in the set of quaternions $\mathbb{H}$ with respect of multiplication is the element $1 + 0 \cdot i + 0 \cdot j + 0 \cdot k$, which is called the identity and is denoted by $\mathbf{1}$.

✳ Theorem 10.3.

Quaternions form an associative non-commutative ring with identity, i.e., $\mathbb{H}$ is an Abelian group with respect to addition and all elements $q, q_1, q_2, q_3 \in \mathbb{H}$ satisfy the conditions:

1. $(q_1q_2)q_3 = q_1(q_2q_3)$
2. $(q_1 + q_2)q_3 = q_1q_3 + q_2q_3$
3. $q_1(q_2 + q_3) = q_1q_2 + q_1q_3$
4. $q \cdot 1 = 1 \cdot q = q$

Proof.

Let $q_1 = a_1 + a_2i + a_3j + a_4k$, $q_2 = b_1 + b_2i + b_3j + b_4k$, $q_3 = c_1 + c_2i + c_3j + c_4k \in \mathbb{H}$.

1. Let us show that multiplication is associative. We have:

$$(q_1q_2)q_3 = [(a_1 + a_2i + a_3j + a_4k)(b_1 + b_2i + b_3j + b_4k)](c_1 + c_2i + c_3j + c_4k) =$$

$$= [(a_1b_1 - a_2b_2 - a_3b_3 - a_4b_4) + (a_1b_2 + a_2b_1 + a_3b_4 - a_4b_3)i +$$

$$+ (a_1b_3 - a_2b_4 + a_3b_1 + a_4b_2)j + (a_1b_4 + a_2b_3 - a_3b_2 + a_4b_1)k](c_1 + c_2i + c_3j + c_4k) =$$

$$= [(a_1b_1 - a_2b_2 - a_3b_3 - a_4b_4)c_1 - (a_1b_2 + a_2b_1 + a_3b_4 - a_4b_3)c_2 -$$

$$- (a_1b_3 - a_2b_4 + a_3b_1 + a_4b_2)c_3 - (a_1b_4 + a_2b_3 - a_3b_2 + a_4b_1)c_4] +$$

$$+ [(a_1b_1 - a_2b_2 - a_3b_3 - a_4b_4)c_2 + (a_1b_2 + a_2b_1 + a_3b_4 - a_4b_3)c_1 +$$

$$+ (a_1b_3 - a_2b_4 + a_3b_1 + a_4b_2)c_4 - (a_1b_4 + a_2b_3 - a_3b_2 + a_4b_1)c_3]i +$$

$$+ [(a_1b_1 - a_2b_2 - a_3b_3 - a_4b_4)c_3 - (a_1b_2 + a_2b_1 + a_3b_4 - a_4b_3)c_4 +$$

$$+ (a_1b_3 - a_2b_4 + a_3b_1 + a_4b_2)c_1 + (a_1b_4 + a_2b_3 - a_3b_2 + a_4b_1)c_2]j +$$

$$+ [(a_1b_1 - a_2b_2 - a_3b_3 - a_4b_4)c_4 + (a_1b_2 + a_2b_1 + a_3b_4 - a_4b_3)c_3 -$$

$$- (a_1b_3 - a_2b_4 + a_3b_1 + a_4b_2)c_2 + (a_1b_4 + a_2b_3 - a_3b_2 + a_4b_1)c_1]k$$

and

$$q_1(q_2q_3) = (a_1 + a_2i + a_3j + a_4k)[(b_1 + b_2i + b_3j + b_4k)(c_1 + c_2i + c_3j + c_4k)] =$$

$$= (a_1 + a_2i + a_3j + a_4k)[(b_1c_1 - b_2c_2 - b_3c_3 - b_4c_4) + (b_1c_2 + b_2c_1 + b_3c_4 - b_4c_3)i +$$

$$+ (b_1c_3 - b_2c_4 + b_3c_1 + b_4c_2)j + (b_1c_4 + b_2c_3 - b_3c_2 + b_4c_1)k] =$$

$$= [a_1(b_1c_1 - b_2c_2 - b_3c_3 - b_4c_4) - a_2(b_1c_2 + b_2c_1 + b_3c_4 - b_4c_3) -$$

$$- a_3(b_1c_3 - b_2c_4 + b_3c_1 + b_4c_2) - a_4(b_1c_4 + b_2c_3 - b_3c_2 + b_4c_1)] +$$

$$+ [a_1(b_1c_2 + b_2c_1 + b_3c_4 - b_4c_3) + a_2(b_1c_1 - b_2c_2 - b_3c_3 - b_4c_4) +$$

$$+a_3(b_1c_4 + b_2c_3 - b_3c_2 + b_4c_1) - a_4(b_1c_3 - b_2c_4 + b_3c_1 + b_4c_2)]i+$$

$$+a_1(b_1c_3 - b_2c_4 + b_3c_1 + b_4c_2) - a_2(b_1c_4 + b_2c_3 - b_3c_2 + b_4c_1)+$$

$$+a_3(b_1c_1 - b_2c_2 - b_3c_3 - b_4c_4) + a_4(b_1c_2 + b_2c_1 + b_3c_4 - b_4c_3)]j+$$

$$+[a_1(b_1c_4 + b_2c_3 - b_3c_2 + b_4c_1) + a_2(b_1c_3 - b_2c_4 + b_3c_1 + b_4c_2)-$$

$$-a_3(b_1c_2 + b_2c_1 + b_3c_4 - b_4c_3) + a_4(b_1c_1 - b_2c_2 - b_3c_3 - b_4c_4)]k,$$

hence, $(q_1q_2)q_3 = q_1(q_2q_3)$.

2. Let us show the distributivity of multiplication of quaternions, with respect to addition:

$$(q_1+q_2)q_3 = [(a_1+a_2i+a_3j+a_4k)+(b_1+b_2i+b_3j+b_4k)](c_1+c_2i+c_3j+c_4k) =$$

$$= [(a_1 + b_1) + (a_2 + b_2)i + (a_3 + b_3)j + (a_4 + b_4)k](c_1 + c_2i + c_3j + c_4k) =$$

$$= [(a_1 + b_1)c_1 - (a_2 + b_2)c_2 - (a_3 + b_3)c_3 - (a_4 + b_4)c_4]+$$

$$+[(a_1 + b_1)c_2 + (a_2 + b_2)c_1 + (a_3 + b_3)c_4 - (a_4 + b_4)c_3]i+$$

$$+[(a_1 + b_1)c_3 - (a_2 + b_2)c_4 + (a_3 + b_3)c_1 + (a_4 + b_4)c_2]j+$$

$$+[(a_1 + b_1)c_4 - (a_2 + b_2)c_3 - (a_3 + b_3)c_2 + (a_4 + b_4)c_1]k =$$

$$= (a_1c_1 - a_2c_2 - a_3c_3 - a_4c_4) + (b_1c_1 - b_2c_2 - b_3c_3 - b_4c_4)+$$

$$+(a_1c_2 + a_2c_1 + a_3c_4 - a_4c_3)i + (b_1c_2 + b_2c_1 + b_3c_4 - b_4c_3)i+$$

$$+(a_1c_3 - a_2c_4 + a_3c_1 + a_4c_2)j + (b_1c_3 - b_2c_4 + b_3c_1 + b_4c_2)j+$$

$$+(a_1c_4 + a_2c_3 - a_3c_2 + a_4c_1)k + (b_1c_4 + b_2c_3 - b_3c_2 + b_4c_1)k =$$

$$= q_1q_3 + q_2q_3.$$

Other conditions can be proved analogously. □

In addition to the operations of addition and multiplication, which are internal operations on the set of quaternions, we can determine the external operation that is multiplication of quaternions by real numbers.

Multiplication of quaternions by real numbers is defined as follows: For a real number $\lambda \in \mathbb{R}$ and a quaternion $q = a_1 + a_2i + a_3j + a_4k \in \mathbb{H}$, we define:

$$\lambda q = \lambda(a_1 + a_2i + a_3j + a_4k) = (\lambda a_1) + (\lambda a_2)i + (\lambda a_3)j + (\lambda a_4)k \quad (10.5)$$

✳ **Theorem 10.4.**

The set of quaternions $\mathbb{H}$ forms a four-dimensional vector space over the field of real numbers $\mathbb{R}$ with basis $\{1, i, j, k\}$, i.e., $\dim_{\mathbb{R}}\mathbb{H} = 4$ and:

1. $\lambda(q_1 + q_2) = \lambda q_1 + \lambda q_2$
2, $(\lambda\mu)q = \lambda(\mu q)$
3. $(\lambda + \mu)q = \lambda q + \mu q$.

for all $q, q_1, q_2, q_3 \in \mathbb{H}$ and all $\lambda, \mu \in \mathbb{R}$.

Proof.

Let $q = q_1 = a_1 + a_2i + a_3j + a_4k, q_2 = b_1 + b_2i + b_3j + b_4k \in \mathbb{H}$. Then:

1. $\lambda(q_1 + q_2) = \lambda(a_1 + a_2i + a_3j + a_4k + b_1 + b_2i + b_3j + b_4k) = \lambda a_1 + \lambda a_2i + \lambda a_3j + \lambda a_4k + \lambda b_1 + \lambda b_2i + \lambda b_3j + \lambda b_4k = \lambda(a_1 + a_2i + a_3j + a_4k) + \lambda(b_1 + b_2i + b_3j + b_4k) = \lambda q_1 + \lambda q_2$.

2. $(\lambda\mu)q = (\lambda\mu)(a_1 + a_2i + a_3j + a_4k) = (\lambda\mu)a_1 + (\lambda\mu)a_2i + (\lambda\mu)a_3j + (\lambda\mu)a_4 = \lambda(\mu a_1 + \mu a_2i + \mu a_3j + \mu a_4k) = \lambda(\mu q)$.

3. $(\lambda + \mu)q = (\lambda + \mu)(a_1 + a_2i + a_3j + a_4k) = (\lambda + \mu)a_1 + (\lambda + \mu)a_2i + (\lambda + \mu)a_3j + (\lambda + \mu)a_4k = \lambda(a_1 + a_2i + a_3j + a_4k) + \mu(a_1 + a_2i + a_3j + a_4k) = \lambda q + \mu q$.

$\square$

Similar to complex numbers, we can determine the operation of conjugation for quaternions. For a quaternion $q = a_1 + a_2i + a_3j + a_4k \in \mathbb{H}$, we define the **conjugate** of q as follows:

$$q^* = a_1 - a_2i - a_3j - a_4k \tag{10.6}$$

This operation satisfies the conditions:

$$(q^*)^* = q \tag{10.7}$$

$$(q_1 + q_2)^* = q_1^* + q_2^* \tag{10.8}$$

$$(q_1q_2)^* = q_2^*q_1^* \tag{10.9}$$

for all $q, q_1, q_2 \in \mathbb{H}$.

Really, we have:

$$(q_1 + q_2)^* = [(a_1 + a_2i + a_3j + a_4k) + (b_1 + b_2i + b_3j + b_4k)]^* =$$

$$= [(a_1 + b_1) + (a_2 + b_2)i + (a_3 + b_3)j + (a_4 + b_4)k]^* =$$

$$= [(a_1 + b_1) - (a_2 + b_2)i - (a_3 + b_3)j - (a_4 + b_4)k] =$$

$$= [(a_1 - a_2i - a_3j - a_4k) + (b_1 - b_2i - b_3j - b_4k)] = q_1^* + q_2^*$$

From (10.4), we obtain:

$$(q_1q_2)^* = [(a_1 + a_2i + a_3j + a_4k)(b_1 + b_2i + b_3j + b_4k)]^* =$$

$$= (a_1b_1 - a_2b_2 - a_3b_3 - a_4b_4) - (a_1b_2 + a_2b_1 + a_3b_4 - a_4b_3)i -$$

$$-(a_1b_3 - a_2b_4 + a_3b_1 + a_4b_2)j - (a_1b_4 + a_2b_3 - a_3b_2 + a_4b_1)k$$

and

$$q_2^*q_1^* = (b_1 + b_2i + b_3j + b_4k)^*(a_1 + a_2i + a_3j + a_4k)^* =$$

$$= (b_1 - b_2i - b_3j - b_4k)(a_1 - a_2i - a_3j - a_4k) =$$

$$= (b_1a_1 - b_2a_2 - b_3a_3 - b_4a_4) + (-b_1a_2 - b_2a_1 - b_3a_4 + b_4a_3)i +$$
$$+ (-b_1a_3 - b_2a_4 - b_3a_1 + b_4a_2)j + (-b_1a_4 + b_2a_3 + b_3a_2 - b_4a_1)k =$$
$$= (a_1b_1 - a_2b_2 - a_3b_3 - a_4b_4) - (a_1b_2 + a_2b_1 + a_3b_4 - a_4b_3)i -$$
$$- (a_1b_3 - a_2b_4 + a_3b_1 + a_4b_2)j - (a_1b_4 + a_2b_3 - a_3b_2 + a_4b_1)k.$$

Hence, $(q_1q_2)^* = q_2^* q_1^*$.

So, if $q_1 = q_2 = q$, then

$$qq^* = q^*q = a_1^2 + a_2^2 + a_3^2 + a_4^2 \geq 0 \qquad (10.10)$$

is a real number which is equal to 0 if and only if $q = \mathbf{0}$.

Thus, we can define the norm of a quaternion as the product of quaternion and its conjugate:

$$N(q) = qq^* \qquad (10.11)$$

which satisfies the conditions:
1. $N(q) = 0 \Leftrightarrow q = \mathbf{0}$
2. $N(q_1q_2) = N(q_1)N(q_2)$
3. $N(\mathbf{1}) = 1$
for all $q, q_1, q_2 \in \mathbb{H}$.

Really, from associativity of quaternions and taking into account equalities (10.9) and (10.11), we obtain:

$$N(q_1q_2) = (q_1q_2)(q_1q_2)^* = (q_1q_2)(q_2^* q_1^*) = q_1(q_2q_2^*)q_1^* = N(q_1)N(q_2).$$

If a quaternion $q \neq \mathbf{0}$, then the element $qq^* = N(q) \neq 0$ as well. Therefore, for each non-zero quaternion q, we can define a quaternion of the form

$$q^{-1} = \frac{q^*}{N(q)} \qquad (10.12)$$

which is the inverse to the quaternion q, since

$$qq^{-1} = \frac{qq^*}{N(q)} = \frac{N(q) \cdot 1}{N(q)} = 1$$

and

$$q^{-1}q = \frac{q^*q}{N(q)} = \frac{N(q) \cdot 1}{N(q)} = 1$$

Therefore, $\mathbb{H}$ is a non-commutative associative ring with identity in which each non-zero element has the inverse, i.e., $\mathbb{H}$ is a division ring.

10.2. Octonions-Cayley's Octaves

In 1843, while generalizing the concept of quaternions, John Thomas Graves (1806-1870), a long-time friend of Hamilton, was led to the construction of a new class of numbers which were called by octaves, now known as octonions. Graves sent the results of their thinking to Hamilton. New invented numbers had more unusual properties because multiplication of these numbers was neither commutative nor associative. In his letter to Hamilton, Graves wrote the following:

"*In this system (of quaternions) there is something which does not give rest to me. So far I have not clear look in what degree we have right freely to create image numbers and supply them into unusual properties...*

If thanks your alchemy you can create three pounds of golds why should this be limited?"

Unfortunately, Graves' article was never published, since Hamilton's diligence was not sufficient, and he was too busy studying quaternions. At the end of 1845 Arthur Cayley rediscovered the octonions, independently on Graves, based on the works of Hamilton. Since then they come to history as **Cayley numbers**.

✳ Definition 10.5.

Octonions (or **octaves**) are elements of an 8-dimensional vector space over the field of real numbers $\mathbb{R}$ of the form:

$$x = a_1 e_1 + a_2 e_2 + \cdots + a_7 e_7 + a_8 e_8 \qquad (10.13)$$

where $a_i \in \mathbb{R}$ and $e_1, e_2, \ldots, e_7, e_8$ are basis elements for which multiplication is defined by the following table:

·	e_1	e_2	e_3	e_4	e_5	e_6	e_7	e_8
e_1	e_1	e_2	e_3	e_4	e_5	e_6	e_7	e_8
e_2	e_2	$-e_1$	e_4	$-e_3$	e_6	$-e_5$	$-e_8$	e_7
e_3	e_3	$-e_4$	$-e_1$	e_2	e_7	e_8	$-e_5$	$-e_6$
e_4	e_4	e_3	$-e_2$	$-e_1$	e_8	$-e_7$	e_6	$-e_5$
e_5	e_5	$-e_6$	$-e_7$	$-e_8$	$-e_1$	e_2	e_3	e_4
e_6	e_6	e_5	$-e_8$	e_7	$-e_2$	$-e_1$	$-e_4$	e_3
e_7	e_7	e_8	e_5	$-e_6$	$-e_3$	e_4	$-e_1$	$-e_2$
e_8	e_8	$-e_7$	e_6	e_5	$-e_4$	$-e_3$	e_2	$-e_1$

Table 10.2

The set of octonions is denoted by $\mathbb{O}$:

$$\mathbb{O} = \{a_1 e_1 + a_2 e_2 + \cdots + a_7 e_7 + a_8 e_8 \ : \ a_i \in \mathbb{R}\}$$

Since each basis element $e_i \in \mathbb{O}$, for $i = 2, \ldots, 7, 8$ satisfies the condition $e_i^2 = -1$, elements e_i are often called **image units** of octonions. From Table 10.2, the following properties for image units of octonions follow:

1) $e_i e_j = e_j e_i = e_j$, $e_1 e_1 - e_1$
2) $e_i e_j = -e_j e_i$ for $i, j \neq 1$, $i \neq j$;

3) if $e_i e_j = e_k$, then $e_{i+1} e_{j+1} = e_{k+1}$
4) $e_i e_{i+1} = e_{i+3}$ for $i \neq 1$
5) if $e_i e_j = e_k$, then $e_{2i} e_{2j} = e_{2k}$.

The property 2) shows that multiplication of the basis elements is non-commutative. The property 3) shows the cyclicality of indexes of image units, and property 5) shows the reduplication of indexes.

In order to understand and remember the rule of multiplication for image units of octonions, the so-called Fano plane can be used (see Figure 10.2).

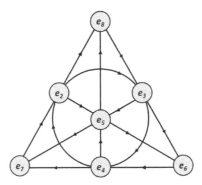

Figure 10.2: Fano plane.

This plane contains seven lines and seven points corresponding to seven standard basis elements of octonions. Three lines form the sides of the triangle, the next three lines are perpendiculars drawn from each vertex of the triangle, while the extra seventh line forms a circle connecting the middles of the triangle's sides. So, each line contains three points. The lines are directed, so the direction in which we move between points is important. For example, assuming that a_1, a_2, a_3 denote three ordered points belonging to the same line, we can determine products:

$$(a_1 a_2 = a_3) \wedge (a_2 a_1 = -a_3) \wedge (a_2 a_3 = a_1) \wedge (a_3 a_2 = -a_1) \wedge (a_3 a_1 = a_2) \wedge (a_1 a_3 = -a_2)$$

On the set of octonions $\mathbb{O}$, the operation of addition can be introduced in the following form:

$$x + y = (a_1 + b_1)e_1 + (a_2 + b_2)e_2 + \cdots + (a_8 + b_8)e_8 \qquad (10.14)$$

for all octonions $x = a_1 e_1 + a_2 e_2 + \cdots + a_8 e_8$, $y = b_1 e_1 + b_2 e_2 + \cdots + b_8 e_8$.

The neutral element, with respect to addition, in the set of octonions is the element $0 \cdot e_1 + 0 \cdot e_2 + \cdots + 0 \cdot e_8$, which is called the **zero element** and is denoted by $\mathbf{0}$. The octonion $y = -a_1 e_1 - a_2 e_2 - \cdots - a_8 e_8 \in \mathbb{O}$ is the inverse element to an octonion $x = a_1 e_1 + a_2 e_2 + \cdots + a_8 e_8$, since $x + y = \mathbf{0}$, and is denoted by $-x$.

Since addition of octonions is reduced to addition of their real coordinates, we immediately obtain the following theorem:

✳ Theorem 10.6.
Octonions form an additive Abelian group, i.e., all elements $u, v, w \in \mathbb{O}$ satisfy the conditions:
1) $u + w = w + u$
2) $(u + v) + w = u + (v + w)$
3) $u + \mathbf{0} = \mathbf{0} + u = u$
4) $u + (-u) = (-u) + u = \mathbf{0}$

Similar to quaternions in the set of octonions, we can also introduce the operation of multiplication of octonions by real numbers in the following way:

$$\lambda x = \lambda(a_1 e_1 + a_2 e_2 + \cdots + a_8 e_8) = (\lambda a_1)e_1 + (\lambda a_2)e_2 + \cdots + (\lambda a_8)e_8 \quad (10.15)$$

for each real number λ and an octonion $x = a_1 e_1 + a_2 e_2 + \cdots + a_7 e_7 + a_8 e_8$.

✳ Theorem 10.7.
The set of octonions $\mathbb{O}$ form an eight-dimensional linear space over the field of real numbers $\mathbb{R}$ with basis $\{e_1, e_2, \ldots, e_7, e_8\}$, i.e., $\dim_\mathbb{R} \mathbb{O} = 8$ and the following conditions hold:
1. $\lambda(x_1 + x_2) = \lambda x_1 + \lambda x_2$
2, $(\lambda\mu)x = \lambda(\mu x)$
3. $(\lambda + \mu)x = \lambda x + \mu x$.
for all $x, x_1, x_2, x_3 \in \mathbb{O}$ and all $\lambda, \mu \in \mathbb{R}$.

The proof of this theorem is exactly the same as the proof of Theorem 10.4.

Multiplication on the basis elements of octonions can be extended to all octonions. The neutral element, with respect to multiplication, in the set of octonions is the element $1 \cdot e_1 + 0 \cdot e_2 + \cdots + 0 \cdot e_8$, which is called the identity of $\mathbb{O}$ and is denoted by $\mathbf{1}$. Multiplication of octonions is not commutative, since, for example:

$$e_4 = e_2 e_3 \neq e_3 e_2 = -e_4.$$

Moreover, multiplication of octonions is not associative, since, for example:

$$e_4(e_2 e_8) = e_4 e_7 = e_6,$$

but

$$(e_4 e_2)e_8 = e_3 e_8 = -e_6.$$

Meanwhile, octonions satisfy the weaker versions of associativity:
1) $u(uw) = (uu)w$ - left alternativity
2) $u(ww) = (uw)w$ - right alternativity
for all $u, w \in \mathbb{O}$.
The proof of this fact will be given later in Section 10.5.

For octonions, we can also define the notion of conjugation. The **conjugate** of an octonion $x = a_1e_1 + a_2e_2 + \cdots + a_8e_8$ is an octonion $x^* = a_1e_1 - a_2e_2 - \cdots - a_8e_8$. It is easy to verify that the conjugation is an involution of octonions, i.e., it satisfies the basic properties:

$$(x^*)^* = x \tag{10.16}$$

$$(x_1 + x_2)^* = x_1^* + x_2^* \tag{10.17}$$

$$(x_1x_2)^* = x_2^*x_1^* \tag{10.18}$$

for all $x, x_1, x_2 \in \mathbb{O}$ and

$$xx^* = x^*x = a_1^2 + a_2^2 + a_3^2 + a_4^2 + a_5^2 + a_6^2 + a_7^2 + a_8^2 \geq 0 \tag{10.19}$$

and $xx^* = \mathbf{0}$ if and only if $x = \mathbf{0}$.

Therefore, we can define the norm of an octonion x as the product of x with its conjugate x^*:

$$N(x) = xx^* \tag{10.20}$$

which satisfies the properties:
1. $N(x) = 0 \iff x = \mathbf{0}$
2. $N(\mathbf{1}) = 1$.

Thus, for each non-zero octonion x, we can define the inverse octonion:

$$x^{-1} = \frac{x^*}{N(x)} \tag{10.21}$$

since

$$xx^{-1} = \frac{xx^*}{N(x)} = \frac{N(x) \cdot \mathbf{1}}{N(x)} = \mathbf{1}$$

and

$$x^{-1}x = \frac{x^*x}{N(x)} = \frac{N(x) \cdot \mathbf{1}}{N(x)} = \mathbf{1}$$

10.3. Algebras and their Properties

❋ **Definition 10.8.**

An **algebra** over a field K (or a K-algebra) is an algebraic structure $(A, K, +, \bullet, \circ)$ with two internal binary operations:

$$+ : A \times A \to A,$$

$$\bullet : A \times A \to A,$$

named by addition and multiplication, respectively, and one external operation:

$$\circ : K \times A \to A,$$

satisfying the conditions:

1) $(A, +, \bullet)$ is a ring;
2) $(A, K, +, \circ)$ is a vector space over the field K;
3) $\lambda \circ (a \bullet b) = (\lambda \circ a) \bullet b = a \bullet (\lambda \circ b)$ for all $a, b \in A$, and for all $\lambda \in K$.

An algebra A is called **commutative** if the operation of multiplication is commutative:

$$a \bullet b = b \bullet a \quad \text{for all } a, b \in A.$$

An algebra A is called **associative** if the operation of multiplication is associative:

$$a \bullet (b \bullet c) = (a \bullet b) \bullet c \quad \text{for all } a, b, c \in A.$$

So, a commutative algebra $(A, +, \bullet)$ is a commutative ring, while an associative algebra $(A, +, \bullet)$ is an associative ring.

✳ **Definition 10.9.**

An algebra A is called an **algebra with identity** if $(A, +, \bullet)$ is a ring with identity, i.e., if $A \neq \{0\}$ and there is a neutral element with respect to multiplication, $1 \in A$ such that

$$a \bullet 1 = 1 \bullet a = a \quad \text{for all } a \in A$$

Each vector subspace $B \subseteq A$ closed with respect to multiplication $\bullet$ in A (i.e., $B \bullet B \subseteq B$) is called a **subalgebra** of A.

If A is an algebra with identity 1, then $1 \in B$ for each subalgebra B.

✳ **Definition 10.10.**

A **basis** of a K-algebra A is a basis of a vector space A over K.

A **dimension** of a K-algebra is a dimension of a vector space A over the field K and is denoted by $\dim_K A$.

If $\dim_K A < \infty$, then an algebra A is called **finite dimensional**.

✶ **Examples 10.11.**

1. A field K is a K-algebra and $\dim_K K = 1$.

2. If K, L are fields and $K \subset L$, then L is a commutative associative K-algebra with identity and $\dim_K L = [L : K]$.

3. The ring of polynomials $K[x]$ over a field K with operations of addition and multiplication of polynomials is a commutative, associative algebra with identity and $\dim_K K[x] = \infty$.

4. The ring of polynomials $K[x_1, x_2, \ldots, x_n]$ in n variables over a field K, with operations of addition and multiplication of polynomials, is a commutative, associative algebra with identity and $\dim_K K[x_1, x_2, \ldots, x_n] = \infty$.

5. The ring of quadratic matrices $M_n(K)$, where K is a field, with operations of addition and multiplication of matrices, is a non-commutative, associative algebra with identity and $\dim_K M_n(K) = n^2$. The basis of this algebra consists of matrices $\mathbf{E}_{ij} \in M_n(K)$, where

$$\mathbf{E}_{ik}\mathbf{E}_{rj} = \begin{cases} \mathbf{E}_{ij}, & \text{if } k = r \\ \mathbf{0}, & \text{otherwise} \end{cases}$$

6. The ring of quadratic matrices $M_n(A)$, where A is an algebra over a field K, with operations of addition and multiplication of matrices, is a finite dimensional algebra and $\dim_K M_n(A) = n^2 \dim_K(A)$.

7. Quaternions $(\mathbb{H}, \mathbb{R}, +, \cdot)$ form an algebra over the field of real numbers $\mathbb{R}$, and it is called the **algebra of quaternions**. This algebra is a non-commutative, associative algebra with identity and $\dim_{\mathbb{R}} \mathbb{H} = 4$.

8. Let V be a vector space over a field K and $\text{End}_K(V)$ be a set of all endomorphisms of V. In the set $\text{End}_K(V)$, we can define the zero endomorphism:

$$0_V : V \to V, \quad \text{where } 0_V(u) = 0 \text{ for all } u \in V$$

and the identity endomorphism:

$$1_V : V \to V, \quad \text{where } 1_V(u) = u \text{ for all } u \in V$$

In the set $\text{End}_K(V)$, we can define two internal operations: addition:

$$\sigma + \tau : V \to V, \quad \text{where } (\sigma + \tau)(u) = \sigma(u) + \tau(u)$$

multiplication:

$$\sigma \circ \tau : V \to V, \quad \text{where } (\sigma \circ \tau)(u) = \sigma(\tau(u))$$

for all endomorphisms $\sigma, \tau \in \text{End}_K(V)$.

With respect to these two operations, $\text{End}_K(V)$ is an associative ring with identity. In the set $\text{End}_K(V)$, we can also define an external operation:

Multiplication of endomorphisms by scalars:

$$x\sigma : V \to V, \quad (x\sigma)(u) = x\sigma(u)$$

for all $x \in K$ and $\sigma \in \mathrm{End}_K(V)$, satisfying the conditions:

$x(\sigma + \tau) = x\sigma + x\tau,$

$x(\sigma \circ \tau) = (x\sigma) \circ \tau = \sigma \circ (x\tau)$

for all $\sigma, \tau \in \mathrm{End}_K(V)$ and all $x \in K$.

Therefore, $\mathrm{End}_K(V)$ is an algebra over a field K which is called the **algebra of endomorphisms** of a vector space V.

✳ **Definition 10.12.**
   An algebra $(A, K, +, \bullet, \circ)$ is called **alternative** if multiplication satisfies the conditions:
$$a \bullet (a \bullet b) = (a \bullet a) \bullet b$$
$$a \bullet (b \bullet b) = (a \bullet b) \bullet b$$
for all $a, b \in A$.

✳ **Example 10.13.**
   Considering the aforementioned in Section 10.2, it follows that the set of octonions forms an algebra over the field of real numbers $\mathbb{R}$, which is called the **algebra of octonions**. This algebra is non-commutative and non-associative with identity. In Section 10.5, we will show that the algebra of octonions is alternative.

✳ **Definition 10.14.**
   A **Lie algebra** is a vector space A over a field K with a binary operation which is called the **Lie bracket** or the **commutator**:
$$[\,,\,] : A \times A \to A,$$
satisfying the conditions:

1. bilinearity:

$$[ax + by, z] = a[x, z] + b[y, z] \text{ for all } a, b \in K \text{ and all } x, y, z \in A \tag{10.22}$$

2. Jacobi identity:

$$[x, [y, z]] + [z, [x, y]] + [y, [z, x]] = 0 \text{ for all } x, y, z \in A \tag{10.23}$$

3.

$$[x, x] = 0 \text{ for all } x \in A \tag{10.24}$$

Note, that from the first and third conditions, the condition of antisymmetry follows:
$$[x, y] = -[y, x] \tag{10.25}$$

Really:

$$0 = [x+y, x+y] = [x,x] + [x,y] + [y,x] + [y,y] = [x,y] + [y,x].$$

Inversely, if char $K \neq 2$, then, from equality (10.25), the condition (10.24) follows.

A Lie algebra was named in honer of the outstanding Norwegian mathematician, **Marius Sophus Lie** (1842-1899).

Two elements x, y of a Lie algebra A are called **commutative** if $[x, y] = 0$. A Lie algebra A is called **commutative** if

$$[x,y] = 0, \quad \text{for all } x, y \in A.$$

✱ Example 10.15.
Let $A = \mathbb{R}^3$ be a vector space over the field of real numbers with orthogonal basis $\{i, j, k\}$ and the operation:
$$[a, b] = a \times b$$

which is the vector product of elements $a, b \in A$. If

$$a = a_1 i + a_2 j + a_3 k$$

$$b = b_1 i + b_2 j + b_3 k$$

Marius Sophus Lie
(1842-1899)

then the vector product is defined by:

$$[a, b] = \det \begin{pmatrix} i & j & k \\ a_1 & a_2 & a_3 \\ b_1 & b_2 & b_3 \end{pmatrix}.$$

From the properties of determinants, we immediately obtain that the following conditions hold:
1) antisymmetry:
$$[a, b] = -[b, a]$$

2) bilinearity:
$$[\alpha a + \beta b, c] = \alpha[a, c] + \beta[b, c]$$

3) Jacobi identity:

$$[a, [b, c]] + [c, [a, b]] + [b, [c, a]] = 0 \quad \text{for all } a, b, c, \in A$$

for all $\alpha, \beta \in \mathbb{R}$, and all $a, b, c \in A$.
Therefore, A is a Lie algebra.

✳ Examples 10.16.

1. Each vector space V with Lie bracket:

$$[u, w] = 0 \text{ for all } u, w \in V$$

   is a Lie algebra, which is commutative.

2. Let A be an associative algebra with operations of addition $+$ and multiplication $\bullet$. We introduce a new operation called the **commutator**:

$$[a, b] = a \bullet b - b \bullet a \text{ for all } a, b \in A.$$

   If A is a commutative algebra, then $[a, b] = 0$. If A is not a commutative algebra, then A is a Lie algebra with respect to commutator.

3. If V is a vector space over a field K and $A = \text{End}_K V$, then we can introduce the operation:

$$[x, y] = x \circ y - y \circ x, \text{ for all } x, y \in A,$$

   where $\circ$ is a composition of linear mappings. With respect to this operation, A is a Lie algebra, which is denoted $A = \text{gl}(V)$, and it is called the **general linear Lie algebra**.

4. The algebra of matrices $A = M_n(\mathbb{R})$ over the field of real numbers $\mathbb{R}$ is a Lie algebra with respect to operation:

$$[\mathbf{X}, \mathbf{Y}] = \mathbf{XY} - \mathbf{YX}$$

   for all $\mathbf{X}, \mathbf{Y} \in A$ and is denoted by $A = \mathbf{gl}(n, K)$.

✳ Definition 10.17.

Let A be an algebra over a field K. A **derivation** is a K-linear mapping:

$$\partial : A \to A$$

satisfying the conditions:

    1) $\partial(\alpha p + \beta q) = \alpha \partial(p) + \beta \partial(q)$
    2) $\partial(pq) = \partial(p)q + p\partial(q)$
    for all $p, q \in A$ and all $\alpha, \beta \in K$.

The condition 2) is called the **Leibnitz product rule**.

✳ Example 10.18.

The set of all derivations $\text{Der}(A)$ of an algebra A is a Lie algebra with respect to the operation:

$$[\partial_1, \partial_2] = \partial_1 \partial_2 - \partial_2 \partial_1.$$

Let A be an n-dimensional algebra over a field K with basis $\{e_1, e_2, \ldots, e_n\}$. Then, each element $a \in A$ can be uniquely written in the form:

$$a = \sum_{i=1}^{n} x_i e_i$$

where $x_i \in K$, $i = 1, 2, \ldots, n$.

The product of elements of A is completely defined by multiplication of basis elements:

$$e_i e_j = \sum_{k=1}^{n} \gamma_{ij}^{k} e_k,$$

where $\gamma_{ij}^{k} \in K$.

If $a = \sum_{i=1}^{n} x_i e_i$ and $b = \sum_{i=1}^{n} y_i e_i$, then $ab = \sum_{i,j,k=1}^{n} (x_i y_j) \gamma_{ij}^{k} e_k$. The structure of an algebra A is completely defined by the choose of n^3 elements $\gamma_{ij}^{k} \in K$, for $i, j, k \in \{1, 2, \ldots, n\}$. Elements γ_{ij}^{k} are called **structural constants** of the algebra A.

By direct verification, it is not difficult to prove (which we leave to the Reader as an exercise) the following theorem:

✳ **Theorem 10.19.**

- An algebra A is commutative if and only if $e_i e_j = e_j e_i$.

- An algebra A is associative if and only if $(e_i e_j) e_k = e_i (e_j e_k)$.

✳ **Definition 10.20.**
A **homomorphism** of K-algebras A and B is a mapping:

$$\varphi : A \to B$$

satisfying the following conditions:
  1) $\varphi(\lambda a) = \lambda \varphi(a)$
  2) $\varphi(a + b) = \varphi(a) + \varphi(b)$
  3) $\varphi(ab) = \varphi(a)\varphi(b)$
  for all $\lambda \in K$ and all $a, b \in A$.

If A and B are algebras with identities 1_A and 1_B, respectively, then $\varphi(1_A) = 1_B$.

Note, that a homomorphism of algebras is simultaneously a homomorphism of rings and vector spaces.

▌ ✳ **Definition 10.21.**
A bijective homomorphism of K-algebras is called an **isomorphism** of K-algebras.

Similar to the case of rings, we will use the following denotations:

- $\mathrm{Hom}_K(A, B)$ is the set of all homomorphisms $\varphi : A \to B$.

- $\mathrm{End}_K(A) = \mathrm{Hom}_K(A, A)$ is the set of all endomorphisms of A.

It is easy to verify that $\mathrm{End}_K(A)$ is an K-algebra over a field K.

Let A be a K-algebra and $a \in A$. We define a mapping $R_a : A \to A$ such that $R_a(x) = xa$, for all $x \in A$. Then, R_a satisfies the following conditions:

- $R_a(x + y) = R_a(x) + R_a(y)$ for all $x, y \in A$,

- $R_a(\lambda x) = \lambda R_a(x)$ for all $x \in A$ and all $\lambda \in K$.

Therefore, R_a is a linear operator on the algebra A called the **right regular representation** of A.

Analogously, we can introduce the **left regular representation** L_a of A: $L_a : A \to A$ such that $L_a(x) = ax$ for all $a, x \in A$.

We show that

- $R_{a+b} = R_a + R_b$,

- $R_{\lambda a} = \lambda R_a$.

Really, $R_{a+b}(x) = x(a + b) = xa + xb = R_a(x) + R_b(x)$ and $R_{\lambda a}(x) = x(\lambda a) = \lambda(xa) = \lambda R_a(x)$ for all $x \in A$ and all $\lambda \in K$.

If A is an associative algebra, then

$$R_{ab} = x(ab) = (xa)b = R_b(xa) = R_b R_a(x)$$

for all $a, b, x \in A$. Therefore, $R_{ab} = R_b R_a$. If A is an algebra with identity 1, then $R_1(x) = x$, i.e., $R_1 = Id$ is the identity operation.

Since $R_a(1) = 1 \cdot a = a$ and $R_b(1) = 1 \cdot b = b$, $R_a \neq R_b$ if $a \neq b$.

Analogously we can show that:

- $L_{a+b} = L_a + L_b$

- $L_{\lambda a} = \lambda L_a$.

- $L_1 = Id$

- $L_a \neq L_b$ if $a \neq b$.

If A is an associative algebra, then $L_{ab} = L_a L_b$.

Therefore, the sets of operators $X = \{R_a \ : \ a \in A\}$ and $Y = \{L_a \ : \ a \in A\}$ form subalgebras in the algebra of endomorphisms $\mathrm{End}_K(A)$ of an associative algebra A over a field K.

✳ **Theorem 10.22. (Cayley)**
Each associative algebra A with identity over a field K is isomorphic to a subalgebra $\mathrm{End}_K(A)$ of linear transformations of A over K.

Proof.
Let $Y = \{L_a \ : \ a \in A\}$. From the reasoning above, it follows that Y is a subalgebra of $\mathrm{End}_K(A)$. Consider the mapping $f : A \to Y$, where $f(a) = L_a$ for all $a \in A$.

Since $f(a + b) = L_{a+b} = L_a + L_b = f(a) + f(b)$ and $f(ab) = L_{ab} = L_a L_b = f(a)f(b)$, $f(\lambda a) = L_{\lambda a} = \lambda L_a = \lambda f(a)$, $f(1) = L_1 = Id$, it follows that f is a homomorphism of algebras which is obviously an epimorphism. Since $L_a \neq L_b$ for $a \neq b$, f is a monomorphism, so f is an isomorphism. □

10.4. Division Algebras. Algebras with Involution. Composition Algebras

In this section, we consider some interesting classes of algebras with particular properties.

✳ **Definition 10.23.**
An algebra A is called a **division algebra** if each of equations: $ax = b$ and $ya = b$, where $a, b \in A$, $a \neq 0$, has a solution.

✳ **Lemma 10.24.**
If A is a finite dimensional associative algebra with identity, then the following conditions are equivalent:
1) A is a division algebra.
2) An algebra A has no zero divisors, i.e., each equality $ab = 0$ implies $a = 0$ or $b = 0$ where $a, b \in A$.

Proof.
1) $\Rightarrow$ 2). Suppose that A is a division algebra. Then, each equation of the form $ax = 1$ and $ya = 1$, where $0 \neq a \in A$, has a solution. In this case, an ideal aA, where $a \in A$, contains the identity 1. Then, by Lemma 4.28, $aA = A = Aa$. Consider the right regular representation: $R_a : A \to A$, where $R_a(x) = xa$ for all $x \in A$. Since $Aa = A$, R_a is an epimorphism. From theorem A.25, it follows that

$$\dim A = \dim \mathrm{Ker}(R_a) + \dim \mathrm{Im}(R_a).$$

Since $\dim A = \dim \mathrm{Im}(R_a)$, we obtain that $\dim \mathrm{Ker}(R_a) = 0$, hence, $\mathrm{Ker}(R_a) = 0$, i.e., R_a is an monomorphism. This means that, if $R_a(x) = xa = 0$, then $x = 0$. Since a is an arbitrary element of A, we obtain that the algebra A has no zero divisors.

2) $\Rightarrow$ 1). Suppose that A has no zero divisors. Let $a \in A$, $a \neq 0$. Consider right ideals of A, which form increasing chain of ideals:

$$aA \supseteq a^2 A \supseteq a^3 A \supseteq \cdots \supseteq a^n A \supseteq a^{n+1} A \supseteq \cdots$$

Since $\dim A < \infty$, there exists an integer n such that $a^n A = a^{n+1} A$. Therefore, for a given element $b \in A$ there is an element $x \in A$ such that $a^n b = a^{n+1} x$. Hence, $a^n (b - ax) = 0$. Since A has no zero divisors and $a \neq 0$, we obtain that $b - ax = 0$, i.e., the equation $ax = b$ has a solution.

Analogously, it can be proved that an equation $ya = b$ has a solution. $\square$

✳ **Lemma 10.25.**

If A is a finite dimensional associative algebra with the identity $\mathbf{1}$, then the following conditions are equivalent:
1) A is a division algebra.
2) Each element of A is invertible, i.e., for each element $a \in A$ there exists an element $a^{-1} \in A$ such that

$$aa^{-1} = a^{-1}a = \mathbf{1}.$$

Proof.

1) $\Rightarrow$ 2). Suppose that A is a division algebra. Then, an equation $ax = \mathbf{1}$ has one solution and $ya = \mathbf{1}$ has one solution. Hence, $y = y(ax) = (ya)x = x$. Hence, $x = y = a^{-1}$.

2) $\Rightarrow$ 1). Consider an equation $ax = b$, $a \neq 0$. Since there is an invertible element a^{-1}, $x = a^{-1}(ax) = a^{-1}b$. Analogously, an equation $ya = b$, $a \neq 0$ has a solution $y = (ya)a^{-1} = ba^{-1}$. $\square$

✳ **Definition 10.26.**

An algebra A over the field $\mathbb{R}$ is called a **∗-algebra** (or an **algebra with involution**) if there is a linear mapping

$$* : A \to A$$

satisfying the conditions:
1) $(x + y)^* = x^* + y^*$
2) $(\lambda x)^* = \lambda x^*$
3) $(x^*)^* = x$
4) $(xy)^* = y^* x^*$
for all $x, y \in A$ and all $\lambda \in \mathbb{R}$.
If additionally
5) $x^* = x$
for all $x \in A$, then A is called a **real ∗-algebra**.

✳ Examples 10.27.

1. The algebra of real numbers $\mathbb{R}$ is a real $*$-algebra with trivial involution $x^* = x$ for all $x \in \mathbb{R}$.

2. The algebra of complex numbers $\mathbb{C}$ is a $*$-algebra with involution

$$(a + bi)^* = a - bi$$

   for all $a + bi \in \mathbb{C}$.

3. The algebra of quaternions $\mathbb{H}$ is a $*$-algebra with involution

$$(a + bi + cj + dk)^* = a - bi - cj - dk$$

   for all $a + bi + cj + dk \in \mathbb{H}$.

4. The algebra of octonions $\mathbb{O}$ is a $*$-algebra with involution

$$(a_1 e_1 + a_2 e_2 + \cdots + a_7 e_7 + a_8 e_8)^* = a_1 e_1 - a_2 e_2 - \cdots - a_7 e_7 - a_8 e_8$$

   for all $a_1 e_1 + a_2 e_2 + \cdots + a_7 e_7 + a_8 e_8 \in \mathbb{O}$.

Recall that a mapping $q : A \to \mathbb{R}$ is a **quadratic form** if

$$q(\lambda x) = \lambda^2 q(x) \quad \text{for all } \lambda \in \mathbb{R} \text{ and all } x \in A. \tag{10.26}$$

A function $\beta : A \times A \to \mathbb{R}$ given by:

$$\beta(x, y) = 1/2[q(x + y) - q(x) - q(y)] \tag{10.27}$$

for all $x, y \in A$ is called a **binary form**. From equalities (10.26) and (10.27), we immediately obtain that $q(0) = 0$ and $\beta(x, y)$ is a symmetric binary form. From (10.27) we also obtain that $\beta(x, x) = 1/2[q(2x) - q(x) - q(x)] = q(x)$, i.e.,

$$q(x) = \beta(x, x) \tag{10.28}$$

A quadratic form $q(x)$ and the corresponding binary form $\beta(x, y)$ are called **associated**. A quadratic form $q(x)$ is called **non-singular** if its associated binary form $\beta(x, y)$ is non-singular, i.e., $\beta(x, y) = 0$ for each $y \in A$ implies that $x = 0$.

✳ Definition 10.28.

A non-zero finite dimensional algebra A (not necessary associative) over a field F is called a **composition algebra** if there is a mapping:

$$N : A \to F$$

which is a non-singular quadratic form satisfying the condition of multi-plicity:

$$N(xy) = N(x)N(y) \tag{10.29}$$

In this case, the quadratic form $N(x)$ is called the **norm** (or **algebraic norm**) of A.

In a particular case, if $F = \mathbb{R}$ is the field of real numbers, a composition algebra is sometimes called a **real normed algebra**.

A composition algebra with identity is called a **Hurwitz algebra**.

If A is a composition algebra with identity, then, from (10.29), it follows that $N(1) = 1$. Really, $N(x) = N(x \cdot 1) = N(x)N(1)$. In particular, $N(1) = N(1)N(1)$, hence, $N(1) = 0$ or $N(1) = 1$. If $N(1) = 0$, then $N(x) = 0$ for all $x \in A$. Then, for the associated binary form $\beta(x, y)$, we obtain that $\beta(1, y) = 1/2[N(1 + y) - N(1) - N(y)] = 0$ for all $y \in A$. Since the binary form $\beta(x, y)$ is non-singular, we obtain that $1 = 0$, which is the case only if A is a trivial algebra. If algebra is non-trivial, then $N(1) = 1$.

In our consideration in this chapter, we will consider composition algebras with identity over the field of real numbers.

❋ Examples 10.29.

1. The algebra of real numbers $\mathbb{R}$ is a composition algebra, since $N(x) = x^2$ is a non-singular quadratic form and $N(xy) = N(x)N(y)$.

2. The algebra of complex numbers $\mathbb{C}$ is a composition algebra, since $N(x) = xx^* = a^2 + b^2$, for $x = a + bi$, is a non-singular quadratic form and $N(xy) = N(x)N(y)$.

3. The algebra of quaternions $\mathbb{H}$ is a composition algebra, since $N(x) = xx^* = a^2 + b^2 + c^2 + d^2$, for $x = a + bi + cj + dj$, is a non-singular quadratic form and $N(xy) = N(x)N(y)$.

10.5. Cayley-Dickson Construction

In the previous sections, we considered the algebras of quaternions and octonions, which are some generalizations of real and complex numbers. Note that these algebras form an increasing chain of algebras:

$$\mathbb{R} \subset \mathbb{C} \subset \mathbb{H} \subset \mathbb{O}$$

such that $\dim_{\mathbb{R}} \mathbb{C} = 2$, $\dim_{\mathbb{R}} \mathbb{H} = 4 = 2\dim_{\mathbb{R}} \mathbb{C}$, $\dim_{\mathbb{R}} \mathbb{O} = 8 = 2\dim_{\mathbb{R}} \mathbb{H}$.

It would be very interesting to have a construction of division algebras $\mathbb{R}$, $\mathbb{C}$, $\mathbb{H}$, $\mathbb{O}$ which explains why each of these algebras contains the other one, in what way they are interconnected, why the algebra of quaternions is not commutative and the algebra of octonions is not associative. Moreover, it would be also very interesting to have a construction which gives an infinity chain of algebras, each with twice the dimension of the previous one. In fact, such a construction exists and is called the **Cayley-Dickson construction**.

While studying quaternions, Hamilton noted that a complex number $a+bi$ can be considered as a pair (a, b) of real numbers with corresponding operations.

Consider a set of pairs of real numbers:

$$X = \{(a, b) \ : \ a, b \in \mathbb{R}\}.$$

We can define the operations of addition, multiplication, multiplication by a scalar and conjugation on this set by the follows:

$$(a, b) + (c, d) = (a + c, b + d),$$

$$(a, b) \cdot (c, d) = (ac - bd, ad + bc),$$

$$\lambda(a, b) = (\lambda a, \lambda b)$$

$$(a, b)^* = (a, -b).$$

for all $\lambda, a, b, c, d \in \mathbb{R}$.

It is easy to verify (which we leave to the Reader as an exercise) that all axioms of an $*$-algebra are satisfied. In this way, X turns out to be a $*$-algebra over the real numbers, and the identity in X is the element $1 = (1, 0)$. At that time, X is a two-dimensional $*$-algebra over the real numbers with the basis $\{(1, 0), (0, 1)\}$.

Consider a mapping $f : \mathbb{C} \to X$ given by $f(a + ib) = (a, b)$. By this mapping, a real number a corresponds to the pair $(a, 0)$, the image identity i corresponds to the pair $(0, 1)$, and the number i^2 corresponds the pair $(0, 1)(0, 1) = (-1, 0)$, i.e., the integer -1. It is easy to verify that the mapping f is an isomorphism of the algebra of complex numbers $\mathbb{C}$ to the algebra X, while the restriction of the mapping f to the real numbers $f_{\mathbb{R}} : \mathbb{R} \to X$ given by $f_{\mathbb{R}}(a) = (a, 0)$ is an embedding of the field of real numbers into the algebra X. Therefore, the algebra X is isomorphic to the field of complex numbers $\mathbb{C}$, and the algebra of the real numbers $\mathbb{R}$ is a subalgebra of X with $\dim_{\mathbb{R}} X = 2$.

Assuming that a, b, c, d are real numbers and $x^* = x$ for all real numbers x, the operations of multiplication and conjugation on X can be written in the following form:

$$(a, b) \cdot (c, d) = (ac - bd, ad + bc) = (ac - db, da + bc) = (ac - d^*b, da + bc^*)$$

and

$$(a, b)^* = (a^*, -b)$$

In a similar way, we can obtain the algebras of quaternions, octonions and other algebras using the Cayley-Dickson construction.

✳ Definition 10.30.

The **general Cayley-Dickson construction** is an iterative process of generating new algebras from old ones, starting from the initial K-algebra $A_0 = K$, so that we get a sequence of K-algebras with involution $*$:

$$A_0 = K \subset A_1 \subset A_2 \subset \cdots \subset A_{n-1} \subset A_n \subset A_{n+1} \subset \cdots$$

where $A_{n+1} = A_n \times A_n$ is a K-algebra with involution $*$ and the operations defined as follows:

1) addition:

$$(a, b) + (c, d) = (a + c, b + d)$$

2) multiplication by a scalar:

$$\lambda(a, b) = (\lambda a, \lambda b)$$

3) multiplication:

$$(a, b)(c, d) = (ac + \gamma_n d^* b, da + bc^*),$$

4) conjugation:

$$(a, b)^* = (a^*, -b)$$

for all $a, b, c, d \in A_n$, where (γ_n) is a sequence of non-zero elements of K.

An algebra A_{n+1} which is generated by this process is called the **Cayley-Dickson algebra** or the **Cayley-Dickson extension** of A_n.

Note, that in the general Cayley-Dickson construction, a K-algebra A_{n+1} contains A_n as a subalgebra via the embedding $a \mapsto (a, 0)$ for all $a \in A_n$. Moreover, $\dim_K A_{n+1} = 2 \dim_K A_n$.

It is easy to show that each Cayley-Dickson extension A of a $*$-algebra B is a K-algebra with involution $*$ defined in Definition 10.30, i.e., $(a, b)^* = (a^*, -b)$ for all $a, b \in B$, if $\lambda^* = \lambda$ for all $\lambda \in K$. To this aim, we need to verify all conditions of definition 10.26:

1. $[(a, b) + (c, d)]^* = (a + c, b + d)^* = [(a + c)^*, -(b + d)] = (a^* + c^*, -b - d) = (a^*, -b) + (c^*, -d) = (a, b)^* + (c, d)^*$

2. $[\lambda(a, b)]^* = (\lambda a, \lambda b)^* = ((\lambda a)^*, -\lambda b) = (\lambda a^*, -\lambda b) = \lambda(a^*, -b) = \lambda(a, b)^*$

3. $[(a, b)^*]^* = (a^*, -b)^* = (a^{**}, -(-b)) = (a, b)$

4. $[(a, b)(c, d)]^* = [ac - d^* b, da + bc^*]^* = [(ac - d^* b)^*, -(da + bc^*)] = [(c^* a^* - b^* d), (-da - bc^*)] = (c^*, -d)(a^*, -b) = (c, d)^*(a, b)^*$

for all $a, b \in \mathbb{C}$ and all $\lambda \in K$.

In the classical case, which will be considered in this section, $A_0 = K = \mathbb{R}$ is the field of real numbers and $\gamma_n = -1$ for all n. Such a construction is called the **Cayley-Dickson construction**.

Other cases with $\gamma_n \in \{1, 0, -1\}$ will be considered in the next sections.

As a corollary of the Cayley-Dickson construction, we obtain the following theorem whose proof can be obtained from Definition 10.30 by straightforward verification.

❋ Theorem 10.31.
Let an algebra B be the Cayley-Dickson extension of an algebra A with involution $*$ over the field of real numbers $\mathbb{R}$. Then:
1) B is commutative if and only if A is commutative and $a^* = a$ for all $a \in A$.
2) B is associative if and only if A is commutative and associative.

❋ Example 10.32.
We show that the algebra of quaternions is the Cayley-Dickson extension of $*$-algebra of complex numbers $\mathbb{C}$ with $\gamma = -1$.

Let us apply the Cayley-Dickson construction to the complex numbers with $\gamma = -1$. Consider the set of pairs of complex numbers, i.e., the cartesian product:

$$X = \{(a, b) \; : \; a, b \in \mathbb{C}\}.$$

In this set, we introduce the operations of addition, multiplication, multiplication by a scalar and conjugation as follows:

1) addition:
$$(a, b) + (c, d) = (a + c, b + d),$$

2) multiplication:
$$(a, b)(c, d) = (ac - d^*b, da + bc^*)$$

3) multiplication by a scalar:
$$\lambda(a, b) = (\lambda a, \lambda b)$$

4) conjugation:
$$(a, b)^* = (a^*, -b)$$

for all $a, b, c, d \in \mathbb{C}$ and all $\lambda \in \mathbb{R}$.

Note, that each quaternion $x = a_1 + a_2 i + a_3 j + a_4 k$ can be written in the following form:

$$x = a_1 + a_2 i + a_3 j + a_4 k = (a_1 + a_2 i) + (a_3 j + a_4 ij) = (a_1 + a_2 i) + (a_3 + a_4 i)j = a + bj,$$

where $a = a_1 + a_2 i, b = a_3 + a_4 i \in \mathbb{C}$, $j^2 = -1$. So, each quaternion $x = a + bj$ corresponds in a natural way to the pair of complex numbers (a, b), where $a, b \in \mathbb{C}$.

Consider a mapping $f : \mathbb{H} \to X$ given by:

$$f(x) = f(a_1 + a_2 i + a_3 j + a_4 k) = f((a_1 + a_2 i) + (a_3 + a_4 i)j) = f(a + bj) = (a, b),$$

where $a = a_1 + a_2 i, b = a_3 + a_4 i \in \mathbb{C}$.

By this mapping, a complex number a corresponds to a pair $(a, 0)$, and the image element j corresponds to a pair $(0, 1)$, so the number j^2 corresponds to a pair $(0, 1)(0, 1) = (-1, 0)$, i.e., the number -1.

We show that the mapping f is an isomorphism of the algebra of quaternions $\mathbb{H}$ to the algebra X, while the restriction of the mapping f on the complex numbers $f_{\mathbb{C}} : \mathbb{C} \to X$, given by $f_{\mathbb{C}}(a) = (a, 0)$, is a natural embedding of the complex numbers into X. Therefore, the algebra X is isomorphic to the algebra of quaternions $\mathbb{H}$, and the algebra of the complex numbers is a subalgebra of the algebra X with $\dim_{\mathbb{C}} X = 2$.

Let $x = a_1 + a_2 i + a_3 j + a_4 k = a + bj$, $y = b_1 + b_2 i + b_3 j + b_4 k = c + dj$, then

1. $f(x + y) = f((a + bj) + (c + dj)) = f((a + c) + (b + d)j) = (a + c, b + d) = (a, b) + (c, d) = f(x) + f(y)$,

2. $f(\lambda x) = f(\lambda(a + bj)) = f(\lambda a + (\lambda b)j) = (\lambda a, \lambda b) = \lambda(a, b) = \lambda f(x)$,

3. $f(xy) = f((a_1 + a_2 i + a_3 j + a_4 k)(b_1 + b_2 i + b_3 j + b_4 k)) = f((a_1 b_1 - a_2 b_2 - a_3 b_3 - a_4 b_4) + (a_1 b_2 + a_2 b_1 + a_3 b_4 - a_4 b_3)i + ((a_1 b_3 - a_2 b_4 + a_3 b_1 + a_4 b_2) + (a_1 b_4 + a_2 b_3 - a_3 b_2 + a_4 b_1)i)j) = f((ac - d^* b) + (da + bc^*)j) = (ac - d^* b, da + bc^*) = (a, b)(c, d) = f(x)f(y)$.

Moreover, $f(\mathbf{1}) = (1, 0)$ is the identity of X. At the time, f is a homomorphism of algebras. It is obvious that f is an epimorphism, since $f(a + bj) = (a, b)$ for each element $(a, b) \in X$. The homomorphism f is also a monomorphism, since $f(a + bj) = (a, b) = (0, 0)$ implies $a = b = 0$, i.e., $a + bj = \mathbf{0}$. So, f is an isomorphism. Therefore, we have:

$$f(\mathbf{1}) = (1, 0), \quad f(j) = (0, 1), \quad f(i) = (i, 0),$$

$$f(k) = f(ij) = f(i)f(j) = (i, 0)(0, 1) = (0, i).$$

In this way, X is a four-dimensional algebra over the field of real numbers with basis:

$$\{(1, 0), (i, 0), (0, 1), (0, i)\}.$$

Since the conjugate element for a quaternion

$$x = a_1 + a_2 i + a_3 j + a_4 k = (a_1 + a_2 i) + (a_3 + a_4 i)j = a + bj$$

is a quaternion:

$$x^* = a_1 - a_2 i - a_3 j - a_4 k = (a_1 - a_2 i) - (a_3 + a_4 i)j = a^* - bj$$

we obtain that

$$[f(x)]^* = (a, b)^* = (a^*, -b) = f(a^* - bj) = f(x^*).$$

Therefore, f is an isomorphism of $*$-algebras X and $\mathbb{H}$. From Theorem 10.31, it follows that $\mathbb{H}$ is an associative algebra.

> ✳ **Definition 10.33.**
> A $*$-algebra with identity over the field of real numbers $\mathbb{R}$, satisfying the conditions:
> 1) $x + x^* \in \mathbb{R}$ for all $x \in A$,
> 2) $0 < xx^* = x^*x \in \mathbb{R}$ for all $x \in A$ and $x \neq 0$
> is called a **nicely normed $*$-algebra**.

✳ **Example 10.34.**
The algebras of real numbers, complex numbers and quaternions are nicely normed $*$-algebras.

Using Definitions 10.30 and 10.33, we obtain the following theorem whose proof can be obtained by direct verifying.

✳ **Theorem 10.35.**
Let $*$-algebra A be a Cayley-Dickson extension of $*$-algebra B over the field of real numbers $\mathbb{R}$ with $\gamma = -1$. Then:

1. An algebra A is nicely normed $*$-algebra if and only if an algebra B is nicely normed $*$-algebra.

2. An algebra A is alternative and nicely normed $*$-algebra if and only if an algebra B is associative and nicely normed $*$-algebra.

✳ **Example 10.36.**
We show that the algebra of octonions $\mathbb{O}$ is the Cayley-Dickson extension of the $*$-algebra of quaternions $\mathbb{H}$, which is isomorphic to the $*$-algebra $X = \mathbb{H} \times \mathbb{H}$.

Since the table of multiplication for the basis elements e_1, e_2, e_3, e_4 is exactly the same as for quaternions $1, i, j, k$, using Table 10.2, each octonion $x = a_1e_1 + a_2e_2 + \cdots + a_8e_8$ can be written in the form:

$$x = (a_1e_1 + a_2e_2 + a_3e_3 + a_4e_4) + (a_5e_1 + a_6e_2 + a_7e_3 + +a_8e_4)e_5 = (a + be_5),$$

where $a = a_1e_1 + a_2e_2 + a_3e_3 + a_4e_4$, $b = a_5e_1 + a_6e_2 + a_7e_3 + +a_8e_4$ and $e_5^2 = -1$.

Consider now the set Y of elements of the form $a + be$, where $a, b \in \mathbb{H}$, $e^2 = -1$, and the operations of addition, multiplication by a scalar, multiplication, and conjugation are defined as follows:

$$(a + be) + (c + de) = (a + c) + (b + d)e$$

$$\lambda(a + be) = \lambda a + \lambda be$$

$$(a + be)(c + de) = (ac - d^*b) + (da + bc^*)e$$

$$(a + be)^* = (a^* - be)$$

for all $a, b, c, d \in \mathbb{H}$, and all $\lambda \in \mathbb{R}$.

Then, elements $1, i, j, k, e, ie, je, ke$ form a basis of Y over the field of real numbers and the mapping

$$\varphi : \mathbb{O} \to Y$$

such that $\varphi(a + be_5) = a + be$ is an isomorphism of algebras. In this case, $\varphi(e_1) = 1$, $\varphi(e_2) = i$, $\varphi(e_3) = j$, $\varphi(e_4) = k$, $\varphi(e_5) = e$, $\varphi(e_6) = ie$, $\varphi(e_7) = je$, $\varphi(e_8) = ke$. So, we can identify the algebras $\mathbb{O}$ and Y.

Let $X = \mathbb{H} \times \mathbb{H}$ be the Cayley-Dickson extension of the $*$-algebra of quaternions $\mathbb{H}$. Then, we can consider the mapping $f : X \to Y$ such that $f(a, b) = a + be$ for all $(a, b) \in X$. In particular, $f(1, 0) = 1$, $f(i, 0) = i$, $f(j, 0) = j$, $f(k, 0) = k$, $f(0, 1) = e$, $f(0, i) = ie$, $f(0, j) = je$, $f(0, k) = ke$. Comparing the conditions of operations on X, from Definition 10.30 and conditions of operations on X, we obtain that f is an isomorphism of $*$-algebras.

In this way, we obtain that $\mathbb{O} \cong Y \cong X$, i.e., the algebra of octonions is isomorphic to the Cayley-Dickson extension of the $*$-algebra of quaternions $\mathbb{H}$.

Since $\mathbb{H}$ is an associative nicely normed $*$-algebra, from theorem 10.35, it follows that the algebra of octonions $\mathbb{O}$ is an alternative nicely normed $*$-algebra.

✳ Theorem 10.37.

Let an algebra A be the Cayley-Dickson extension of a composition $*$-algebra B over the field of real numbers $\mathbb{R}$. If A is a nicely normed $*$-algebra, then the mapping $N : A \to [0, \infty)$ given by $N(x) = xx^*$ defines a non-singular quadratic form.

Proof.

The function $N(x) = xx^*$ is a quadratic form, since $N(\lambda x) = (\lambda x)(\lambda x)^* = (\lambda x)(\lambda x^*) = \lambda^2(xx^*) = \lambda^2 N(x)$ and the function $\beta(x, y) = \dfrac{1}{2}[N(x + y) - N(x) - N(y)] = \dfrac{1}{2}[(x + y)(x + y)^* - xx^* - yy^*] = \dfrac{1}{2}(xy^* + yx^*)$ is a symmetric binary form.

We show that $\beta(x, y)$ is a non-singular binary form. Let B be a composition $*$-algebra with algebraic norm $N_1(x_1) = x_1 x_1^*$, which is a non-singular quadratic form, and associated binary form $\beta_1(x_1, y_1) = \dfrac{1}{2}[N_1(x_1 + y_1) - N_1(x_1) - N_1(y_1)]$, where $x_1, y_1 \in B$. Let A be the Cayley-Dickson extension of $*$-algebra B and $x = (x_1, x_2)$, $y = (y_1, y_2) \in A$, where $x_1, x_2, y_1, y_2 \in B$. Then, $N(x) = xx^* = (x_1, x_2)(x_1, x_2)^* = (x_1, x_2)(x_1^*, -x_2) = (x_1 x_1^* +$

$x_2 x_2^*, -x_2 x_1 + x_2 x_1) = ((x_1 x_1^* + x_2 x_2^*, 0) = (N_1(x_1) + N_1(x_2), 0)$. Analogously, $N(y) = yy^* = (N_1(y_1) + N_1(y_2), 0)$, and $N(x+y) = (x+y)(x+y)^* = (N_1(x_1+y_1) + N_1(x_2+y_2), 0)$. Therefore, $\beta(x, y) = 1/2(N_1(x_1+y_1) + N_1(x_2+y_2) - N_1(x_1) - N_1(x_2) - N_1(y_1) - N_1(y_2), 0) = 1/2[(N_1(x_1 + y_1) - N_1(x_1) - N_1(y_1)) + (N_1(x_2 + y_2) - N_1(x_2) - N_1(y_2)), 0] = (\beta_1(x_1, y_1) + \beta_1(x_2, y_2), 0)$.

Assume that $\beta(x, y) = 0$ for all $y \in A$. Then, $\beta_1(x_1, y_1) + \beta_1(x_2, y_2) = 0$ for all $y_1, y_2 \in B$. If $y_2 = 0$, then $\beta_1(x_1, y_1) = 0$ for all $y_1 \in B$, since $\beta_1(x_2, 0) = 0$. From the assumption that $\beta_1(x_1, y_1)$ is a non-singular form, we obtain that $x_1 = 0$. Analogously, if $y_1 = 0$, then $\beta_1(x_2, y_2) = 0$ for all $y_2 \in B$. Hence, $x_2 = 0$. At this time $x = 0$, which means that $\beta(x, y)$ is a non-singular binary form, therefore, $N(x) = xx^*$ is a non-singular quadratic form. $\square$

✳ Theorem 10.38.
Let an algebra A be a Cayley-Dickson extension of a composition $*$-algebra B over the field of real numbers $\mathbb{R}$. If A is an associative finite dimensional $*$-algebra which is nicely normed, then A is both a composition and division algebra.

Proof.
1) We show that the function $N : A \to [0, \infty)$ given by $N(x) = xx^*$ defines an algebraic norm in A. By Theorem 10.37, the function N is a non-singular quadratic form. We show that the function N is multiplicative. Since A is associative we have:

$$N(xy) = (xy)(xy)^* = (xy)(y^* x^*) = x(yy^*)x^* = xN(y)x^* = N(x)N(y),$$

for all $x, y \in A$. At the time A is a composition $*$-algebra.

2) We show that A is a division algebra. Since $*$-algebra A is a nicely normed, $N(x) = xx^* > 0$ for any non-zero element $x \in A$. Therefore, each non-zero element $x \in A$ has the inverse element of the form:

$$x^{-1} = \frac{x^*}{N(x)} \tag{10.30}$$

such that $xx^{-1} = x^{-1}x = 1$ is the identity of A. By Lemma 10.25, A is a division algebra. $\square$

✳ Example 10.39.
Since $\mathbb{H}$ is the Cayley-Dickson extension of the algebra of complex numbers, the algebra of quaternions $\mathbb{H}$ is associative and nicely normed $*$-algebra, by Theorem 10.35. So, by Theorem 10.38, the algebra $\mathbb{H}$ is both a division and composition algebra with algebraic norm $N(x) = xx^* = a^2 + b^2 + c^2 + d^2$ for $x = a + bi + cj + dk$.

To show that octonions also form a composition algebra, we will use the next important theorem proved by E. Artin.

> **✻ Theorem 10.40. (E. Artin)**
> An algebra is alternative if and only if each its subalgebra generated by any two elements is associative.

Proof. We give only brief outline of the proof. The more complete proof of this theorem can be found in [16], [19] and [22].

⇒. The proof of this theorem can be performed by induction. Suppose that A is an alternative algebra over a field F. To prove that $F[x, y]$ is an associative algebra for all $x, y \in A$, it suffices to show that $x^i(y^j x^k) = (x^i y^j) x^k$ and $y^i(x^j y^k) = (y^i x^j) y^k$ for all $i, j, k \in \mathbb{N}$.

Since A is an alternative algebra, all $x, y \in A$ satisfies the equalities: $x^2 y = x(xy)$ and $yx^2 = (yx)x$ for all $x, y \in A$. The first equality implies that $[(x+y)(x+y)]y = (x+y)[(x+y)y]$, i.e.,

$$x^2 y + (yx)y + (xy)y + y^3 = x(xy) + y(xy) + xy^2 + y^3,$$

and taking into account the main equalities of alternativity, we obtain that $(yx)y = y(xy)$, and analogously $(xy)x = x(yx)$.

We show first that subalgebra $F[x]$ is associative. Substituting $y = x$ in the last equality, we obtain $x^2 x = xx^2$. Defining $x^{i+1} = xx^i$ for all $i \in \mathbb{Z}^+$, it is easy to prove by induction that $x^i x = xx^i$ and $x^{i+j} = x^i x^j$ for all $i, j \in \mathbb{Z}^+$. Therefore, a subalgebra $F[x]$ is associative. Then again, by induction, we can easy show that $x^i(x^j y^k) = x^{i+j} y^k$ and then $(x^i y^k) x^j = x^i(y^k x^j)$.

⇐. This is obvious. ☐

> **✻ Theorem 10.41.**
> Let an algebra A be the Cayley-Dickson extension of a composition *-algebra B over the field of real numbers $\mathbb{R}$. If A is an alternative finite dimensional *-algebra which is nicely normed, then A is both a composition and division algebra.

Proof.

1) First we show that A is a composition algebra, i.e., the function $N : A \to [0, \infty)$ given by $N(x) = xx^*$ defines an algebraic norm in A. By Theorem 10.37, the function N define a non-singular quadratic form. So, it only needs to show that N is a multiplicative function. Since A is a nicely normed algebra, $x + x^* \in \mathbb{R}$. Then, we can define:

$$\mathrm{Re}(x) = \frac{x + x^*}{2} \in \mathbb{R}, \quad \mathrm{Im}(x) = \frac{x - x^*}{2} \in A.$$

So $x = \dfrac{x + x^*}{2} + \dfrac{x - x^*}{2}$. For each two elements $x, y \in A$, consider a subalgebra in A generated by two elements:

$$M(x, y) = \langle \mathrm{Im}(x), \mathrm{Im}(y) \rangle.$$

By Theorem 10.31, $M(x, y)$ is an associative algebra. Since elements $x, y, x^*, y^* \in M(x, y)$, we have

$$N(xy) = (xy)(xy)^* = (xy)(y^* x^*) = x(yy^*)x^* = xN(y)x^* = N(x)N(y),$$

so A is a composition $*$-algebra.

2) Since A is a nicely normed $*$-algebra, from definition 10.33, it follows that $N(x) = 0 \Leftrightarrow x = 0$. If $xy = 0$, then $0 = N(xy) = N(x)N(y)$. Hence, $N(x) = 0$ or $N(y) = 0$, therefore, $x = 0$ or $y = 0$. So, A has no zero divisors. For any two elements $a, b \in A$, consider equations $ax = b$ and $ya = b$ and the algebra $M(a, b)$ considered above in part 1. Then, $M(a, b)$ is an associative algebra which has no divisors of zero. So, by Lemma 10.24, $M(a, b)$ is a division algebra and these equations have solutions. So, A is also a division algebra. □

✳ Example 10.42.

Since the algebra of octonions $\mathbb{O}$ is the the Cayley-Dickson extension of the algebra of quaternions $\mathbb{H}$, which is associative and nicely normed $*$-algebra, it is an alternative and nicely normed $*$-algebra, by Theorem 10.35(1). So, by Theorem 10.41, the algebra $\mathbb{O}$ is both a division and composition algebra with algebraic norm

$$N(x) = xx^* = a_1^2 + a_2^2 + a_3^2 + a_4^2 + a_5^2 + a_6^2 + a_7^2 + a_8^2 \tag{10.31}$$

for any element $x = a_1 e_1 + a_2 e_2 + \cdots + a_8 e_8 \in \mathbb{O}$, which is a non-singular quadratic form. Therefore, $N(1) = 1$ and $N(x) = 0$ if and only if $x = 0$.

For any non-zero element $x \in \mathbb{O}$, the inverse element is defined as follows:

$$x^{-1} = \frac{x^*}{N(x)},$$

since $xx^{-1} = x^{-1}x = 1$.

✳ Example 10.43.

Consider the Cayley-Dickson extension of the $*$-algebra of octonions $\mathbb{O}$, i.e.,

$$\mathbb{S} = \{(a, b) \ : a, b \in \mathbb{O}\}$$

with operations defined by the follows:

1) addition:
$$(a, b) + (c, d) = (a + c, b + d),$$

2) multiplication:

$$(a, b)(c, d) = (ac - bd^*, ad + bc^*),$$

3) multiplication by a scalar:

$$\lambda(a, b) = (\lambda a, \lambda b),$$

4) conjugation:
$$(a, b)^* = (a^*, -b).$$

for all $a, b, c, d \in \mathbb{O}$

The obtained algebra $\mathbb{S}$ is called the **algebra of sedenions**. From Theorems 10.31 and 10.35, it follows that it is a nicely normed $*$-algebra, which is non-commutative, non-associative and non-alternative. $\mathbb{S}$ is a 16-dimentional algebra over the field of real numbers: $\dim_{\mathbb{R}} \mathbb{S} = 16$. Each element $x \in \mathbb{S}$ can be written as $x = \sum_{i=0}^{15} a_i e_i$, where $a_i \in \mathbb{R}$ and e_i, for $i = 0, 1, \ldots, 15$, are basis elements of $\mathbb{O}$ over $\mathbb{R}$. The basis elements e_i satisfy the conditions:

1) $e_0 e_i = e_i e_0 = e_i$ for all i
2) $e_i e_i = -e_0$ for all $i \neq 0$
3) $e_i e_j = -e_j e_i$ for all $i, j \neq 0$ and $i \neq j$.

So the element e_0 is the identity of $\mathbb{S}$.

Since $\mathbb{S}$ is an algebra with involution $*$, we can define the function $N : \mathbb{S} \to \mathbb{R}$ given as $N(x) = xx^* \geq 0$ for all $x \in \mathbb{S}$, moreover $N(x) = 0 \iff x = 0$. However, N is not an algebraic norm because $N(xy) \neq N(x)N(y)$ for all $x, y \in \mathbb{S}$.

Nevertheless, for each non-zero element $x \in \mathbb{S}$ it can be defined the inverse element:

$$x^{-1} = \frac{x^*}{N(x)},$$

such that $xx^{-1} = x^{-1}x = e_0$.

However, the algebra $\mathbb{S}$ is not a division algebra since it has zero divisors. For example, $(e_1 + e_{13})(e_2 - e_{14}) = 0$.

Thus, we can summarize the main properties of algebras considered in this section:

- The algebra of real numbers $\mathbb{R}$ is a commutative associative nicely normed division algebra with trivial involution.

- The algebra of complex numbers $\mathbb{C}$ is a commutative associative nicely normed division $*$-algebra and $\dim_{\mathbb{R}} \mathbb{C} = 2$.

- The algebra of quaternions $\mathbb{H}$ is a non-commutative associative nicely normed division $*$-algebra and $\dim_{\mathbb{R}} \mathbb{H} = 4$.

- The algebra of octonions $\mathbb{O}$ is an alternative composition division $*$-algebra and $\dim_{\mathbb{R}} \mathbb{O} = 8$.

- The algebra of sedenions $\mathbb{S}$ is a composition $*$-algebra and $\dim_{\mathbb{R}} \mathbb{S} = 16$.

The following theorems, which we give for completeness without proofs, shows the exclusivity of these algebras.

✳ **Theorem 10.44 (Frobenius, 1878).**
There exist up to isomorphism exactly three finite dimensional associative division algebras over the field of real numbers: $\mathbb{R}$, $\mathbb{C}$ and $\mathbb{H}$.

The next three theorems are the generalizations of the Frobenius Theorem for composition and alternative algebras. The Hurwitz Theorem was proved in 1878 by Adolf Hurwitz (1859-1919), while it was published only in 1923.

✳ **Theorem 10.45 (Hurwitz, 1878, 1923).**
There exist up to isomorphism exactly four finite dimensional composition division algebras over the field of real numbers: $\mathbb{R}$, $\mathbb{C}$, $\mathbb{H}$ and $\mathbb{O}$ of dimension 1, 2, 4, 8, respectively.

✳ **Theorem 10.46 (Zorn, 1933).**
There exist up to isomorphism exactly four finite dimensional alternative division algebras over the field of real numbers: $\mathbb{R}$, $\mathbb{C}$, $\mathbb{H}$ and $\mathbb{O}$ of corresponding dimensions 1, 2, 4, 8.

✳ **Theorem 10.47. (M. Kervaire, R. Bott - J. Milnor).**
All finite dimensional division algebras over the field of real numbers have dimensions 1, 2, 4 or 8.

10.6. Dual Numbers and Double Numbers

As was shown in the previous section, the complex numbers give an important example of two-dimensional Cayley-Dickson algebra over real numbers with $\gamma = -1$. In this section, we consider two another important examples of Cayley-Dickson algebras of dimension two over the field of real numbers with $\gamma \in \{1, 0\}$.

✳ **Definition 10.48.**
The Cayley-Dickson extension C_+ of the field of real numbers with $\gamma = 1$ is called the algebra of **double numbers** or **hyporbolic complex numbers**.

So, the algebra of double numbers

$$C_+ = \{(a, b) \ : \ a, b \in \mathbb{R}\} \tag{10.32}$$

is a set of pairs of real numbers with operations of addition, multiplication, multiplication by a scalar and conjugation on this set given as follows:

$$(a, b) + (c, d) = (a + c, b + d),$$

$$(a, b) \cdot (c, d) = (ac + bd, ad + bc),$$

$$\lambda(a, b) = (\lambda a, \lambda b)$$

$$(a, b)^* = (a, -b).$$

for all $\lambda, a, b, c, d \in \mathbb{R}$.

By Theorem 10.31, C_+ is a commutative and associative algebra with involution $*$. The zero element of C_+ is $\mathbf{0} = (0,0)$ and the identity element is $\mathbf{1} = (1,0)$. Consider the element $\varepsilon = (0,1)$. Then, each double number (a,b) can be expressed in the form $a + b\varepsilon = a(1,0) + b(0,1)$ and $\varepsilon^2 = (0,1)(0,1) = (1,0)$, i.e., $\varepsilon^2 = \mathbf{1}$ is the identity element of C_+. So, C_+ can be considered as a two-dimensional algebra over the real numbers with the basis $\{\mathbf{1}, \varepsilon\}$ and the Cayley table for multiplication of basis elements has the form:

$\cdot$	$\mathbf{1}$	ε
$\mathbf{1}$	$\mathbf{1}$	ε
ε	ε	$\mathbf{1}$

Table 10.3

Thus, we can define C_+ as follows:

$$C_+ = \{a + b\varepsilon \ : \ \varepsilon^2 = \mathbf{1}, \ a, b \in \mathbb{R}\} \tag{10.33}$$

Then, the conjugate in C_+ to an element $z = a + b\varepsilon$ is defined as $z^* = (a + b\varepsilon)^* = a - b\varepsilon$ and

$$N(z) = zz^* = (a + b\varepsilon)(a - b\varepsilon) = a^2 - b^2$$

is the algebraic norm of z, since it satisfies the condition of multiplicity $N(z_1 z_2) = N(z_1)N(z_2)$. However, since the quadratic form $N(z)$ is not positive definite, C_+ is not a nicely normed algebra, i.e., N does not satisfy the condition $N(z) = 0 \Leftrightarrow z = 0$. Since the quadratic form $N(z) = a^2 - b^2$ has the hyperbolic type, the double numbers are also called **hyperbolic complex numbers**. The algebra C_+ has an infinite number of elements with $N(z) = 0$, all of them have the form $z = a \pm a\varepsilon$. All these numbers are divisors of zero in C_+. Hence, the algebra of double numbers C_+ is not a division algebra. Nevertheless, any double number with $N(z) \neq 0$ has a unique inverse

$$z^{-1} = \frac{z^*}{N(z)}$$

such that $zz^{-1} = z^{-1}z = 1$.

Consider two double numbers: $e_- = (1 - \varepsilon)/2$ and $e_+ = (1 + \varepsilon)/2$. Then, $e_-^2 = e_-$, $e_+^2 = e_+$, i.e., e_- and e_+ are **idempotents** of C_+. Moreover, they satisfy the following conditions $e_- + e_+ = 1$ and $e_-e_+ = e_+e_- = 0$. Therefore, e_- and e_+ are **orthogonal idempotents**. Each double number $z = a + b\varepsilon \in C_+$ can be written as $xe_- + ye_+$, where $x = a - b$, $y = a + b$. Then, addition and multiplication in C_+ are reduced to addition and multiplication on double numbers by coordinatewise:

$$(xe_- + ye_+) + (x_1e_- + y_1e_+) = (x + x_1)e_- + (y + y_1)e_+$$

$$(xe_- + ye_+)(x_1e_- + y_1e_+) = (xx_1)e_- + (yy_1)e_+$$

Taking these equalities into account, it is easy to show that C_+ is isomorphic to the direct sum of two fields of real numbers $\mathbb{R} \oplus \mathbb{R}$. For this reason, the double numbers are also called **split complex numbers**.

If $z = a + b\varepsilon$ and $N(z) = 1$, then we can set $a = \cosh(t)$ and $b = \sinh(t)$, and z can be expressed in the form:

$$z = e^{\varepsilon t} = \cosh(t) + \varepsilon \sinh(t) \tag{10.34}$$

Now consider the another Cayley-Dickson extension of real numbers with $\gamma = 0$.

✳ Definition 10.49.
   The Cayley-Dickson extension C_0 of the field of real numbers with $\gamma = 0$ is called the algebra of **dual numbers** or **parabolic complex numbers**.

Therefore, the algebra of dual numbers

$$C_0 = \{(a, b) \ : \ a, b \in \mathbb{R}\} \tag{10.35}$$

is a set of pairs of real numbers with operations of addition, multiplication, multiplication by a scalar and conjugation on this set as follows:

$$(a, b) + (c, d) = (a + c, b + d),$$

$$(a, b) \cdot (c, d) = (ac, ad + bc),$$

$$\lambda(a, b) = (\lambda a, \lambda b)$$

$$(a, b)^* = (a, -b)$$

for all $\lambda, a, b, c, d \in \mathbb{R}$.

By Theorem 10.31, C_0 is a commutative and associative algebra with involution $*$. The zero element of C_0 is $\mathbf{0} = (0, 0)$ and the identity element is $\mathbf{1} = (1, 0)$. Consider the element $\varepsilon = (0, 1)$. Then, each double number (a, b) can be written in the form $a + b\varepsilon = a(1, 0) + b(0, 1)$ and $\varepsilon^2 = (0, 1)(0, 1) = (0, 0) = \mathbf{0}$, i.e., the zero element of C_0. Hence, C_0 can be considered as a two-dimensional algebra over the real numbers with the basis $\{\mathbf{1}, \varepsilon\}$ and the Cayley table for multiplication of basis elements has the form:

$\cdot$	$\mathbf{1}$	ε
$\mathbf{1}$	$\mathbf{1}$	ε
ε	ε	$\mathbf{0}$

Table 10.4

Thus, we can define C_0 as follows:

$$C_0 = \{a + b\varepsilon \; : \; \varepsilon^2 = \mathbf{0}, \; a, b \in \mathbb{R}\} \tag{10.36}$$

Then, the conjugation in C_0 for an element $z = a + b\varepsilon$ can be defined as $z^* = (a + b\varepsilon)^* = a - b\varepsilon$ and

$$N(z) = zz^* = (a + b\varepsilon)(a - b\varepsilon) = a^2$$

is the algebraic norm of z, since it satisfies the condition of multiplicity $N(z_1 z_2) = N(z_1)N(z_2)$. However, since the quadratic form $N(z)$ is not positive definite, it does not satisfy the condition $N(z) = 0 \Leftrightarrow z = 0$. Since the quadratic form $N(a + b\varepsilon) = a^2$ has the parabolic type, the dual numbers are also called **parabolic complex numbers**. The algebra C_0 has an infinite number of elements with $N(z) = 0$, all of them have the form $z = \pm b\varepsilon$. All these numbers are divisors of zero in C_0. Hence, the algebra of dual numbers C_0 is not a division algebra. Nevertheless, any dual number with $N(z) \neq 0$ has a unique inverse

$$z^{-1} = \frac{z^*}{N(z)}$$

such that $zz^{-1} = z^{-1}z = \mathbf{1}$.

Let $z = a + b\varepsilon$, where $\varepsilon^2 = 0$, and $N(z) = 1$. Since $\varepsilon^n = 0$ for all $n > 1$, from Taylor's series of the exponent function we get:

$$z = e^{\varepsilon t} = 1 + \varepsilon t \tag{10.37}$$

10.7. Clifford Algebras. Grassmann Algebras

In this section, we consider some important examples of finite dimensional associative algebras, namely, Clifford algebras and Grassmann algebras. They can be considered as some generalizations of Cayley-Dickson algebras.

✳ **Definition 10.50.**

Let V be an n-dimensional vector space over a field K with $\mathrm{char}K \neq 2$. Suppose that a quadratic form $Q(v)$ is given and corresponding bilinear symmetric form is defined as follows

$$< u, v > = \frac{1}{2}[Q(u + v) - Q(u) - Q(v)] \tag{10.38}$$

for all $u, v \in K$.

An associative K-algebra $A = Cl(V, Q)$ which freely generated by elements $v \in V$ modulo the relation

$$uv + vu = -2 < u, v > \cdot \mathbf{1}, \tag{10.39}$$

where $\mathbf{1}$ is the identity of $Cl(V, Q)$, is called the **Clifford algebra**.

The equality (10.39) is called the **fundamental identity** of $Cl(V, Q)$.

If $u = v$, then $v^2 = -Q(v) \cdot \mathbf{1}$ for all $v \in V$.

If $\dim V = n$ and $\{e_1, e_2, \ldots, e_n\}$ is a basis of V, then the set of elements

$$\{e_{i_1} e_{i_2} \ldots e_{i_k} : 1 \leq i_1 < i_2 < \cdots < i_k \leq n \text{ for } 0 \leq k \leq n\}$$

forms the basis of the Clifford algebra $Cl(V, Q)$. If $k = 0$, then the empty element is the identity of this algebra and denoted by $\mathbf{1}$.

$$\dim Cl(V, Q) = \sum_{k=0}^{n} \binom{n}{k} = 2^n.$$

An arbitrary element of algebra $Cl(V, Q)$ has the form:

$$a = \sum_{k=0}^{n} \sum_{i_1 < i_2 < \cdots < i_k} \gamma_{i_1 i_2 \ldots i_k} e_{i_1} e_{i_2} \ldots e_{i_k},$$

where $\gamma_{i_1 i_2 \ldots i_k} \in K$.

Elements of the Clifford algebra $Cl(V, Q)$ are called the **Clifford numbers**.

If a basis $\{e_1, e_2, \ldots, e_n\}$ of a vector space V is orthogonal, i.e., $\langle e_i, e_j \rangle = 0$ for all $i \neq j$, then the equality (10.39) has the form:

$$e_i e_j = -e_j e_i, \quad \text{for } i \neq j.$$

Let $K = \mathbb{R}$ and $V = \mathbb{R}^n$. In this case, the quadratic form $Q(v)$ can be reduced to the form:

$$Q(v) = v_1^2 + \cdots + v_p^2 - v_{p+1}^2 - \cdots - v_{p+q}^2. \tag{10.40}$$

In this case, the Clifford algebra defined by (10.40) is called the **real Clifford algebra** and is denoted by $Cl(p, q, n, \mathbb{R})$. If the form $Q(v)$ is non-singular, then $n = p + q$.

If the basis of V is orthogonal, then we have the following equalities:

$$e_i e_j = -e_j e_i, \quad \text{for } i \neq j;$$
$$e_i^2 = 1, \quad \text{for } i = 1, \ldots, p;$$
$$e_j^2 = -1, \quad \text{for } j = p+1, \ldots, p+q;$$
$$e_k^2 = 0, \quad \text{for } k = p+q+1, \ldots, n$$

✳ Examples 10.51.

I. Suppose $Q(v)$ is a non-singular form and $p = 0$, $q = n$. Then, we have a Clifford algebra: $Cl(0, n, n, \mathbb{R})$, which is an associative algebra over the real numbers. If $\{e_1, e_2, \ldots, e_n\}$ is an orthogonal basis of V, then

$$e_i e_j = -e_j e_i, \quad \text{for all } i \neq j;$$
$$e_i^2 = -1, \quad \text{for } i = 1, \ldots, n$$

1. If $n = 0$, then $Cl(0, n, n, \mathbb{R}) \cong \mathbb{R}$ is the algebra of real numbers.

2. If $n = 1$ and $\{e_1\}$ is the basis of V, then $e_1^2 = -1$ and $\{1, e_1\}$ is the basis of the Clifford algebra $Cl(0, 1, 1, \mathbb{R})$. Therefore,

$$Cl(0, 1, 1, \mathbb{R}) = \{a + e_1 b \ : \ a, b \in \mathbb{R}, \ e_1^2 = -1\} \cong \mathbb{C}$$

is the algebra of complex numbers.

3. If $n = 2$ and $\{e_1, e_2\}$ is the basis of V, then $e_1^2 = -1$, $e_2^2 = -1$, $e_1 e_2 = -e_2 e_1$ and $\{1, e_1, e_2, e_{12} = e_1 e_2\}$ is the basis of the Clifford algebra $Cl(0, 2, 2, \mathbb{R})$. Since $e_{12}^2 = (e_1 e_2)^2 = (e_1 e_2)(e_1 e_2) = -(e_1 e_2)(e_2 e_1) = -1$, the Clifford algebra is defined as follows:

$$Cl(0, 2, 2, \mathbb{R}) = \{a + e_1 b + c e_2 + d e_{12} \ : \ a, b, c, d \in \mathbb{R}, \ e_1^2 = e_2^2 = e_{12}^2 = -1,$$

$$e_1 e_2 = -e_2 e_1\} \cong \mathbb{H}$$

is the algebra of quaternions, where $e_1 = i$, $e_2 = j$ and $e_{12} = k$.

4. If $n = 3$ and $\{e_1, e_2, e_3\}$ is the basis of V, then the following equalities hold:
$$e_i^2 = -1, \ e_i e_j = -e_j e_i, \ i, j = 1, 2, 3$$
and $\{1, e_1, e_2, e_3, e_{12} = e_1 e_2, e_{13} = e_1 e_3, e_{23} = e_2 e_3, e_{123} = e_1 e_2 e_3\}$ is the basis of the Clifford algebra $Cl(0, 3, 3, \mathbb{R}) \cong \mathbb{H} \oplus \mathbb{H}$, which is called the **Pauli algebra** or the algebra of **complex quaternions**.

5. $Cl(0, 4, 4, \mathbb{R}) \cong M_2(\mathbb{H})$

6. $Cl(0, 5, 5, \mathbb{R}) \cong M_4(\mathbb{C})$

7. $Cl(0, 6, 6, \mathbb{R}) \cong M_8(\mathbb{R})$

8. $Cl(0, 7, 7, \mathbb{R}) \cong M_8(\mathbb{R}) \oplus M_8(\mathbb{R})$

9. $Cl(0, 8, 8, \mathbb{R}) \cong M_{16}(\mathbb{R})$

II. Suppose $Q(v)$ is a non-singular form and $p \neq 0$, $q = n - p$.

1. Let $n = 1$, $p = 1$, $q = 0$ and $\{e_1\}$ be a basis of the vector space V. Then, $\{1, e_1\}$ is the basis of the Clifford algebra $Cl(1, 0, 1, \mathbb{R})$ with $e_1^2 = 1$. Therefore,

$$Cl(1, 0, 1, \mathbb{R}) = \{a + e_1 b \ : \ a, b \in \mathbb{R}, \ e_1^2 = 1\} \cong C_+$$

is the algebra of hyperbolic numbers (or double numbers).

2. Let $n = 2$, $p = 2$, $q = 0$ and $\{e_1, e_2\}$ be a basis of the vector space V. Then, $e_1^2 = e_2^2 = 1$ and $e_1 e_2 = -e_2 e_1$ and $\{1, e_1, e_2, e_{12} = e_1 e_2\}$ is the basis of the Clifford algebra $Cl(1, 0, 1, \mathbb{R})$ with the following table of multiplication:

·	**1**	e_1	e_2	e_{12}
1	**1**	e_1	e_2	e_{12}
e_1	e_1	**1**	e_{12}	e_2
e_2	e_2	$-e_{12}$	**1**	$-e_1$
e_{12}	e_{12}	$-e_2$	e_1	**-1**

Table 10.5

3. Let $n = 2$, $p = 1$, $q = 1$ and $\{e_1, e_2\}$ be a basis of the vector space V. Then, $e_1^2 = 1$, $e_2^2 = -1$, $e_1 e_2 = -e_2 e_1$ and $\{1, e_1, e_2, e_{12} = e_1 e_2\}$ is the basis of the Clifford algebra $Cl(1, 0, 1, \mathbb{R})$ with the following table of multiplication:

·	**1**	e_1	e_2	e_{12}
1	**1**	e_1	e_2	e_{12}
e_1	e_1	**1**	e_{12}	e_2
e_2	e_2	$-e_{12}$	**-1**	e_1
e_{12}	e_{12}	$-e_2$	$-e_1$	**1**

Table 10.6

Let $K = \mathbb{C}$ and $V = \mathbb{C}^n$. In this case, the quadratic form $Q(v)$ can be reduced to the form:

$$Q(v) = v_1^2 + \cdots + v_p^2 \qquad (10.41)$$

and the Clifford algebra defined by (10.41) is called the **complex Clifford algebra** and denoted by $Cl(p, n, \mathbb{C})$. If the form $Q(v)$ is non-singular, then $n = p$.

✻ Examples 10.52.

1. $Cl(0, 0, \mathbb{C}) \cong \mathbb{C}$

2. $Cl(1, 1, \mathbb{C}) \cong \mathbb{C} \oplus \mathbb{C}$

3. $Cl(2, 2, \mathbb{C}) \cong M_2(\mathbb{C})$

Consider now the particular example of Clifford algebras when the quadratic form is singular, which are called Grassmann algebras.

✻ Definition 10.53.

Let V be a vector space over a field K with basis $\{e_1, e_2, \ldots, e_n\}$. The **Grassmann algebra** $\Gamma_n = \Gamma(e_1, e_2, \ldots, e_n)$ is an associative algebra with identity 1 generated by elements $\{e_1, e_2, \ldots, e_n\}$ with relations

$$e_i e_j = -e_j e_i, \quad \text{for all } i \neq j;$$

$$e_i^2 = 0, \quad \text{for } i = 1, \ldots, n$$

✳ Examples 10.54.

Consider real Grassmann algebras, i.e., $K = \mathbb{R}$.

1. $\Gamma_0 = \mathbb{R}$.

2. $\Gamma_1 = Cl(0,0,1,\mathbb{R})$.

   $\Gamma_1 = \Gamma(e_1)$ with basis $\{1, e_1\}$ and the relation $e_1^2 = 0$. Therefore,

   $$Cl(0,0,1,\mathbb{R}) = \{a + e_1 b \ : \ a, b \in \mathbb{R}, \ e_1^2 = 0\} \cong C_-$$

   is the algebra of parabolic numbers (or dual numbers).

3. $\Gamma_2 = Cl(0,0,2,\mathbb{R})$.

   $\Gamma_2 = \Gamma(e_1, e_2)$ with basis $\{1, e_1, e_2, e_{12} = e_1 e_2\}$ and relations $e_1^2 = e_2^2 = 0$, $e_1 e_2 = -e_2 e_1$ and the table of multiplications:

$\cdot$	1	e_1	e_2	e_{12}
1	1	e_1	e_2	e_{12}
e_1	e_1	**0**	e_{12}	**0**
e_2	e_2	$-e_{12}$	**0**	**0**
e_{12}	e_{12}	**0**	**0**	**0**

   Table 10.7

10.8. Exercises

1. Find $g + h$, gh, hg, g^*, h^*, $N(g)$, $N(h)$ and g^{-1}, h^{-1} where g, h are quaternions of the form:

   $$g = 1 - 2i - 3j + 4k, \ h = 2 + 5i - j - k$$

2. Show that the equation $x^2 = -1$ has an infinite number of solutions in the algebra of quaternions.

3. In the set of quaternions $\mathbb{H}$, consider a subset $G = \{\pm 1, \pm i, \pm j, \pm k\}$. Prove that G is a group with respect to multiplication.

4. Show that the set of quaternions of the form $G = \{q \in \mathbb{H} \ : \ N(q) = 1\}$ forms a multiplicative subgroup in $\mathbb{H}$.

5. Solve equations: $qx = h$ $yq = h$ in the algebra of quaternions, if

   $$q = -1 + 4i - j + 4k, \ \ h = 4 - 2j - k.$$

6. Find $g + h$, gh, hg, g^*, h^*, $N(g)$, $N(h)$ and g^{-1}, h^{-1}, where g, h are octonions of the form:

   $$g = 2e_1 + 3e_2 - 3e_5 + 5e_6 - 2e_7, \ \ h = e_1 + 3e_2 + 2e_3 + e_7.$$

7. Solve equations: $qx = h$ and $yq = h$ in the algebra of octonions, if

$$g = 2e_1 + 3e_2 + 5e_4 - 2e_7, \quad h = e_1 - 2e_2 + 2e_5 + e_6.$$

8. Prove that, in the algebra $GL(n, \mathbb{R})$, the matrices of the form $\begin{pmatrix} a & -b \\ b & a \end{pmatrix}$ form a subalgebra which is isomorphic to the field of complex numbers.

9. Prove that, in the algebra $M_2(\mathbb{C})$, the matrices of the form $\begin{pmatrix} a + bi & c + di \\ -c + di & a - bi \end{pmatrix}$, where $a, b, c, d \in \mathbb{R}$ and $i^2 = -1$, form a subalgebra which is isomorphic to the algebra of quaternions over the field of real numbers.

10. Prove that a zero-divisor in an associative algebra can not be an invertible element.

11. Prove that, if $f : A \to B$ is an epimorphism of K-algebras, where K is a field, and A is an associative algebra, then B is also an associative algebra.

References

[1] Baez, J. 2002. The octonions. Bull. Amer. Math. Soc. 39: 145–205.

[2] Baez, J. and J. Huerta. 2 May 2016. Division algebras and supersymmetry. arXiv/0909.0551.

[3] Clifford, W.K. 1882. On the classification of geometric algebras. pp. 397–401. *In*: Tucker R. (ed.). Mathematical Papers, Macmillian, London.

[4] Conway, J.H. and D.A. Smith. 2003. On Quaternions and Octonions: Their Geometry, Arithmetic, and Symmetry, A K Peters, Massachusetts.

[5] Davenport, C.M. 1996. A commutative hypercomplex algebra with associated function theory. pp. 213–227. *In*: Abamowicz, R., P. Lounesto and J.M. Parra [eds.]. Clifford Algebras with Numeric and Symbolic Computations, Boston, MA: Birkhuser.

[6] Delanghe, R. 2001. Clifford analysis: History and perspective. Computational Methods and Function Theory 1(1): 107–153.

[7] Dickson, L.E. 1919. On quaternions and their generalization and the history of the eight square theorem. Ann. Math. 20: 155–171.

[8] Franchini, S., G. Vassallo and F. Sorbello. 2010. A brief introduction to Clifford algebra. Università degli Studi di Palermo, Technical Report N. 2/ 2010, Palermo.

[9] Franssens, G. 2009. Introduction to Clifford analysis. pp. 1–19. *In*: Gürlebeck, K. and C. Könke [eds.]. 18th International Conference on the Application of Computer Science and Mathematics in Architecture and Civil Engineering, Weimer, Germany.

[10] Hamilton, W.R. 1844. On quaternions; or an a new system of imagniaries in algebra. London, Edinburg, and Dublin Philosophical Magazine and Journal of Science 25(3): 489–495.

[11] Hestenes, D. and G. Sobczyk. 1992. Clifford Algebra to Geometric Calculus, Kluwer, Dordrecht.

[12] Hurwitz, A. 1919. Vorlesungen ber die Zahlentheorie der Quaternionen, Berlin: J. Springer.

[13] Joyner, D., R. Kreminski and J. Turisco. 2004. Applied Abstract Algebra, The Johns Hopkins University Press.

[14] Judson, Th.W. 1994. Abstract Algebra. Theory and Applications, PWS Publishing Company, Michigan.

[15] Kantor, I.L. and A.S. Solodovnikov. 1989. Hypercomplex Numbers: An Elementary Introduction to Algebras. New York: Springer-Verlag.

[16] Kurosh, A.G. 1963. Lectures on General Algebra, Chelsea Pub. Co, New York.

[17] Oneto, A. 2002. Alternative real division algebras of finite dimension. Divulgaciones Matemáticas 10(2): 161–169.

[18] Rochon, D. and Shapiro, M. 2004. On algebraic properties of bicomplex and hyperbolic numbers. An. Univ. Oradea Fasc. Mat. 11: 71–110.

[19] Schafer, R.D. 1952. Representations of alternative algebras. Trans. Amer. Math. Soc. 72: 1–17.

[20] Schafer, R.D. 1954. On the algebras formed by the Cayley-Dickson process. Amer. J. Math. 76: 435–446.

[21] Schafer, R.D. 1966. An introduction to nonassociative algebra, Pure and Applied Mathematics, Academic Press, New York, London, Vol. 22.

[22] Zhevlakov, K.A., A.M. Slin'ko, I.P. Shestakov and A.I. Shirshov. 1982. Rings that are Nearly Associative, Academic Press.

Chapter 11

Applications of Quaternions and Octonions

"If you have a garden and a library, you have everything you need."
Marcus Tullius Cicero

"To solve math problems, you need to know the basic mathematics before you can start applying it."
Catherine Asaro

Quaternions, octonions and other hypercomplex numbers have a lot of different applications, not only in mathematics, but also in other fields of physics and technics. For example, the algebra of quaternions is used to solve important practical problems of inertial navigation and kinematic motion of solids. The outstanding Scottish physician and mathematician J.C. Maxwell used quaternions to develop and write the equations of electrodynamics in the monograph "A Treatise on Electricity & Magnetism", published in 1873. At the present time, Grassmann and Clifford algebras are more commonly used for this. The algebra of quaternions has a wide range of applications, from computer graphic video games to air traffic control systems. Octonions now have an increasing importance in physics, in particular, in the theory of elementary particles, supersymmetry and the theory of strings.

This chapter presents some applications of quaternions and octonions in number theory and computer graphics. In particular, in the first section, we apply quaternions and octonions in order to prove the identity on representing a product of two n-squares sums of integers as a sum of n squares of some integers. The next sections show the connection of Fermat's theorem on two-square sum with Gaussian integers and the connection of Lagrange's theorem, which states that each natural number can be written as a sum of four squares, with some kinds of quaternions. At the end of this chapter, we show the use of quaternions to write rotations in the three-dimensional real space.

11.1. Square Sum Identities

The problem of representing a product of two n squares sums of integers as a sum of n squares of some integers originated in the early days of the history of mathematics. For $n = 1$, this problem is trivial and follows from commutativity and associativity of integers:

$$a^2 b^2 = (ab)^2.$$

For the case $n = 2$, the formula that expresses the product of two sums of two squares of integers as a sum of two squares of some integers was known already in the VII century and was discovered by the Indian mathematician Brahmagupta (598-668). This formula was also discovered by Leonard Fibonacci and appeared in 1225 in his book "Liber quadratorum". Therefore, this identity is now known as the **Brahmagupta-Fibonacci identity**.

✳ **Theorem 11.1 (Brahmagupta-Fibonacci two-square identity).**
The product of two sums of two squares of integers can be also written as a sum of two squares of integers, i.e.,

$$(a_1^2 + b_1^2)(a_2^2 + b_2^2) = (a_1 a_2 - b_1 b_2)^2 + (a_1 b_2 + a_2 b_1)^2. \qquad (11.1)$$

Proof.
The equality (11.1) follows from the simple transformations:

$$(a_1 a_2 - b_1 b_2)^2 + (a_1 b_2 + a_2 b_1)^2 = a_1^2 a_2^2 - 2a_1 a_2 b_1 b_2 + b_1^2 b_2^2 + a_1^2 b_2^2 + 2a_1 b_2 a_2 b_1 + a_2^2 b_1^2 =$$

$$= (a_1^2 + b_1^2)(a_2^2 + b_2^2).$$

This equality can also be easily proved using complex numbers. Since the norm of complex numbers is a multiplicative function,

$$N(a_1 + b_1 i) N(a_2 + b_2 i) = N[(a_1 + b_1 i)(a_2 + b_2 i)] = N[(a_1 a_2 - b_1 b_2) + (a_1 b_2 + a_2 b_1)i],$$

hence, we obtain:

$$(a_1^2 + b_1^2)(a_2^2 + b_2^2) = (a_1 a_2 - b_1 b_2)^2 + (a_1 b_2 + a_2 b_1)^2.$$

$\square$

In the case of four squares, the analogous equality was obtained by Leonhard Euler in 1748.

✳ **Theorem 11.2 (Euler's four-square identity).**
The product of two sums of four squares of integers can be also written as a sum of four squares of integers.

Proof.

Euler's original proof is rather difficult. Now, to prove this theorem, we use quaternions.

Let $q_1 = a_1 + b_1 i + c_1 j + d_1 k$, $q_2 = a_2 + b_2 i + c_2 j + d_2 k \in \mathbb{H}$ and $a_p, b_p, c_p, c_p \in \mathbb{Z}$ for $p = 1, 2$. Then, $q = q_1 q_2 = a + bi + cj + dj$, where

$$a = a_1 b_1 - a_2 b_2 - a_3 b_3 - a_4 b_4,$$

$$b = a_1 b_2 + a_2 b_1 + a_3 b_4 - a_4 b_3,$$

$$c = a_1 b_2 + a_2 b_1 + a_3 b_4 - a_4 b_3,$$

$$d = a_1 b_4 + a_2 b_3 - a_3 b_2 + a_4 b_1.$$

It is obvious that $a, b, c, d \in \mathbb{Z}$. By multiplicity of the norm of quaternions, we have:

$$N(q) = N(q_1 q_2) = N(q_1) N(q_2), \tag{11.2}$$

where $N(q_1) = a_1^2 + b_1^2 + c_1^2 + d_1^2$, $N(q_2) = a_2^2 + b_2^2 + c_2^2 + d_2^2$, $N(q) = a^2 + b^2 + c^2 + d^2$.

Hence, from (11.2), it follows that

$$(a_1^2 + b_1^2 + c_1^2 + d_1^2)(a_2^2 + b_2^2 + c_2^2 + d_2^2) = a^2 + b^2 + c^2 + d^2. \tag{11.3}$$

□

The identity of eight squares was first discovered by Carl Ferdinand Degen in 1818 and then it was independently rediscovered by John Thomas Graves in 1843 and Arthur Cayley in 1845.

✳ **Theorem 11.3 (Degen's eight-square identity).**
The product of two sums of eight squares of integers can also be written as a sum of eight squares of integers.

Proof.

We perform the proof of this theorem analogously to the previous theorem, this time using octonions. Consider octonions with integer coordinates:

$$x = a_1 e_1 + a_2 e_2 + \cdots + a_7 e_7 + a_8 e_8,$$

$$y = b_1 e_1 + b_2 e_2 + \cdots + b_7 e_7 + b_8 e_8.$$

Then, their product is also an octonion with integer coordinates:

$$xy = c_1 e_1 + c_2 e_2 + \cdots + c_7 e_7 + c_8 e_8.$$

Since the norm of octonions is a multiplicative function, we have:

$$N(xy) = N(x)N(y), \tag{11.4}$$

where

$$N(x) = a_1^2 + a_2^2 + \cdots + a_7^2 + a_8^2,$$

$$N(y) = b_1^2 + b_2^2 + \cdots + b_7^2 + b_8^2,$$
$$N(xy) = c_1^2 + c_2^2 + \cdots + c_7^2 + c_8^2.$$

Then, from equality (11.4), it follows that

$$(a_1^2 + a_2^2 + a_3^2 + a_4^2 + a_5^2 + a_6^2 + a_7^2 + a_8^2)(b_1^2 + b_2^2 + b_3^2 + b_4^2 + b_5^2 + b_6^2 + b_7^2 + b_8^2) =$$
$$= (c_1^2 + c_2^2 + c_3^2 + c_4^2 + c_5^2 + c_6^2 + c_7^2 + c_8^2). \tag{11.5}$$

$\square$

As it turned out later, the analogous identity for arbitrary number of summands is not possible. In 1898, Hurwitz proved the following theorem:

> ❊ **Theorem 11.4 (Hurwitz, 1898).**
> The identity of the form:
>
> $$(a_1^2 + a_2^2 + \cdots + a_n^2)(b_1^2 + b_2^2 + \cdots + b_n^2) = c_1^2 + c_2^2 + \cdots + c_n^2 \tag{11.6}$$
>
> with the c_i bilinear functions of the a_i, b_i is possible only for $n \in \{1, 2, 4, 8\}$.

11.2. Gaussian Integers

In this section, we consider a particular subset of complex numbers and study their most important properties.

> ❊ **Definition 11.5.**
> A complex number of the form $a + ib$, where $a, b \in \mathbb{Z}$, is called a **Gaussian integer**.

We write the set of Gaussian integers as:

$$\mathbb{Z}[i] = \{a + bi \ : \ a, b \in \mathbb{Z}\}. \tag{11.7}$$

It is obvious that if $a + bi, c + di \in \mathbb{Z}[i]$, then:
1) $(a + bi) + (c + di) = (a + c) + (b + d)i$
2) $(a + bi)(c + di) = (ac - bd) + (ac + bd)i$
3) $(a + bi)^* = \overline{a + bi} = a - bi$
4) $1 = 1 + 0 \cdot i$
5) $0 = 0 + 0 \cdot i$
are also Gaussian integers, i.e., $\mathbb{Z}[i]$ is a subring of the ring of complex numbers with respect to operations of addition and multiplication. Since in the set of complex numbers there are no zero divisors, $\mathbb{Z}[i]$ is an integral domain.

For each Gaussian number $a + bi$, we can define its norm:

$$N(a + bi) = (a + bi)(a + bi)^* = (a + bi)(a - bi) = a^2 + b^2 \in \mathbb{N} \tag{11.8}$$

satisfying the properties:
1) $N(x) = 0 \Leftrightarrow x = 0$
2) $N(1) = 1$
3) $N(xy) = N(x)N(y)$
for all $x, y \in \mathbb{Z}[i]$.

For a complex number $a + bi$, there is the notion of its module, defined as $|a + bi| = \sqrt{a^2 + b^2}$. It is obvious that $N(a + bi) = |a + bi|^2$.

The group of invertible elements of the ring $\mathbb{Z}[i]$ is denoted by $U(\mathbb{Z}[i])$. We show that

$$U(\mathbb{Z}[i]) = \{z \in \mathbb{Z}[i] \ : \ N(z) = 1\} = \{1, -1, i, -i\}.$$

Really, if $\alpha = a + bi \in U(\mathbb{Z}[i])$, then there is an element $\beta \in \mathbb{Z}[i]$ such that $\alpha\beta = 1$. Then, $N(\alpha\beta) = N(\alpha)N(\beta) = 1$. Since $N(\alpha), N(\beta) \in \mathbb{Z}^+$, then $N(\alpha) = 1$, i.e., $a^2 + b^2 = 1$. Hence, $(a = \pm 1) \wedge (b = 0)$ or $(a = 0) \wedge (b = \pm 1)$, so we have only 4 possibilities: $z = \pm 1; \pm i$.

In the set of complex numbers, we can always divide one complex number by another non-zero complex number. Unfortunately, in the set of Gaussian integers, this is not the case. For example, if $\alpha = 10 + 3i$, $\beta = 2 + 5i$, then

$$\frac{\alpha}{\beta} = \frac{10 + 3i}{2 + 5i} = \frac{(10 + 3i)(2 - 5i)}{(2 + 5i)(2 - 5i)} = \frac{35 - 44i}{29} = \frac{35}{29} - \frac{44}{29}i \notin \mathbb{Z}[i].$$

✳ Theorem 11.6.
If $\alpha, \beta \in \mathbb{Z}[i]$ and $\beta|\alpha$, then $N(\beta)|N(\alpha)$ in $\mathbb{Z}$.

Proof.
If $\beta|\alpha$, then there is a number $\gamma \in \mathbb{Z}[i]$ such that $\alpha = \beta\gamma$. Hence, $N(\alpha) = N(\beta\gamma) = N(\beta)N(\gamma)$, i.e., $N(\beta)|N(\alpha)$ in $\mathbb{Z}$.   □

✳ Corollary 11.7.
The norm of a Gaussian integer α is even if and only if α is divided by $1 + i$.

Proof.
If $(1 + i)|\alpha$, then $\alpha = \beta(1 + i)$, hence, $N(\alpha) = N(\beta)N(1 + i) = 2N(\beta)$.
Inversely, let $\alpha = m + ni$ and $N(\alpha) = m^2 + n^2 \equiv 0 \,(\text{mod } 2)$. Hence, $m \equiv n \,(\text{mod } 2)$. Then, $\alpha = m + ni = (1 + i)(u + vi)$, where $u = (m + n)/2$ and $v = (n - m)/2$. Moreover, $u, v \in \mathbb{Z}$, since $m \equiv n \,(\text{mod } 2)$.   □

For Gaussian integers, we have the theorem similar to the theorem of division with remainder for integers.

✳ Theorem 11.8.
For all $\alpha, \beta \in \mathbb{Z}[i]$ and $\beta \neq 0$, there are $\gamma, \rho \in \mathbb{Z}[i]$ such that

$$\alpha = \gamma\beta + \rho, \quad \text{where } 0 \leq N(\rho) < N(\beta). \qquad (11.9)$$

Proof.

Let $\alpha, \beta \in \mathbb{Z}[i]$ and $\beta \neq 0$. We write:

$$\frac{\alpha}{\beta} = \frac{\alpha\bar{\beta}}{\beta\bar{\beta}} = \frac{\alpha\bar{\beta}}{N(\beta)} = \frac{m + ni}{N(\beta)},$$

where $\alpha\bar{\beta} = m + ni$. Since $m, n, N(\beta) \in \mathbb{Z}$, there exist integers q_1, q_2, r_1, r_2 such that

$$m = q_1 N(\beta) + r_1, \quad n = q_2 N(\beta) + r_2, \text{ and } 0 \leq |r_1|, |r_2| \leq \frac{1}{2} N(\beta). \quad (11.10)$$

So,

$$\frac{\alpha}{\beta} = \frac{(q_1 N(\beta) + r_1) + (q_2 N(\beta) + r_2)i}{N(\beta)} = q_1 + q_2 i + \frac{r_1 + r_2 i}{N(\beta)} =$$

$$= q_1 + q_2 i + \frac{r_1 + r_2 i}{\beta\bar{\beta}} \quad (11.11)$$

Let $\gamma = q_1 + q_2 i$. Then, from the equality (11.11), we obtain:

$$\alpha = \beta\gamma + \frac{r_1 + r_2 i}{\bar{\beta}} \quad (11.12)$$

We write $\rho = \alpha - \beta\gamma \in \mathbb{Z}[i]$. Then, taking (11.10) and (11.12) into account, we obtain:

$$N(\rho) = N(\frac{r_1 + r_2 i}{\bar{\beta}}) = \frac{r_1^2 + r_2^2}{N(\beta)} \leq \frac{(1/4)N^2(\beta) + (1/4)N^2(\beta)}{N(\beta)} = \frac{1}{2} N(\beta).$$

Thus, $\alpha = \beta\gamma + \rho$, where $\gamma, \rho \in \mathbb{Z}[i]$ and $0 \leq N(\rho) < N(\beta)$. $\qquad \square$

∗ Examples 11.9.

1. Divide with remainder the number $\alpha = 11 + 10i$ by the number $\beta = 4 + i$. We have $N(\beta) = 17$ and compute:

$$\frac{\alpha}{\beta} = \frac{\alpha\bar{\beta}}{N(\beta)} = \frac{(11 + 10i)(4 - i)}{17} = \frac{54 + 29i}{17} =$$

$$= \frac{(51 + 3) + (34 - 5)i}{17} = 3 + 2i + \frac{3 - 5i}{17}.$$

Note, that we write $54 = 3 \cdot 17 + 3$ and $29 = 2 \cdot 17 - 5$ in such forms that $|r_1| = |3| < (1/2) \cdot 17$ and $|r_2| = |-5| < (1/2) \cdot 17$. Then,

$$\alpha = \beta(3 + 2i) + \rho, \text{ where } \rho = \frac{(3 - 5i)}{17}\beta = \frac{(3 - 5i)(4 + i)}{17} = \frac{17 - 17i}{17} = 1 - i.$$

Therefore, $11 + 10i = (4 + i)(3 + 3i) + (1 - i)$ and $N(1 - i) = 2 < 17 = N(4 + i)$.

2. Divide with remainder the number $\alpha = 37 + 2i$ by the number $\beta = 11 + 2i$. Now, we have $N(\beta) = 125$ and compute:

$$\frac{\alpha}{\beta} = \frac{\alpha\bar{\beta}}{N(\beta)} = \frac{(37 + 2i)(11 - 2i)}{125} = \frac{411 - 52i}{125} =$$

$$= \frac{(3 \cdot 125 + 36) + (-52)i}{125} = 3 + \frac{36 - 52i}{125}.$$

We note that $|r_1| = |36| < (1/2) \cdot 125$ and $|r_2| = |-52| < (1/2) \cdot 125$. Then, $\alpha = 3\beta + \rho$, where

$$\rho = \frac{(36 - 52i)}{125}\beta = \frac{(36 - 52i)(11 + 2i)}{125} = \frac{500 - 500i}{125} = 4 - 4i.$$

Therefore, $37 + 2i = 3(11 + 2i) + (4 - 4i)$ and $N(4 - 4i) = 32 < 125 = N(11 + 2i)$. Note, that we also have: $37 + 2i = (11 + 2i)(3 - i) + (2 - 7i)$ and $N(2 - 7i) = 53 < 125 = N(11 + 2i)$, which means that the choice of elements q, r from the equality (11.9) is not unique.

Since $\mathbb{Z}[i]$ is an integral domain, from Theorem 11.8, we get the following corollary:

❋ Corollary 11.10.
$\mathbb{Z}[i]$ is an Euclidian ring.

Now, from this corollary and Theorem 4.69, we immediately get the following corollary:

❋ Corollary 11.11.
The ring $\mathbb{Z}[i]$ is a principal ideal domain.

Similar to an arbitrary ring, in the ring of Gaussian integers, we can define prime elements and irreducible elements.

✳ Definition 11.12.
  A non-zero Gaussian integer z is called **prime** if it satisfies the conditions:
  1) $z \in \mathbb{Z}[i] \backslash U(\mathbb{Z}[i])$, i.e., an element z is not invertible.
  2) $z|(\alpha\beta)$ implies that $z|\alpha$ or $z|\beta$ for all $\alpha, \beta \in \mathbb{Z}[i]$.
  A non-zero Gaussian integer $z \in \mathbb{Z}[i]$ is called **irreducible** in $\mathbb{Z}[i]$ if $z = \alpha\beta$ implies that $\alpha \in U(\mathbb{Z}[i])$ or $\beta \in U(\mathbb{Z}[i])$. Otherwise, z is called **decomposable**.

From Corollary 11.11 and Theorem 4.82, we get the following corollary:

❋ Corollary 11.13.
The ring $\mathbb{Z}[i]$ is a unique factorization domain.

It is obvious that, if a Gaussian integer $z \in \mathbb{Z}[i]$ is decomposable and $z = \alpha\beta$, then $N(z) = N(\alpha)N(\beta)$. Hence, we get the sufficient condition for a number $z \in \mathbb{Z}[i]$ to be irreducible.

❋ Theorem 11.14.
If $z \in \mathbb{Z}[i]$ and $N(z)$ is irreducible in $\mathbb{Z}$, then z is irreducible in $\mathbb{Z}[i]$.

From Corollary 11.13 and Theorem 4.81, we get the following corollary.

> ✳ **Corollary 11.15.**
> Each irreducible Gaussian integer is a prime Gaussian integer, i.e., each number which is not a prime Gaussian integer is decomposable in $\mathbb{Z}[i]$.

From this corollary and Theorem 11.14, we obtain the following theorem:

> ✳ **Theorem 11.16.**
> If $z \in \mathbb{Z}[i]$ and $N(z)$ is irreducible in $\mathbb{Z}$, then z is prime in $\mathbb{Z}[i]$.

✳ **Examples 11.17.**

1. The number $z = 4 + i$ is irreducible in $\mathbb{Z}[i]$ because $N(z) = 17$ is irreducible in $\mathbb{Z}$. From Corollary 11.11, it also follows that $z = 4 + i$ is prime in $\mathbb{Z}[i]$.

2. The number $z = 2$ is irreducible in $\mathbb{Z}$ and decomposable in $\mathbb{Z}[i]$, since $2 = (1+i)(1-i)$ and $N(1+i) = N(1-i) = 2$, $N(2) = 4$, i.e., elements $(1+i), (1-i) \notin U(\mathbb{Z}[i])$.

11.3. Fermat's Theorem on Sums of Two Squares

In this section, we prove the important Fermat Theorem that answers the question: "When is a prime a sum of two squares?". As a corollary from this theorem, we will obtain a criterion that shows when a Gaussian integer is prime, which is equivalent to the fact that this number is irreducible in $\mathbb{Z}[i]$.

> ✳ **Lemma 11.18 (Lagrange's Lemma).**
> For each prime $p \in \mathbb{Z}$ of the form $p = 4n + 1$, there exists an integer $m \in \mathbb{Z}$ such that $p|(m^2 + 1)$.

Proof.

Let $p = 4n + 1$. Consider the set of all representatives of the group $\mathbb{Z}_p^*$ of the form

$$S = \{-2n, -2n + 1, \ldots, -1, 1, 2, \ldots, 2n - 1, 2n\}.$$

Then, $|\mathbb{Z}_p^*| = |S| = p - 1 = 4n$. Since $p - x \equiv -x \,(\mathrm{mod}\, p)$ for all $x \in \mathbb{Z}$, from Wilson's Theorem 7.15, it follows that:

$$-1 \equiv (4n)! \equiv (2n)!(2n + 1)(2n + 2)\ldots(4n - 1)(4n) \equiv$$

$$\equiv (2n)!(-2n)(-2n + 1)\ldots(-2)(-1) \equiv (2n)!(-1)^{2n}(2n)! \equiv [(2n)!]^2 \,(\mathrm{mod}\, p).$$

Therefore,

$$m^2 + 1 \equiv 0 \,(\mathrm{mod}\, p),$$

where $m = (2n)!$ ∎

✳ **Theorem 11.19 (Fermat's Theorem).**
A prime $p \in \mathbb{Z}$ is a sum of two squares of integers if and only if

$$p = 2 \quad \text{or} \quad p \equiv 1 \,(\text{mod}\,4). \qquad (11.13)$$

Proof.
$\Rightarrow$. If $p = 2$ then $p = 1^2 + 1^2$.

Let $p > 2$ and $p \equiv 1 \,(\text{mod}\,4)$. By Lagrange's Lemma 11.18 for a prime $p = 4n+1$, there is an integer m such that $p|(m^2+1)$, i.e., there is an integer q such that $m^2 + 1 = pq$.

Since $m^2 + 1 = (m+i)(m-i) \in \mathbb{Z}[i]$, we get that $(m+i)(m-i) = pq$. Nevertheless, p divides neither $m+i$ nor $m-i$. Really:

1) If $p|(m+i)$, then $m+i = pz$ for some $z \in \mathbb{Z}[i]$. However $\dfrac{m}{p} + \dfrac{1}{p}i \notin \mathbb{Z}[i]$. Hence, p is not a prime element in $\mathbb{Z}[i]$, i.e., p is decomposable in $\mathbb{Z}[i]$. Therefore, $p = \alpha\beta$, where $\alpha, \beta \in \mathbb{Z}[i]$. So, $p^2 = pp^* = (\alpha\beta)(\alpha\beta)^* = N(\alpha)N(\beta)$ because $p = p^* = (\alpha\beta)^*$. Since $\alpha, \beta \notin U(\mathbb{Z}[i])$ and $p > 2$, we get that $p = N(\alpha) = N(\beta)$. If $\alpha = a + bi$, then $p = a^2 + b^2$.

2) If $p|(m-i)$, the proof is similar.

$\Leftarrow$. Let $p = a^2 + b^2$. If $|a| = |b| = 1$, then $p = 2$.

Suppose that $|a|, |b| > 1$. Without loss of generality, we can assume that $a, b > 1$. If $a \equiv b \equiv 0 \,(\text{mod}\,2)$, then $a^2 + b^2 \equiv 0 \,(\text{mod}\,4)$, i.e., $4|p$. If $a \equiv b \equiv 1 \,(\text{mod}\,2)$ then $a^2 + b^2 \equiv 2 \,(\text{mod}\,4)$, i.e., $2|p$. So, there remains only the case when $a \equiv 0 \,(\text{mod}\,2)$ and $b \equiv 1 \,(\text{mod}\,2)$. Then, $p = a^2 + b^2 \equiv 1 \,(\text{mod}\,4)$.
□

✳ **Corollary 11.20.**
If $p \in \mathbb{Z}$ is a prime and $p \equiv 1 \,(\text{mod}\,4)$, then p is decomposable in $\mathbb{Z}[i]$.

✳ **Theorem 11.21.**
A prime $p \in \mathbb{Z}$ is decomposable in $\mathbb{Z}[i]$ if and only if p is a sum of two squares.

Proof.
Let $p \in \mathbb{Z}$ be irreducible in $\mathbb{Z}$ and decomposable in $\mathbb{Z}[i]$. Then, $p = \alpha\beta$, where $\alpha, \beta \in \mathbb{Z}[i]$. Hence, $p^2 = pp^* = (\alpha\beta)(\alpha\beta)^* = N(\alpha)N(\beta)$. Therefore, $p = N(\alpha) = N(\beta)$. If $\alpha = a + bi$, then $p = a^2 + b^2$.

Inversely, if $p = a^2 + b^2$, then, by Theorem 11.19 and Corollary 11.20, p is decomposable in $\mathbb{Z}[i]$. □

✳ **Corollary 11.22.**
A natural number m is irreducible in $\mathbb{Z}[i]$ if and only if m is a prime and $m \equiv 3 \,(\text{mod}\,4)$.

Now, we can summarize previous results in the form of the following theorem, which gives the description of all irreducible Gaussian integers by means of prime integers.

✳ **Theorem 11.23.**

An irreducible Gaussian integer has one of the form:

- $p \in \mathbb{Z}$ is a prime and $p \equiv 3 \,(\mathrm{mod}\,4)$.

- $1 + i$

- a Gaussian integer $a + bi$ or $a - bi$ if $a^2 + b^2$ is a prime and $a^2 + b^2 \equiv 1 \,(\mathrm{mod}\,4)$.

or a Gaussian number that is associated to a Gaussian integer mentioned above.

11.4. Lagrange's Four-Square Theorem

In the previous section, we showed that not any prime integer can be written as a sum of two squares. However, it turns out that any positive integer can be written as a sum of four squares. This is an outstanding theorem proved by Lagrange in 1770 that is some analog of Fermat's Theorem 11.19. While to prove the Fermat Theorem we used Gaussian integers, to prove the Lagrange theorem, we will use two kinds of quaternions, namely Lipschitz quaternions and Hurwitz quaternions. For both kinds of quaternions, their norm is an integer.

✳ **Definition 11.24.**

    Lipschitz quaternions are quaternions of the form $a+bi+cj+dk \in \mathbb{H}$ where $a, b, c, d \in \mathbb{Z}$. The set of Lipschitz quaternions will be denoted by:

$$\mathbf{L} = \{a + bi + cj + dk \in \mathbb{H} \ : \ a, b, c, d \in \mathbb{Z}\}.$$

Hurwitz quaternions are quaternions of the form $a + bi + cj + dk \in \mathbb{H}$, where $(a, b, c, d \in \mathbb{Z}) \vee (a, b, c, d \in \mathbb{Z} + \frac{1}{2})$. The set of Hurwitz quaternions will be denoted by:

$$\mathbf{Hu} = \{a + bi + cj + dk \in \mathbb{H} \ : \ (a, b, c, d \in \mathbb{Z}) \vee (a, b, c, d \in \mathbb{Z} + \frac{1}{2})\}.$$

It is obvious that $N(q) \in \mathbb{N}$ for all $q \in \mathbf{L}$ and all $q \in \mathbf{Hu}$. It is easy to show that Hurwitz quaternions $\mathbf{Hu}$ form a subalgebra in the algebra of quaternions $\mathbb{H}$, while Lipschitz quaternions $\mathbf{L}$ form a subalgebra in the algebra of Hurwitz quaternions $\mathbf{Hu}$.

❋ **Definition 11.25.**
     A quaternion $q \in \mathbb{H}$ is called a **unit** if $N(q) = 1$.

There are 8 units in the set of Lipschitz quaternions:

$$\pm 1, \pm i, \pm j, \pm k$$

and there are 24 units in the set of Hurwitz quaternions:

$$\pm 1, \pm i, \pm j, \pm k, \frac{1}{2}(\pm 1 \pm i \pm j \pm k).$$

It is easy to prove the theorem:

❋ **Theorem 11.26.**

- Units in the set of Lipschitz quaternions form a group of invertible elements $U(\mathbf{L})$, with respect to multiplication.

- Units in the set of Hurwitz quaternions form a group of invertible elements $U(\mathbf{Hu})$, with respect to multiplication.

❋ **Theorem 11.27. (Division with remainder)**
For all $a, b \in \mathbf{Hu}$ and $b \neq 0$, there exist elements $q, r \in \mathbf{Hu}$ such that

$$a = qb + r, \text{ where } 0 \leq N(r) < N(b) \tag{11.14}$$

and there exist elements $x, t \in \mathbf{Hu}$ such that

$$a = bx + t, \text{ where } 0 \leq N(t) < N(b). \tag{11.15}$$

Proof.
     Since $\mathbb{H}$ is a division ring, for all $a, b \in \mathbf{Hu}$ and $b \neq 0$, there exists an element $ab^{-1} \in \mathbb{H}$. Exactly like the proof of Theorem 11.8, we can choose a Lipschitz quaternion $q \in \mathbf{L}$ such that $s = ab^{-1} - q = x + yi + zj + uk$ and $|x|, |y|, |z|, |u| \leq \frac{1}{2}$. Then, $N(s) = x^2 + y^2 + z^2 + u^2 \leq \frac{1}{4} + \frac{1}{4} + \frac{1}{4} + \frac{1}{4} = 1$.
     Consider separately two cases:
     1) $N(s) < 1$. Let $r = sb = a - qb$, i.e., $a = qb + r$. Then,

$$N(r) = N(s)N(b) < N(b).$$

     2) $N(s) = 1$. Let $s = \omega = \frac{1}{2}(\pm 1 + \pm i \pm j \pm k) \in U(\mathbb{Z}[i])$. Then, $ab^{-1} = q + \omega = h \in \mathbf{Hu}$. Hence, $a = hb + 0$ and $N(0) = 0 < N(b)$. □

❋ **Corollary 11.28.**
Each right (left) ideal I of the ring $A = \mathbf{Hu}$ is a principal ideal of the form bA (Ab) for some $b \in I$.

Proof.

Let I be a right ideal of the ring $A = \mathbf{Hu}$. We choose a non-zero element b in I for which $N(b)$ is minimal. Let $a \in I$. Since $b \neq 0$, from Theorem 11.27, it follows that there are elements $q, r \in \mathbf{Hu}$ such that $a = bq + r$, where $0 \leq N(r) < N(b)$. Since $a, b \in I$, $r = a - bq \in I$ and $0 \leq N(r) < N(b)$. From the choice of the element $b \in I$, it follows that $N(r) = 0$. This means that $r = 0$, i.e., $a = bq$. At the time I is a right ideal of the ring $A = \mathbf{Hu}$ generated by the element b, i.e.,

$$I = bA = \{bq \; : \; q \in A\}.$$

In the case of a left ideal, the proof is similar. $\qquad\square$

✳ Lemma 11.29.

For each $q \in \mathbf{Hu} \backslash \mathbf{L}$, there is an element $\delta \in U(\mathbf{Hu})$ of the form $\frac{1}{2}(\pm 1 + \pm i \pm j \pm k)$ such that $N(\delta) = 1$ and $q\delta \in \mathbf{L}$.

Proof.

Let $q = \frac{1}{2}(a + bi + cj + dk) \in \mathbf{Hu} \backslash \mathbf{L}$. Since a, b, c, d are odd integers, we can choose $\omega_i = \pm 1$ for $i = 1, 2, 3, 4$ such that:

$$a = 4x + \omega_1, \; b = 4y + \omega_2, \; c = 4z + \omega_3, \; d = 4v + \omega_4.$$

Then,

$$q = 2(x + yi + zj + vk) + \frac{1}{2}(\omega_1 + \omega_2 i + \omega_3 j + \omega_4 k) = 2h + \omega,$$

where $\omega = \frac{1}{2}(\omega_1 + \omega_2 i + \omega_3 j + \omega_4 k)$ and $N(\omega) = 1$. Then,

$$q\omega^* = 2h\omega^* + \omega\omega^* \in \mathbf{L}.$$

Therefore, we have that $q\delta \in \mathbf{L}$ and $N(\delta) = N(\omega^*) = N(\omega) = 1$, where $\delta = \omega^* \in U(\mathbf{Hu})$. $\qquad\square$

✳ Lemma 11.30 (Lagrange).

If $p > 2$ is a prime, then there are integers t, m such that

$$t^2 + m^2 + 1 \equiv 0 \,(\mathrm{mod}\, p) \tag{11.16}$$

Proof.

Note that $a^2 \equiv b^2 \,(\mathrm{mod}\, p)$ if and only if $a \equiv \pm b \,(\mathrm{mod}\, p)$. Therefore, the set

$$X = \{0^2, 1^2, \dots, ((p-1)/2)^2\}$$

contains $(p+1)/2$ different remainders mod p and the set

$$Y = \{-(1+x) \; : \; x \in X\}$$

also contains $(p+1)/2$ different remainders mod p. The sum of elements of these sets is equal to

$$|X| + |Y| = (p+1)/2 + (p+1)/2 = p+1.$$

Since the number of different remainders mod p is exactly p, $X \cap Y \neq \emptyset$. Hence, there are elements $x \in X$ and $y \in Y$ such that $x \equiv y \pmod{p}$. This means that there are integers t, m such that $x = t^2, y = -(1 + m^2)$ and so $t^2 + m^2 + 1 \equiv 0 \pmod{p}$. □

✴ Lemma 11.31.
If $p > 2$ is a prime, then there is an element $q \in \mathbf{L}$ such that $p|N(q)$ and p does not divide q.

Proof.
From Lemma 11.30, it follows that there are integers t, m such that $t^2 + m^2 + 1 \equiv 0 \pmod{p}$. Let $q = 1 + ti + mj$. Then, $N(q) = t^2 + m^2 + 1$. So, $p|N(q)$, i.e., there is a positive integer k such that $kp = (1 + ti + mj)(1 - ti - mj)$. However, p does not divide $q = 1 + ti + mj$ because the quaternion $\dfrac{1}{p} + \dfrac{t}{p}i + \dfrac{m}{p}$ is not a Hurwitz quaternion. □

✴ Definition 11.32.
A non-zero Hurwitz quaternion $q \in \mathbf{Hu}$ is called **irreducible in Hu**, if from the equality $q = \alpha\beta$, where $\alpha, \beta \in \mathbf{Hu}$, it follows that $(\alpha \in U(\mathbf{Hu})) \vee (\beta \in U(\mathbf{Hu}))$, i.e., α or β is equal to $\pm 1, \pm i, \pm j, \pm k, \frac{1}{2}(\pm 1 \pm i \pm j \pm k)$. Otherwise, q is called **decomposable in Hu**.

If $\alpha, \beta \in \mathbf{Hu}$, then we say that $\beta|\alpha$, if there is $\varphi \in \mathbf{Hu}$ such that $\alpha = \beta\varphi$ or $\alpha = \varphi\beta$.

✴ Lemma 11.33.
If $p > 2$ is a prime, then p is decomposable in the ring $A = \mathbf{Hu}$.

Proof.
From Lemma 11.31, it follows that there is a quaternion $q = 1 + ti + mj \in \mathbf{L}$ such that $p|N(q)$ and p does not divide q.

I. Consider two left principal ideals Ap, Aq of the ring A. We show that $Ap \neq Ap + Aq$ and $Ap + Aq \neq A$. We perform the proof by contrary.

1) Assume that $Ap = Ap + Aq$. Then, $q \in Ap$, hence, $2q \in \mathbf{L}p$. This means that $p|2$, i.e., we obtain a contradiction.

2) Assume that $A = Ap + Aq$. Then, $1 \in Ap + Aq$, hence, we obtain that there are elements $u, v \in A$ such that $1 = up + vq$, i.e., $vq = 1 - up$. Then,

$$(vq)(vq)^* = (1 - up)(1 - up)^* = (1 - up)(1 - u^*p) =$$

$$= 1 - ((u + u^*)p + (uu^*)p^2.$$

Since the norm of quaternions is a multiplicative function, we have that

$$N(vq) = N(v)N(q) = 1 - (u + u^*)p + (uu^*)p^2.$$

Since $p|N(q)$, $1 - (u + u^*)p + (uu^*)p^2 \equiv 0 \,(\mathrm{mod}\, p)$, we get a contradiction.

II. Since each left ideal of the ring $A = \mathbf{Hu}$ is principal, there is an element $b \in A$ such that $N(b) \neq 1$ and $Ap + Aq = Ab$. Since $Ap \subset Ap + Aq = Ab$ and $Ap \neq Ab$, there is an element $h \in A$ such that $N(h) \neq 1$ and $p = hb$, i.e., p is a decomposable element in $A = \mathbf{Hu}$. □

❋ **Theorem 11.34. (Lagrange's Four-Square Theorem).**
Each natural number is a sum of four squares of integers, i.e., for all $n \in \mathbb{N}$ there are $a, b, c, d \in \mathbb{Z}$ such that:

$$n = a^2 + b^2 + c^2 + d^2. \tag{11.17}$$

Proof.

1. From Euler's Theorem 11.2, we can assume that n is a prime.

2. If $p = 2$, then $2 = 1^2 + 1^2 + 0^2 + 0^2$.

3. If $p > 2$ is a prime, then, from Lemma 11.33, it follows that p is decomposable in $\mathbf{Hu}$, i.e., there are elements $\alpha, \beta \in \mathbf{Hu}$ such that $p = \alpha\beta$. Then, $p = p^* = \beta^*\alpha^*$ and $p^2 = (\alpha\beta)(\beta^*\alpha^*) = N(\alpha)N(\beta)$. Since $p > 2$ is a prime, we obtain that $N(\alpha) = N(\beta) = p$.

If $\alpha = a + bi + cj + dk \in \mathbf{L}$, then $N(\alpha) = a^2 + b^2 + c^2 + d^2$, hence,

$$p = a^2 + b^2 + c^2 + d^2$$

and the proof is complete.

If $\alpha = a + bi + cj + dk \in \mathbf{Hu}\backslash\mathbf{L}$, then, from Lemma 11.29, it follows that there is an element $\delta \in U(\mathbf{Hu})$ such that $N(\delta) = 1$ and $q\delta \in \mathbf{L}$. If $q\delta = x + yi + zj + uk$, where $x, y, z, u \in \mathbb{Z}$, then

$$p = N(q) = N(q)N(\delta) = N(q\delta) = x^2 + y^2 + z^2 + u^2$$

and we are done. □

❋ **Examples 11.35.**

- $23 = 3^2 + 3^2 + 2^2 + 1^2$

- $36 = 5^2 + 3^2 + 1^2 + 1^2$

- $9 = 2^2 + 2^2 + 1^2 + 0^2$

11.5. Trigonometric Form of Quaternions

In this section, we will show that, similar to the case of complex numbers, an arbitrary quaternion can be written in the trigonometric form by means of its module and some angle.

> ✳ **Definition 11.36.**
> A quaternion of the form
>
> $$q = x_1 i + x_2 j + x_3 k, \tag{11.18}$$
>
> where $x_i \in \mathbb{R}$, is called a **pure imaginary quaternion** or a **vector quaternion**.

It is obvious that the set of all pure imaginary quaternions form a subspace in $\mathbb{H}$ which will be denoted by $\mathrm{Im}(\mathbb{H})$.

✳ **Lemma 11.37.**
A quaternion $q \in \mathrm{Im}(\mathbb{H})$ if and only if $q^* = -q$.

Proof.
1) Let $q = x_1 i + x_2 j + x_3 k \in \mathrm{Im}(\mathbb{H})$. Then, $q^* = -x_1 i - x_2 j - x_3 k = -q$.
2) Let $q = x_0 + x_1 i + x_2 j + x_3 k \in \mathbb{H}$ and $q^* = -q$. Then,

$$x_0 - x_1 i - x_2 j - x_3 k = -x_0 - x_1 i - x_2 j - x_3 k,$$

hence, $x_0 = 0$, i.e., $q = x_1 i + x_2 j + x_3 k \in \mathrm{Im}(\mathbb{H})$. □

So, each quaternion $q = x_0 + x_1 i + x_2 j + x_3 k \in \mathbb{H}$ can be written in the form $q = x_0 + \mathbf{u}$, where $x_0 \in \mathbb{R}$ is called the **scalar part** of q and denoted by $\mathrm{Re}(q)$ and $\mathbf{u} = x_1 i + x_2 j + x_3 k \in \mathrm{Im}(\mathbb{H})$ is called the **vector part** of q. It is obvious that $\mathbb{R} \cap \mathrm{Im}(\mathbb{H}) = \emptyset$. Therefore, the algebra of quaternions $\mathbb{H}$ as a vector space over the real numbers $\mathbb{R}$ can be written in the form of the direct sum of subspaces: $\mathbb{H} = \mathbb{R} \oplus \mathrm{Im}(\mathbb{H})$, since, for $q = x_0 + x_1 i + x_2 j + x_3 k$, $h = y_0 + y_1 i + y_2 j + y_3 k \in \mathbb{H}$, we have the operations:
1) $q + h = (x_0 + \mathbf{u}) + (y_0 + \mathbf{v}) = (x_0 + y_0) + (\mathbf{u} + \mathbf{v}) \in \mathbb{R} \oplus \mathrm{Im}(\mathbb{H})$,
2) $\alpha q = \alpha(x_0 + \mathbf{u}) = \alpha x_0 + \alpha \mathbf{u} \in \mathbb{R} \oplus \mathrm{Im}(\mathbb{H})$,
which satisfy all axioms of direct sum of vector subspaces.

Each vector quaternion $q = 0 + x_1 i + x_2 j + x_3 k$ can be identified with the vector (x_1, x_2, x_3) in the three-dimensional vector space $\mathbb{R}^3$. It is easy to verify that the vector subspace $\mathrm{Im}(\mathbb{H})$ is isomorphic to $\mathbb{R}^3$.

Therefore, for vector quaternions $q = 0 + \mathbf{u} = \mathbf{u} = x_1 i + x_2 j + x_3 k$, and $h = 0 + \mathbf{v} = \mathbf{v} = y_1 i + y_2 j + y_3 k \in \mathrm{Im}(\mathbb{H})$, the dot (or scalar) product given by the formula below can be introduced:

$$\langle \mathbf{u}, \mathbf{v} \rangle = x_1 y_1 + x_2 y_2 + x_3 y_3. \tag{11.19}$$

Using isomorphism of vector spaces $\text{Im}(\mathbb{H})$ and $\mathbb{R}^3$, we can also introduce the cross (or vector) product of these quaternions:

$$\mathbf{u} \times \mathbf{v} = \begin{vmatrix} i & j & k \\ x_1 & x_2 & x_3 \\ y_1 & y_2 & y_3 \end{vmatrix} = i \begin{vmatrix} x_2 & x_3 \\ y_2 & y_3 \end{vmatrix} - j \begin{vmatrix} x_1 & x_3 \\ y_1 & y_3 \end{vmatrix} + k \begin{vmatrix} x_1 & x_2 \\ y_1 & y_2 \end{vmatrix} \quad (11.20)$$

In this way, quaternions i, j, k can be identified with orthogonal standard basis $\mathbf{e}_1 = (1,0,0)$, $\mathbf{e}_2 = (0,1,0)$, $\mathbf{e}_3 = (0,0,1)$ of $\mathbb{R}^3$, since $\langle i, j \rangle = \langle j, k \rangle = \langle k, i \rangle = \mathbf{0}$, and $\langle i, i \rangle = \langle j, j \rangle = \langle k, k \rangle = 1$ and $\parallel i \parallel = \parallel j \parallel = \parallel k \parallel = 1$, where $\parallel q \parallel^2 = N(q)$ for a quaternion $q \in \mathbb{H}$.

✳ Example 11.38.

Let $\mathbf{u} = -i + 2j + 5k$, $\mathbf{v} = 2i - 3k \in \text{Im}(\mathbb{H})$. Then,

$$\langle \mathbf{u}, \mathbf{v} \rangle = (-1) \cdot 2 + 2 \cdot 0 + 5 \cdot (-3) = -17.$$

$$\mathbf{u} \times \mathbf{v} = \begin{vmatrix} i & j & k \\ -1 & 2 & 5 \\ 2 & 0 & -3 \end{vmatrix} = -6i + 7j - 4k.$$

✳ Theorem 11.39.

Let $q = x_0 + \mathbf{u}$, $h = y_0 + \mathbf{v} \in \mathbb{R} \oplus \text{Im}(\mathbb{H}) = \mathbb{H}$. Then,

$$qh = (x_0 + \mathbf{u})(y_0 + \mathbf{v}) = (x_0 y_0 - \langle \mathbf{u}, \mathbf{v} \rangle) + (x_0 \mathbf{v} + y_0 \mathbf{u} + \mathbf{u} \times \mathbf{v}) \quad (11.21)$$

Proof.

Let $q = x_0 + x_1 i + x_2 j + x_3 k = x_0 + \mathbf{u}$, $h = y_0 + y_1 i + y_2 j + y_3 k = y_0 + \mathbf{v} \in \mathbb{R} \oplus \text{Im}(\mathbb{H}) = \mathbb{H}$. Then, the left side of equality (11.21) has the following form:

$$gh = (x_0 y_0 - x_1 y_1 - x_2 y_2 - x_3 y_3) + (x_0 y_1 + x_1 y_0 + x_2 y_3 - x_3 y_2)i +$$

$$+(x_0 y_2 - x_1 y_3 + x_2 y_0 + x_3 y_0)j + (x_0 y_3 + x_1 y_2 - x_2 y_1 + x_3 y_0)k.$$

On the other hand,

$$x_0 y_0 - \langle \mathbf{u}, \mathbf{v} \rangle = x_0 y_0 - x_1 y_1 - x_2 y_2 - x_3 y_3 = \text{Re}(gh)$$

$$x_0 \mathbf{v} + y_0 \mathbf{u} + \mathbf{u} \times \mathbf{v} = x_0(y_1 i + y_2 j + y_3 k) + y_0(x_1 i + x_2 j + x_3 k) + i(x_2 y_3 - x_3 y_2) -$$

$$-j(x_1 y_3 - x_3 y_1) + k(x_1 y_2 - x_2 y_1) = (x_0 y_1 + x_1 y_0 + x_2 y_3 - x_3 y_2)i +$$

$$+(x_0 y_2 - x_1 y_3 + x_2 y_0 + x_3 y_0)j + (x_0 y_3 + x_1 y_2 - x_2 y_1 + x_3 y_0)k.$$

Thus, $gh = (x_0 y_0 - \langle \mathbf{u}, \mathbf{v} \rangle) + (x_0 \mathbf{v} + y_0 \mathbf{u} + \mathbf{u} \times \mathbf{v})$. □

From equality (11.21), we get the following corollary:

> **❋ Corollary 11.40.**
> If $q = 0 + \mathbf{u}$, $h = 0 + \mathbf{v} \in \mathrm{Im}(\mathbb{H})$ are vector quaternions, then
>
> $$gh = \mathbf{uv} = -\langle \mathbf{u}, \mathbf{v} \rangle + \mathbf{u} \times \mathbf{v} = -q \cdot h + q \times h \qquad (11.22)$$

Since $\langle \mathbf{u}, \mathbf{v} \rangle = \langle \mathbf{v}, \mathbf{u} \rangle$ and $\mathbf{u} \times \mathbf{v} = -\mathbf{v} \times \mathbf{u}$, from equality (11.22), we get the following:

> **❋ Corollary 11.41.**
> If $q = 0 + \mathbf{u}$, $h = 0 + \mathbf{v} \in \mathrm{Im}(\mathbb{H})$ are vector quaternions, then
>
> $$gh + hg = \mathbf{uv} + \mathbf{vu} = -2\langle \mathbf{u}, \mathbf{v} \rangle = -2\langle g, h \rangle. \qquad (11.23)$$

Let $q = x_0 + \mathbf{u} \in \mathbb{H} \backslash \{0\}$. If $\mathbf{u} \neq 0$, then we can write q in the form:

$$q = \| q \| \left(\frac{x_0}{\| q \|} + \frac{\| \mathbf{u} \|}{\| q \|} \frac{\mathbf{u}}{\| \mathbf{u} \|} \right) = \| q \| \, (b + c\mathbf{n}) \qquad (11.24)$$

where

$$\| \mathbf{u} \|^2 = N(\mathbf{u}), \quad \| q \|^2 = N(q), \quad b = \frac{x_0}{\| q \|}, \quad c = \frac{\| \mathbf{u} \|}{\| q \|}, \quad \mathbf{n} = \frac{\mathbf{u}}{\| \mathbf{u} \|} \qquad (11.25)$$

Therefore, $b^2 + c^2 = \dfrac{1}{N(q)}(x_0^2 + \langle \mathbf{u}, \mathbf{u} \rangle) = \dfrac{1}{N(q)} N(q) = 1$. Taking this equality and (11.25) into account, there exists a real number $\theta \in \mathbb{R}$ such that:

$$b = \cos \theta, \quad c = \sin \theta \qquad (11.26)$$

At the time, a quaternion q can be written as:

$$q = \| q \| \, (\cos \theta + \mathbf{n} \sin \theta) \qquad (11.27)$$

where $\| \mathbf{n} \| = 1$.

> **✳ Definition 11.42.**
> The form (11.27) of a quaternion $q \in \mathbb{H}$ is called the **trigonometric form** of q.
> A quaternion $q \in \mathbb{H}$ with $\| q \| = 1$ is called a **unit quaternion**.

The set of all unit quaternions $\{ q \in \mathbb{H} : \| q \| = 1 \}$ is a sphere of radius one in the 4-space $\mathbb{H}$, and is denoted by S^3.

The S^3 is closed under multiplication, i.e., if $q, h \in S^3$, then $qh \in S^3$ because $\| qh \| = \| q \| \| h \| = 1$. If $q \in S^3$, then $q^* \in S^3$ as well because $\| g^* \| = \| g \| = 1$. Since $q^{-1} = \dfrac{q^*}{N(q)} = q^*$, $q^{-1} \in S^3$. It is obvious that $1 \in S^3$. In this way, S^3 is a group.

If $q = x_0 + x_1 i + x_2 j + x_3 k$ is a unit quaternion, then

$$q = \cos\theta + \mathbf{n}\sin\theta,$$

where $\mathbf{n}$ is a unit vector quaternion. In this case,

$$q^* = \cos\theta - \mathbf{n}\sin\theta = \cos(-\theta) + \mathbf{n}\sin(-\theta)$$

Therefore, we obtain the following theorem:

✳ Theorem 11.43.
If q is a unit quaternion, then it can be expressed in the form:

$$q = \cos\theta + \mathbf{n}\sin\theta, \tag{11.28}$$

and
1) $q^* = \cos\theta - \mathbf{n}\sin\theta = \cos(-\theta) + \mathbf{n}\sin(-\theta)$
2) $q^{-1} = q^*$.
where $\mathbf{n}$ is a unit pure imaginary quaternion.

✳ Examples 11.44.
 1. Write the quaternion $q = -i$ in the trigonometric form.
 We have: $\| q \|^2 = N(q) = N(-i) = 1$, $\| q \| = 1$, $\mathbf{n} = -i$,
 $$b = \frac{x_0}{\| q \|} = 0,\ c = \frac{\| i \|}{\| q \|} = 1.$$
 Therefore, $b = \cos\theta = 0$, $c = \sin\theta = 1$. Hence, $\theta = \pi/2$ and so

$$q = \cos(\pi/2) - \sin(\pi/2)i.$$

 2. Write the quaternion $q = 1 + i$ in the trigonometric form.
 We have: $\| q \|^2 = N(q) = N(1 + i) = 2$, $\| q \| = \sqrt{2}$, $\mathbf{n} = i$,
 $$b = \frac{x_0}{\| q \|} = \frac{1}{\sqrt{2}},\ c = \frac{\| i \|}{\| q \|} = \frac{1}{\sqrt{2}}.$$
 Therefore, $b = \cos\theta = \dfrac{1}{\sqrt{2}}$, $c = \sin\theta = \dfrac{1}{\sqrt{2}}$. Hence, $\theta = \pi/4$ and so

$$q = \sqrt{2}[\cos(\pi/4) + \sin(\pi/4)i].$$

11.6. Rotations and Quaternions

There are three basic transformations used in computer graphics in three-dimensional space $\mathbb{R}^3$: Translations, scaling and rotations. There are a few widespread methods of writing rotations in the space $\mathbb{R}^3$ using:

- Euler's angles

- Rotation matrices

- Quaternions

- Rodrigues parameters and Gibbs vectors

- Cayley-Klein parameters

In this section, we study the use of quaternions to write rotations in $\mathbb{R}^3$.

The use of quaternions to describe a rotation of a vector $\mathbf{p}$ in $\mathbb{R}^3$ around some axis is possible thanks to the mapping:

$$O_q : \operatorname{Im}(\mathbb{H}) \to \operatorname{Im}(\mathbb{H})$$

given by $O_q(p) = qpq^{-1}$, where $q \in \mathbb{H}$ and $p = 0 + \mathbf{p} \in \operatorname{Im}(\mathbb{H})$. If $N(q) = 1$, then $O_q(p) = qpq^{-1} = qpq^*$.

✳ Lemma 11.45.

For any real number c, it holds that $O_q = O_{cq}$.

Proof.

Since the multiplication by a scalar is commutative, for each real number c and each $p \in \operatorname{Im}(\mathbb{H})$, we have:

$$O_{cq}(p) = (cq)p(cq)^{-1} = cqpq^{-1}c^{-1} = cc^{-1}qpq^{-1} = qpq^{-1} = O_q(p).$$

In particular, if $c = -1$, then $(-q)p(-q)^{-1} = qpq^{-1}$, i.e., $O_q = O_{-q}$. □

As we showed in the previous section, a vector $\mathbf{p} = (x, y, z) \in \mathbb{R}^3$ corresponds to a pure imaginary quaternion:

$$p = 0 + xi + yj + zk = 0 + \mathbf{p} \in \operatorname{Im}(\mathbb{H}).$$

Let $\mathbf{n}$ be a vector with $\| \mathbf{n} \| = 1$ that describes an axis around which we perform a rotation.

✳ Theorem 11.46.

A mapping $O_q : \operatorname{Im}(\mathbb{H}) \to \operatorname{Im}(\mathbb{H})$ given by:

$$O_q(p) = qpq^{-1} \tag{11.29},$$

where $q = \| q \| (\cos\theta + \mathbf{n}\sin\theta)$, defines a rotation around the axis $\operatorname{Lin}(\mathbf{n})$ by the angle 2θ in $\mathbb{R}^3$.

Proof.

Since $O_q = O_{cq}$ for each real number c, we can assume that $N(q) = 1$.

Let $p = 0 + \mathbf{p} \in \operatorname{Im}(\mathbb{H})$ and $p' = O_q(p)$ is a quaternion which represents a vector $\mathbf{p}'$ under the rotation. Then,

$$p' = O_q(p) = qpq^{-1} = qpq^*.$$

1) Since $\text{Re}(h) = \frac{1}{2}(h + h^*)$, for all $h \in \mathbb{H}$, and $p \in \text{Im}(\mathbb{H})$, we obtain that $\text{Re}(p) = 0$. Therefore,

$$\text{Re}(p') = \frac{1}{2}[(qpq^*) + (qpq^*)^*] = \frac{1}{2}[qpq^* + qp^*q^*] = \frac{1}{2}[q(p + p^*)q^*] = \text{Re}(p)(qq^*) = 0.$$

Hence, $p' \in \text{Im}(\mathbb{H})$.

2) O_q is a linear mapping that follows from distributivity of multiplication with respect to addition of quaternions.

3) $N(p') = N(qpq^*) = N(q)N(p)N(q^*) = N(qq^*)N(p) = N(p)$. Hence, if $p' = 0 + \mathbf{p}'$, then $\| \mathbf{p}' \| = \| \mathbf{p} \|$. Therefore, O_q is a linear isometry.

4) We show that a mapping O_q performs a rotation about the vector $\mathbf{n}$ as the axis of rotation. It suffices to show that the vector $\mathbf{n}$ is a fixed point of O_q, i.e., it is not changed under O_q.

The vector $\mathbf{n} \in \mathbb{R}^3$ corresponds to the pure imaginary quaternion $h = 0 + \mathbf{n} \in \text{Im}(\mathbb{H})$. It is clear that q and h commute, so $O_q(h) = qhq^{-1} = hqq^{-1} = h$ and $O_q(\alpha h) = \alpha h$ for all $\alpha \in \mathbb{R}$. Hence, all subspace $\text{Lin}(\mathbf{n})$ is a fixed point of O_q. Thus, O_p is a rotation around the axis through the vector $\mathbf{n}$.

5) We show that O_q is a rotation in subspace $\text{Lin}(\mathbf{n})^\perp$ by the angle 2θ. Let us take an arbitrary unit vector $\mathbf{w} \in \text{Lin}(\mathbf{n})^\perp$. Then,

$$\langle \mathbf{w}, \mathbf{n} \rangle = \langle \mathbf{n}, \mathbf{w} \rangle = 0.$$

Let $\mathbf{v} = \mathbf{n} \times \mathbf{w}$. It is obvious that $\mathbf{v} \in \text{Lin}(\mathbf{n})^\perp$, i.e.,

$$\langle \mathbf{v}, \mathbf{n} \rangle = \langle \mathbf{n}, \mathbf{v} \rangle = 0.$$

Using the equality (11.22), we have:

$$\mathbf{n}\mathbf{w} = -\langle \mathbf{n}, \mathbf{w} \rangle + \mathbf{n} \times \mathbf{w} = \mathbf{v},$$

$$\mathbf{w}\mathbf{n} = -\langle \mathbf{w}, \mathbf{n} \rangle + \mathbf{w} \times \mathbf{n} = -\mathbf{v},$$

$$\mathbf{v}\mathbf{n} = -\langle \mathbf{v}, \mathbf{n} \rangle + \mathbf{v} \times \mathbf{n} = \mathbf{w}.$$

Let a vector $\mathbf{w}$ correspond to a pure imaginary quaternion $w = 0 + \mathbf{w} \in \text{Im}(\mathbb{H})$. Then, we obtain

$$O_q(w) = [\cos(\theta) + \sin(\theta)\mathbf{n}]\mathbf{w}[\cos(\theta) - \sin(\theta)\mathbf{n}] =$$

$$= [\cos(\theta)\mathbf{w} + \sin(\theta)\mathbf{v}][\cos(\theta) - \sin(\theta)\mathbf{n}] =$$

$$= [\cos^2(\theta)\mathbf{w} + \sin(\theta)\cos(\theta)\mathbf{v} - \sin(\theta)\cos(\theta)\mathbf{w}\mathbf{n} - \sin^2(\theta)\mathbf{v}\mathbf{n}] =$$

$$= (\cos^2(\theta) - \sin^2(\theta))\mathbf{w} + 2\sin(\theta)\cos(\theta)\mathbf{v} = \cos(2\theta)\mathbf{w} + \sin(2\theta)\mathbf{v}.$$

Therefore, a vector $\mathbf{w}$ which is orthogonal to a vector $\mathbf{n}$ is rotated in a subspace $\text{Lin}(\mathbf{n})^\perp$ by an angle 2θ. Since $\mathbf{w}$ is an arbitrary vector in the subspace $\text{Lin}(\mathbf{n})^\perp$, the proof is completed. $\square$

> ✻ **Definition 11.47.**
> Each rotation about an axis $\mathrm{Lin}(\mathbf{n})$ by an angle 2θ corresponds to a quaternion of the form:
>
> $$q = \cos(\theta) + \sin(\theta)\mathbf{n}, \tag{11.30}$$
>
> with $\| \mathbf{n} \| = 1$, which is called a **rotation quaternion**.

✱ **Example 11.48.**
Find the rotation of the vector $\mathbf{p} = (1,0,0)$ above the axis Oy by the angle π.

The vector $\mathbf{p} = (1,0,0)$ corresponds to the pure imaginary quaternion $p = i$, and we have the rotation around axis Oy through the vector defined by the unit vector quaternion $\mathbf{n} = j$, and the angle of this rotation is $\theta = \pi/2$. Therefore, the rotation quaternion has the form

$$q = \cos(\theta) + \sin(\theta)\mathbf{n} = \cos(\pi/2) + \sin(\pi/2)\mathbf{n} = j.$$

Since $q^* = -j$, at the result of the rotation, we obtain the quaternion

$$p' = O_q(p) = qpq^* = ji(-j) = -i,$$

which corresponds to the vector $\mathbf{p}' = (-1,0,0)$.

✱ **Example 11.49.**
Find the rotation of the vector $\mathbf{p} = (1,1,1)$ above the axis through the vector $\mathbf{n} = (1,-1,1)$ by the angle $\pi/2$.

The vector $\mathbf{p} = (1,1,1)$ corresponds to the vector quaternion $p = i+j+k$, and the axis of rotation corresponds to the vector quaternion $v = i - j + k$ with norm $\| v \| = \sqrt{3}$. Therefore, the corresponding unit vector quaternion is the vector $\mathbf{n} = \sqrt{3}(i - j + k)$, and the angle of rotation $\theta = \pi/4$. Hence, the rotation quaternion has the form

$$q = \cos(\theta) + \sin(\theta)\mathbf{n} = \cos(\pi/4) + \sqrt{3}/3\sin(\pi/4)(i - j + k) =$$

$$= \sqrt{2}/2[1 + \sqrt{3}/3i - \sqrt{3}/3j + \sqrt{3}/3k].$$

Since $q^* = \sqrt{2}/2[1 - \sqrt{3}/3i + \sqrt{3}/3j - \sqrt{3}/3k]$, the result of the rotation of the vector $\mathbf{p}$ is the quaternion

$$p' = O_q(p) = qpq^* =$$
$$= 1/2[1+\sqrt{3}/3i-\sqrt{3}/3j+\sqrt{3}/3k](i+j+k)[1-\sqrt{3}/3i+\sqrt{3}/3j-\sqrt{3}/3k] =$$
$$= \frac{1+2\sqrt{3}}{3}i - \frac{1}{3}j + \frac{1-2\sqrt{3}}{3}k,$$

which corresponds to the vector $\left[\dfrac{1+2\sqrt{3}}{3}, -\dfrac{1}{3}, \dfrac{1-2\sqrt{3}}{3}\right]$ in $\mathbb{R}^3$.

The properties of rotations using quaternions are given in the theorem:

✳ **Theorem 11.50.**

1. The composition of rotations corresponds to multiplication of quaternions:
$$O_h O_q = O_{hq} \tag{11.31}$$
In particular, the n-th multiply repeating rotation described by a quaternion q is a rotation described by a quaternion q^n.

2. The rotation around a vector $\mathbf{n}$ by an angle θ is equivalent to a rotation around a vector $-\mathbf{n}$ by an angle $-\theta$.

3.
$$O_q^{-1} = O_{q^*} = O_{q^{-1}} \tag{11.32}$$

Proof.

1. If h, q are rotation quaternions, then

$$O_h(O_q(p)) = h(qpq^{-1})h^{-1} = (hq)p(q^{-1}h^{-1}) = (hq)p(hq)^{-1} = O_{hq}(p)$$

for any pure imaginary quaternion $p \in \text{Im}(\mathbb{H})$, which means that $O_h O_q = O_{hq}$. In particular, if $h = q$, then $O_q(O_q(p)) = q^2 pq^{-2} = O_{qq}(p)$ and

$$O_q(O_q \ldots (O_q(p)) \ldots) = q^n pq^{-n} = O_{q^n}(p)$$

for any pure imaginary quaternion $p \in \text{Im}(\mathbb{H})$. This means that the nth multiple repeating rotation described by a quaternion q is a rotation described by the quaternion q^n:

$$O_q(O_q \ldots (O_q) \ldots) = O_{q^n}.$$

2. Since $O_q = O_{cq}$ for each real number c, $O_q = O_{-q}$.

3. Let $O_1 = Id$ be the identity rotation described by the identity quaternion $\mathbf{1}$. If $N(q) = 1$, then $q^{-1} = q^*$ and so:

$$O_q O_{q^*}(p) = O_{qq^*}(p) = O_1(p) = p.$$

This means that

$$O_q^{-1} = O_{q^*} = O_{q^{-1}}.$$

□

✳ **Example 11.51.**

Find the rotation of the vector $\mathbf{p} = (0, 1, 1)$ around the axis through the vector $\mathbf{n_1} = (0, 1, 0)$ by the angle $\pi/2$, and then around the axis through the vector $\mathbf{n_2} = (1, 0, 0)$ by the angle $\pi/2$.

Since $2\theta_1 = \pi/2$, $2\theta_2 = \pi/2$, we get that:

$\theta_1 = \pi/4$, $\theta_2 = \pi/4$,

$\mathbf{n}_1 = (0,1,0) = j$, $\mathbf{n}_2 = (1,0,0) = i$,

$p = 0 + \mathbf{p} = j + k$

$q = \cos(\theta_1) + \sin(\theta_1)\mathbf{n}_1 = \cos(\pi/4) + \sin(\pi/4)j = \sqrt{2}/2 + \sqrt{2}/2j$,

$h = \cos(\theta_2) + \sin(\theta_2)\mathbf{n}_2 = \cos(\pi/4) + \sin(\pi/4)i = \sqrt{2}/2 + \sqrt{2}/2i$,

$hq = (\sqrt{2}/2 + \sqrt{2}/2i)\sqrt{2}/2 + \sqrt{2}/2j) = 1/2(1 + i + j + k) = 1/2 + 1/2(i + j + k) = 1/2 + \sqrt{3}/2\mathbf{n}$, where $\mathbf{n} = \dfrac{i + j + k}{\sqrt{3}}$ is the unit vector quaternion.

Since $\cos(\theta) = 1/2$ and $\sin(\theta) = \sqrt{3}/2$, $\theta = 60°$. Therefore, $2\theta = 120°$ is the rotation angle, and the vector $\mathbf{v} = i + j + k$ defines the rotation axis.

$(hq)^{-1} = 1/2(1 - i - j - k) = (hq)^*$.

So $O_{hq}(p) = 1/2(1 + i + j + k)(j + k)(1/2(1 - i - j - k) = i + k$.

If we have a composition of rotations around the same axis defined by a vector $\mathbf{n}$ by an angle 2θ, and next by an angle 2φ, then the first rotation corresponds to the quaternion:

$$q = \| q \| [\cos(\theta) + \sin(\theta)\mathbf{n}],$$

and the second one corresponds to the quaternion:

$$h = \| h \| [\cos(\varphi) + \sin(\varphi)\mathbf{n}],$$

where $\| \mathbf{n} \| = 1$.

Then, the composition of rotations corresponds to the quaternion:

$$hq = \| h \| [\cos(\varphi) + \sin(\varphi)\mathbf{n}] \| q \| [\cos(\theta) + \sin(\theta)\mathbf{n}] =$$

$$= \| h \| \cdot \| q \| \{[\cos(\varphi)\cos(\theta) + \sin(\varphi)\sin(\theta)\mathbf{nn}] + [\cos(\varphi)\sin(\theta) + \sin(\varphi)\cos(\theta)]\mathbf{n}\} =$$

$$= \| h \| \cdot \| q \| \{[\cos(\varphi)\cos(\theta) - \sin(\varphi)\sin(\theta)] + [\cos(\varphi)\sin(\theta) + \sin(\varphi)\cos(\theta)]\mathbf{n}\} =$$

$$= \| h \| \cdot \| q \| [\cos(\varphi + \theta) + \sin(\varphi + \theta)\mathbf{n}].$$

If $\theta = \varphi$, then:

$$q^2 = \| q \|^2 [\cos(2\theta) + \sin(2\theta)\mathbf{n}]. \tag{11.33}$$

Repeating this procedure, we obtain for each natural number k the formula which is similar to Moivre's formula for complex numbers:

$$q^k = \| q \|^k [\cos(k\theta) + \sin(k\theta)\mathbf{n}]. \tag{11.34}$$

If $N(q) = 1$, then for each natural number k we have the formula

$$q^k = \cos(k\theta) + \sin(k\theta)\mathbf{n}. \tag{11.35}$$

Moreover, it can be proved that this formula holds for each real number t:

$$q^t = \cos(t\theta) + \sin(t\theta)\mathbf{n}.$$

From theorem 11.50, it follows that rotations in $\mathbb{R}^3$ form a group which is denoted by $SO(3) = SO(3, \mathbb{R})$.

✳ Theorem 11.52.

Let $G = \{q \in \mathbb{H} \: : \: N(q) = 1\}$. Then, the mapping $F : G \to SO(3, \mathbb{R})$ given by $F(q) = O_q$ is an epimorphism of groups and each rotation is an image of exactly two quaternions from the group G: q and $-q$.

Proof.

Let $g, h \in G$. Then,

$$F(hq) = O_{qh}(p) = (hq)p(hq)^{-1} = h(qpq^{-1})h^{-1} = (F(h) \circ F(q))(p)$$

for all $p \in \text{Im}(\mathbb{H})$. Therefore, $F(hq) = F(h) \circ F(q)$, i.e., F is a homomorphism of groups. It is obvious that F is an epimorphism.

If Id is the identity rotation, then it corresponds to quaternions $\pm 1 \in G$. If ω is a rotation around an axis $\text{Lin}(\mathbf{n})$ by an angle 2θ, then it corresponds to two quaternions:

$$q = \cos(\theta) + \sin(\theta)\mathbf{n}$$

and

$$-q = -\cos(\theta) - \sin(\theta)\mathbf{n} = \cos(\pi + \theta) + \sin(\pi + \theta)\mathbf{n}.$$

□

In summary, we can state the most important advantages that follow from using quaternions for rotations of a vector around an axis in computer graphics:

1. A simple modification and quick connections of rotations.
2. Quick inversions of rotations.
3. Avoidance of expensive normalization.
4. Quaternions occupy less space in the computer memory when compared with matrices.
5. Quick conversion from/to rotation matrices.
6. Smooth interpolation of rotations.

11.7. Exercises

1. Show that the set of complex numbers $T = \{z \in \mathbb{C} \: : \: |z| = 1\}$ forms a multiplicative group with respect to multiplication.

2. Write the following quaternions in the trigonometric form:

   (a) i, j, k,
   (b) $1 + j$

    (c) $1 - k$

    (d) $3 + \sqrt{3}(i - j + k)$

3. Find q^{12} if $q = 3 + 3j$.

4. Find q^{20} if $q = \sqrt{3} + 2i - j + 2k$.

5. In the Euclidian space $E(\mathbb{R}^2)$, write the rotation matrix by $60°$. Find the result of rotation of the point $P(2, -5)$.

6. In the Euclidian space $E(\mathbb{R}^3)$, write the rotation matrix around the axis Oy by $30°$. Find the result of rotation of the point $P(6, -1, -5)$.

7. Find the rotation of the vector $(2, -1, 3)$ around the axis through the vector $(1, 0, 1)$ by the angle $\pi/2$.

8. Find the rotation of the vector $(-5, 0, 4)$ around the axis Ox by the angle $\pi/3$.

9. Find the rotation of the vector $(2, 1, 0)$ around the axis through the vector $(1, 1, 0)$ by the angle $\pi/2$, and then around the axis through the vector $(1, 0, 1)$ by the angle $\pi/3$.

10. Find the rotation of the vector $(1, -1, 3)$ around the axis through the vector $(1, 0, -1)$ by the angle π.

11. Find the rotation of the vector $(3, 1, -1)$ around the axis through the vector $(1, 0, 0)$ by the angle $\pi/2$, and then around the axis through the vector $(0, 1, 0)$ by the angle $\pi/3$.

12. Find the rotation of the vector $(2, -5, 4)$ around the axis through the vector $(2, 0, 1)$ by the angle π, and then around the axis through the vector $(0, 0, 1)$ by the angle $\pi/2$.

References

[1] Coan, B. and Ch. Perng. 2012. Factorization of Hurwitz quaternions. International Mathematical Forum 7(43): 2143–2156.

[2] Conrad, K. The Gaussian Integers, preprint.
http://www.math.uconn.edu/kconrad/blurbs/ugradnumthy/Zinotes.pdf.

[3] Conway, J.H. and D.A. Smith. 2003. On Quaternions and Octonions: Their Geometry, Arithmetic, and Symmetry, A.K. Peters, Massachusetts.

[4] Dickson, L.E. 1919. On quaternions and their generalization and the history of the eight square theorem. Ann. Math. 20: 155–171.

[5] Hardy, G.H. and E.M. Wright. 2008. An Introduction to the Theory of Numbers (6th Ed.), Oxford University Press.

[6] Hurwitz, A. 1898. Über die Composition der quadratischen Formen von beliebig vielen Variabein, Nachr. Ges. Wiss. Göttingen.

[7] Hurwitz, A. 1919. Vorlesungen uber die Zahlentheorie der Quaternionen, Berlin: J. Springer.

[8] Ireland, K. and M. Rosen. 1990. A Classical Introduction to Modern Number Theory (2nd Ed.), Springer.

[9] Joyner, D., R. Kreminski and J. Turisco. 2004. Applied Abstract Algebra, The Johns Hopkins University Press.

[10] Kuipers, J.B. 1999. Quaternions and Rotation Sequences, Prinston University Press.

[11] Niven, I. and H.S. Zuckerman. 1966. Introduction to the Theory of Numbers (2nd Ed.), John Wiley & Sons.

[12] Schafer, R.D. 1966. An introduction to nonassociative algebra. Pure and Applied Mathematics, Academic Press, New York, London, Vol. 22.

Appendix

A.1. Basic Concepts of Set Theory. Relations on Sets

We assume that the Reader knows the basic concepts of set theory, such as a set, a function, operations on sets, cartesian product of sets, and others. We use standard set-theoretical notation. $\emptyset$ denotes the empty set. If x is an element of a set X, we write $x \in X$, otherwise we write $x \notin X$. $A \cup B$ denotes the union of sets A and B, $A \cap B$ the intersection of A and B. Two sets A and B are called disjoint if $A \cap B = \emptyset$. A family is a collection of objects, indexed by some set I, called an index set. A family $\{X_i\}_{i \in I}$ of sets is called pairwise disjoint if $X_i \cap X_j = \emptyset$ for all $i, j \in I$ with $i \neq j$. A pairwise disjoint family of non-empty sets whose union is X is called a **partition** of X.

If A is a finite set, we write $|A|$ for its size, or cardinality. If A is not necessarily a finite set, we also write $\#(A)$ for its cardinality.

We also use standard notation to describe sets. If a subset Y of a set X consists of elements satisfying the condition which is described by the proposition function $P(x)$, we write Y in the following form:

$$Y = \{x \in X \; : \; P(x)\} \quad \text{or} \quad Y = \{x \in X \mid P(x)\}$$

We use standard notation for the most important sets of numbers:
(1) $\mathbb{N} = \{0, 1, 2, 3, \ldots\}$ - the set of natural numbers
(2) $\mathbb{Z} = \{0, \pm 1, \pm 2, \pm 3, \ldots\}$ - the set of integers
(3) $\mathbb{Q} = \{a/b \; : \; a, b \in \mathbb{Z}, \text{ and } b \neq 0\}$ - the set of rational numbers
(4) $\mathbb{R}$ - the set of real numbers
(5) $\mathbb{C} = \{a + bi \; : \; a, b \in \mathbb{R}, \text{ and } i^2 = -1\}$ - the set of complex numbers
(6) $\mathbb{Z}^+ = \{1, 2, 3, \ldots\}$ - the set of positive integers
(7) $\mathbb{Q}^+$ - the set of positive rational numbers
(8) $\mathbb{R}^+$ - the set of positive real numbers

We also use the following logical symbols:
$\wedge$ - the conjunction "and"
$\vee$ - the disjunction "or"
$\Rightarrow$ - the implication "if then"
$\Leftrightarrow$ - the equivalence "if and only if"
$\forall$ - the universal quantifier
$\forall x$ - the quantified variable "for all x"
$\exists$ - the existential quantifier
$\exists x$ - the quantified variable "there exists at least one x"

A binary **relation** on a non-empty set X is a subset $\Re$ of the cartesian product $X \times X$. We use the notation $a\Re b$ to denote that $(a, b) \in \Re$.

A relation $\Re$ on a set X is called:

- **reflexive** if $a\Re a$ for all $a \in X$

- **symmetric** if $a\Re b$ implies $b\Re a$ for all $a, b \in X$

- **antisymmetric** if $(a\Re b) \wedge (b\Re a)$ implies $a = b$ for all $a, b \in X$

- **transitive** if $(a\Re b) \wedge (b\Re c)$ implies $a\Re c$ for all $a, b, c \in X$

- **connex** if $(a\Re b) \vee b\Re a$ for all $a, b \in X$.

A relation $\Re \subseteq X \times X$ is called an **equivalence relation** (or simply an **equivalence**) if it is reflexive, symmetric and transitive. In this case, we often write $a \sim b$ instead of $a\Re b$. Any equivalence relation $\Re$ on a set X is closely connected with the notion of an **equivalence class**. If $\Re$ is an equivalence relation on a set X then the equivalence class of an element $a \in X$ is the set:

$$\Re(a) = [a]_\Re = \{x \in X \; : \; a \sim x\}.$$

If $b \in [a]_\Re$ we say that b is a **representative** of this equivalence class.

A family of all equivalence classes of a set X with respect to an equivalence relation $\Re$ forms a partition of the set X, since for all $a, b \in X$:

- $[a]_\Re \neq \emptyset$

- the set X is a union of all equivalence classes $[a]_\Re$

- $[a]_\Re = [b]_\Re$ if and only if $a\Re b$

- $[a]_\Re \cap [b]_\Re = \emptyset$ if and only if $[a]_\Re \neq [b]_\Re$.

A relation $\Re \subseteq X \times X$ is called a **partial order** or a **partial ordering** if it is reflexive, antisymmetric and transitive. In this case instead of $a\Re b$ it is often written $a \preceq b$. If $a \preceq b$ and $a \neq b$ we will write $a \prec b$. A set X with a partial order $\Re$ is called a **partially ordered set** (or **poset**) and is denoted by $(X, \Re)$.

A partial order $\Re \subseteq X \times X$ is called a **linear order** or a **total order** if it is connex. A poset $(X, \Re)$ is called a **totally ordered set** or a **linearly ordered set**, or a **chain**, if $\Re$ is a linear order on X.

✳ Examples A.1.

1. A set $(\wp(X), \subseteq)$, where $\wp(X)$ is a set of all subsets of a set X, is a partially ordered set for each X.

2. A subset X of the set of natural numbers is a linearly ordered set with respect to common relation $\leq$ for numbers.

In this book, we will use one of the most fundamental axioms in set theory, called the axiom of choice. This axiom was formulated by Zermelo in 1904 and is, therefore, sometimes called Zermelo's axiom of choice.

✳ The Axiom of Choice.
For any indexed family $\Re = \{X_\alpha : \alpha \in I\}$ of non-empty and pairwise disjoint sets X_α there exists a set Y which has exactly one common element y_α with each set $X_\alpha \in \Re$, i.e. $Y \cap X_\alpha = y_\alpha$.

An element a of a partially ordered set $(X, \Re)$ is called **maximal** in X if there is no element $b \in X$ such that $a\Re b$, i.e.

$$(b \in X) \wedge (a\Re b) \Rightarrow (b = a).$$

Analogously, an element a of a partially ordered set $(X, \Re)$ is called **minimal** in X if there is no element $b \in X$ such that $b\Re a$, i.e.

$$(b \in X) \wedge (b\Re a) \Rightarrow (b = a).$$

Let $A \subseteq X$ be a subset of an ordered set $(X, \Re)$. An element $k \in A$ is called the **least element** (corr. **greatest element**) of A if $\forall(x \in A)[k\Re x]$ (corr. $\forall(x \in A)[x\Re k]$).

A poset $(X, \Re)$ is called **well-ordered** if every non-empty subset $A \subseteq X$ has a least element under the ordering relation $\Re$.

An important property of well-ordered sets is the following result, which is equivalent to the axiom of choice and is also known as Zermelo's theorem or the **Well Ordering Theorem**.

✳ Theorem A.2 (Zermelo's Theorem).
Every set can be well-ordered.

A partially ordered set $(X, \Re)$ is called **upper bounded** (corr. **lower bounded**) if there exists an element $a \in X$ such that $\forall(x \in X)[x\Re a]$ (corr. $\forall(x \in X)[a\Re x]$.

We will also use Kuratowski-Zorn's lemma, which is equivalent to the axiom of choice and to Well-Ordering Theorem A.2.

✳ Lemma A.3. (Kuratowski-Zorn's Lemma).
If in a partially ordered set X every linearly ordered subset is upper bounded (corr. lower bounded), then there exists at least one maximal (corr. minimal) element of X.

We will also use the following important statement, which is often called the **Well Ordering Principle** or the **Well-Ordering Property**.

✳ The Well Ordering Principle
Every non-empty set of non-negative integers has a least element.

A function $f : A \to B$ is called an **injection**, or an **injective function**, or a **one-to-one function** if

$$\forall(a_1, a_2 \in A)[f(a_1) = f(a_2) \Longleftrightarrow a_1 = a_2].$$

A function $f : A \to B$ is called a **surjection**, or a **surjective function**, or **onto** if

$$\forall(b \in B)\exists(a \in A)[b = f(a)].$$

A function $f : A \to B$ is called a **bijection** if it is both injective and surjective.

The following important result of set theory, known as the Cantor-Schröder-Bernstein theorem, allows one to compare infinities. It shows that if two cardinalities are less than or equal to each other then they are equal.

✷ **Theorem A.4 (Cantor-Schröder-Bernstein Theorem).**
Let A and B be sets. If there exist injective functions $f : A \to B$ and $g : B \to A$, then there exists a bijection $h : A \to B$.

A.2. Operations on Sets. Algebraic Structures

Let X be a non-empty set. The cartesian product of n copies of the set X is denoted by X^n. An operation on a set X is a special kind of function.

✳ **Definition A.5.**
  A function $f : X^n \to X$ is called an n-ary **operation** (for $n \geq 0$) on a set X. Moreover, we assume that an operation with 0 argument on a set X is an arbitrary, but fixed, element of X.
  If $n = 2$, a function $f : X \times X \to X$ is called a **binary operation** on X.

A binary operation is the simplest and the most frequently used operation in mathematics. Well-known examples of binary operations are addition, subtraction and multiplication on the sets of real and complex numbers. In set theory, the examples of binary operations are union, intersection and difference of sets. A binary operation f is often written using infix notations, such as $x + y$, $x - y$, $x \bullet y$, $x \cdot y$, $x * y$, instead of the form $f(x, y)$. Note, that binary operations written by $+$, $-$, $\cdot$ need not to be arithmetic operations. The meaning of operations will imply from definitions.

✷ **Examples A.6.**

1. Addition of numbers on the set $X = \{-1, 0, 1\}$ is not an operation on X, since $1 + 1 = 2 \notin X$.

2. Multiplication of numbers on the set $T = \{a + b\sqrt{2} + c\sqrt{3} : a, b \in \mathbb{Q}\}$, where $\mathbb{Q}$ is the set of rational numbers, is not an operation on T, since $\sqrt{2} \cdot \sqrt{3} = \sqrt{6} \notin T$.

If $X = \{x_1, x_2, \ldots, x_n\}$ is a finite set, then an operation • on this set is fully defined if for each pair of elements (x_i, x_j) we know the element $x_i \bullet x_j$. In this case, this operation can be given in the form of the **Cayley table**, named after British mathematician Arthur Cayley. In this table, elements of X are written twice: In the first row and in the first column. Furthermore, the result of the operation $x_i \bullet x_j$ is written at the intersection of the i-th row and the j-th column.

✳ Examples A.7.
Let $X = \{a, b, c\}$. Then, an operation • can be described by the following Cayley table:

•	a	b	c
a	a	b	c
b	c	c	b
c	b	b	b

Many operations in different algebraic systems have the same properties (or laws) known from elementary algebra. The names of these laws are standard in mathematics. We will write an arbitrary operation as •.

✳ Definition A.8.
A binary operation • on a set X is called:
1) **commutative** if
$$x \bullet y = y \bullet x \qquad (A.1)$$
for all $x, y \in X$.
2) **associative** if
$$x \bullet (y \bullet z) = (x \bullet y) \bullet z \qquad (A.2)$$
for all $x, y, z \in X$.

Addition and multiplication of real numbers are commutative and associative operations. Union and intersection of sets are also commutative and associative operations.

✳ Definition A.9.
Let • and $+$ be binary operations defined on a set X. We say that the operation • is **left-distributive** over the operation $+$ if
$$x \bullet (y + z) = (x \bullet y) + (x \bullet z)$$
and it **right-distributive** over the operation $+$ if
$$(y + z) \bullet x = (y \bullet x) + (z \bullet x).$$
for all $x, y, z \in X$.
The operation • is **distributive** over the operation $+$ if it is both left-distributive and right-distributive over $+$.

Multiplication of real numbers is distributive over addition. However, addition of real numbers is neither left-distributive nor right-distributive over multiplication. In the set algebra the operation of union is distributive over intersection, and intersection is distributive over union.

The important property of an operation on a set X is the existence of a neutral element with respect to this operation.

> ⁕ **Definition A.10.**
>
> An element $e \in X$ is called a **neutral element** or an **identity element** with respect to an operation $\bullet$ if $e \bullet x = x \bullet e = x$ for every element $x \in X$.

In the set of real numbers, the neutral element with respect to multiplication is 1, and 0 is the neutral element with respect to addition. In the set algebra, the universe set U is the neutral element with respect to intersection of sets, and the empty set $\emptyset$ is the neutral element with respect to union of sets.

Note, that not every operation on a set X has neutral elements, but if they exist there is only one neutral element with respect to the given operation.

> ⁕ **Theorem A.11.**
> For an operation $\bullet$ on a set X, there is at most one neutral element with respect to this operation.

Proof. Suppose that $e_1, e_2 \in X$ are neutral elements with respect to an operation $\bullet$. Then, from definition A.10, it follows that:

$$e_1 \bullet e_2 = e_1 \quad \text{and} \quad e_1 \bullet e_2 = e_2,$$

which implies that $e_1 = e_2$. □

> ⁕ **Definition A.12.**
>
> Let $e \in X$ be a neutral element of X with respect to an operation $\bullet$. An element $y \in X$ is called an **inverse element** to an element $x \in X$ with respect to the operation $\bullet$ if $x \bullet y = y \bullet x = e$.

> ⁕ **Theorem A.13.**
> For an associative operation $\bullet$ on a set X and an element $x \in X$ there is at most one inverse element with respect to this operation.

Proof. Let $e \in X$ be the neutral element with respect to an operation $\bullet$, and elements $y, z \in X$ are inverse elements to $x \in X$, i.e., $x \bullet y = y \bullet x = e$ and $x \bullet z = z \bullet x = e$. Then, it follows from Definition A.10 and Definition A.12, that:

$$(y \bullet x) \bullet z = e \bullet z = z \quad \text{and} \quad y \bullet (x \bullet z) = y \bullet e = y.$$

Since $\bullet$ is an associative operation on X, $(y \bullet x) \bullet z = y \bullet (x \bullet z)$. Hence, $y = z$. □

If for an element $x \in X$ there exists an inverse element with respect to an operation $\bullet$, then x is called an **invertible element** on X. If an operation $\bullet$ is associative in X, then the unique inverse element to an element x is often denoted by x^{-1}.

> ✳ **Definition A.14.**
> Let A and X be non-empty sets. A function $\varphi : A \times X \to X$ is called a **left external binary operation** on X over A. A function $\psi : X \times A \to X$ is called a **right external binary operation** on X over A.

If $\alpha \in A$ and $x \in X$, then $\varphi(\alpha, x)$ is usually written as $\alpha \cdot x$ or αx. So, a left external binary operation on a set X over a set A associates with a pair of elements (α, x), where $\alpha \in A$, $x \in X$, a uniquely defined element $\alpha x \in X$. An element $\alpha \in A$ is called an **operator**, and a set A is called a **set of operators**.

Usually, sets A and X are distinct, but sometimes $A \subset X$ or $X \subset A$. If $A = X$, we get the definition of a binary operation on a set X.

✳ Examples A.15.

1. Let $A = \mathbb{Z}$ be the set of integers, and $X = \mathbb{Q}$ be the set of rational numbers. Then, multiplication of an integer α by a rational number x gives a rational number αx and defines an external operation on the set $\mathbb{Q}$. The set of operators in this case is $\mathbb{Z}$.

2. Let $A = \mathbb{N}$ be the set of natural numbers, and $X = \mathbb{R}$ be the set of all real numbers. Then, multiplication of an integer $n \in A$ by a real number $x \in X$ gives a real number $x^n \in X$ and defines an external binary operation on the set X. The set of operators in this case is $\mathbb{N}$.

3. Let $A = \mathbb{R}$ be the set of real numbers, and X be the set of all functions defined on $(0, 1)$ which values are real numbers. Then, multiplication of a real number $\alpha \in A$ by a function $f \in X$ gives a real function αf and defines an external binary operation on the set X. The set of operators in this case is $\mathbb{R}$.

> ✳ **Definition A.16.**
> We say that an **algebraic structure** is given on a set X if there are given one or more non-empty sets together with a family of operations, such that these sets are closed under these operations and satisfy some axioms. It can be written in the following form:
>
> $$(X, A_1, A_2, \ldots, A_m, f_1, f_2, \ldots, f_n, g_1, g_2, \ldots, g_k),$$
>
> where $X, A_1, A_2, \ldots, A_m$ are non-empty sets, and $f_1, f_2, \ldots, f_n$ are operations defined on X and g_i is a left (or right) external binary operation defined on the set X over A_i for $i = 1, 2, \ldots, k$. Moreover, the priority of operators $f_1, f_2, \ldots, f_n$ and the priority of operators $g_1, g_2, \ldots, g_k$ is not significant.

✳ Examples A.17.

1. Let $\mathbb{Z}$ be the set of integers with operations of addition $+$ and multiplication $\cdot$. The system $(\mathbb{Z}, +, \cdot)$ is an algebraic structure with two binary operations $+$ and $\cdot$. The number 0 is the neutral element with respect to addition and 1 is the neutral element with respect to multiplication.

2. The system $(\mathbb{R}, +, -, \cdot, :)$, where $+, -, \cdot, :$ mean addition, substraction, multiplication and division on real numbers, is not an algebraic structure.

3. Let $(\wp(X), \cup, \cap)$ be the power set with operations of union $\cup$ and intersection $\cap$. This system is an algebraic operation with two binary operations $\cup$ and $\cap$. The empty set $\emptyset$ is the neutral element with respect to union and the set X is the identity element with respect to intersection.

4. The system $(\mathbb{Q}, \mathbb{Z}, +, \cdot, g)$, where $+$ and $\cdot$ mean addition and multiplication of numbers, and g is a transformation defined by $g(a, x) = ax$ for all $a \in \mathbb{Z}$ and all $x \in \mathbb{Q}$, is an algebraic structure.

5. The system $(\mathbb{Q}, \mathbb{N}, +, \cdot, g)$, where $+$ and $\cdot$ mean addition and multiplication of numbers, and g is a transformation defined by $g(n, x) = x^n$ for all $n \in \mathbb{N}$ and all $x \in \mathbb{Q}$, is an algebraic structure.

A.3. Vector Spaces

In this section, we recall the necessary definitions and information from linear algebra which will be used in this book.

✳ Definition A.18.

A **vector space** or (**linear space**) over a field K, with $1 \neq 0$, is an algebraic structure $(V, K, +, \circ)$ with two operations: One binary operation $+$, named addition:

$$+ : V \times V \to V$$

$$(u, w) \mapsto u + w, \quad \forall (u, w \in V)$$

and one external binary operation $\circ$, named multiplication by a scalar:

$$\circ : K \times V \to V$$

$$(\alpha, u) \mapsto \alpha \circ u, \quad \forall (\alpha \in K), \forall (u \in V)$$

which satisfy the conditions:

1. $(V, +)$ is an Abelian group, i.e.

   - the operation $+$ is associative and commutative

- there exists a neutral element with respect to operation $+$
- for any $x \in V$ there exists an inverse element.

2. $1 \circ u = u = u \circ 1$, where 1 is the identity of K

3. $\alpha \circ (\beta \circ u) = (\alpha\beta) \circ u$ - associativity

4. $\alpha \circ (u + v) = \alpha \circ u + \alpha \circ v$ - distributivity of $\circ$ over addition $+$ in V

5. $(\alpha + \beta) \circ u = \alpha \circ u + \beta \circ u$ - distributivity of $\circ$ over addition $+$ in K

for all $u, v \in V$ and all $\alpha, \beta \in K$.

Elements of a vector space V are called **vectors**, and elements of a field K are called **scalars**. The neutral element of an Abelian group $(V, +)$ will be denoted by θ and called the **zero element** of V. So, in each vector space $(V, K, +, \circ)$ the following properties hold:

1. $0 \circ u = \theta$

2. $\alpha \circ \theta = \theta$

3. if $\alpha \circ u = \theta$, then $\alpha = 0$ or $u = \theta$

4. $(-1) \circ u = -u$.

✳ Examples A.19.

1. The set of n-dimensional vectors with coordinates in $\mathbb{R}$:

$$\mathbb{R}^n = \{v = (x_1, x_2, \ldots, x_n) \ : \ x_i \in \mathbb{R}\}$$

together with operations of addition of vectors and multiplication of a vector by a real number is a vector space over $\mathbb{R}$.

2. The set of infinity sequences with coordinates in $\mathbb{R}$:

$$\mathbb{R}^\infty = \{v = (x_1, x_2, \ldots, x_n, \ldots) \ : \ x_i \in \mathbb{R}\}$$

together with operations of addition:

$$(x_1, x_2, \ldots, x_n, \ldots) + (y_1, y_2, \ldots, y_n, \ldots) = (x_1 + y_1, x_2 + y_2, \ldots, x_n + y_n, \ldots)$$

and multiplication by a real number:

$$\alpha \circ (x_1, x_2, \ldots, x_n, \ldots) = (\alpha x_1, \alpha x_2, \ldots, \alpha x_n, \ldots)$$

is a vector space over $\mathbb{R}$.

3. The set of $n \times m$ matrices $M_{n \times m}(K)$ with entries in a field K with addition of matrices and multiplication of a matrix by a scalar from K is a vector space over K.

4. The set $K[x]$ of polynomials over a field K with addition of polynomials and multiplication of a polynomials by a scalar from K is a vector space over K.

5. The set $C[a, b]$ of continuous functions $f : [a, b] \to \mathbf{R}$ with addition of functions:
$$(f + g)(x) = f(x) + g(x)$$
and multiplication of a function by a real number α:
$$(\alpha \circ f)(x) = \alpha f(x)$$
is a vector space over $\mathbb{R}$.

✳ Definition A.20.

A set $U \subseteq V$ is called a **vector subspace** of a vector space V over a field K if
$$a, b \in U \ \Rightarrow\ \alpha a + \beta b \in U$$
for all $\alpha, \beta \in K$.

From this definition, it follows that the zero element $\theta \in V$ also belongs to U, since for an arbitrary vector $a \in U$ we have: $0 \circ a = \theta$. Moreover, if U is a vector subspace of V then $a, b \in U$ implies that $a + b \in U$ and $\alpha \circ a \in U$ for all $\alpha \in K$.

✳ Theorem A.21.
A subset $U \subseteq V$ is a vector subspace of a vector space V if and only if U itself is a vector space with operations defined in V.

Each non-zero vector space V has at least two subspaces: The zero vector space $\{\theta\}$ and itself, i.e., V, which are called **trivial subspaces**.

✳ Example A.22.
The set of vectors of the form $(a, b, 0)$, where a, b are real numbers, is a vector subspace of the vector space $\mathbb{R}^3$.

✳ Definition A.23.

Let V be a vector space over a field K. Let $\alpha_1, \alpha_2, \ldots, \alpha_m \in K$. A vector $v \in V$ of the form:
$$v = \alpha_1 v_1 + \alpha_2 v_2 + \cdots + \alpha_m v_m$$
is called a **linear combination** of vectors $v_1, v_2, \ldots, v_m \in V$. A linear combination is called **trivial** if $\alpha_i = 0$ for all $i = 1, 2, \ldots, m$.

✳ Definition A.24.

A system of vectors $v_1, v_2, \ldots, v_m, \ldots$ of a vector space V over a field K is called **linearly dependent** if there exists a non-trivial linear combination of these vectors which is equal to the zero vector of V, i.e., there are elements $\alpha_1, \alpha_2, \ldots, \alpha_m, \ldots \in K$, not all equal to zero, such that

$$\alpha_1 v_1 + \alpha_2 v_2 + \cdots + \alpha_m v_m + \cdots = \theta.$$

Otherwise a system of vectors $v_1, v_2, \ldots, v_m, \ldots \in V$ is called **linearly independent**.

✴ Examples A.25.

1. The system of vectors $v_1 = (1, 0, -3)$, $v_2 = (2, 0, 0)$, $v_3 = (0, 0, 3)$ is linear dependent because $-2v_1 + v_2 - 2v_3 = \theta$.

2. The system of vectors $e_1 = (1, 0, \ldots, 0)$, $e_2 = (0, 1, \ldots, 0)$, $\ldots$, $e_i = (0, \ldots, 1, \ldots, 0)$ (where 1 is at the i-th place), $\ldots$, $e_n = (0, 0, \ldots, 1)$ of the vector space K^n over a field K is linear independent. Really, an equality $\alpha_1 e_1 + \alpha_2 e_2 + \cdots \alpha_n e_n = \theta$ implies that $(\alpha_1, \alpha_2, \ldots, \alpha_n) = (0, 0, \ldots, 0)$, and so $\alpha_1 = \alpha_2 = \ldots = \alpha_n = 0$.

✳ Definition A.26.

Let V be a vector space and $A \subset V$. A set of all linear combinations of vectors of A is called a **linear shell** of A and is denoted by $\text{Lin}(A)$.

If $B \subseteq V$ and $\text{Lin}(A) = B$, then we say that a set B is generated by a set A. Elements of a set A are said to be **generators** of B.

✳ Definition A.27.

A system of vectors A of a vector space V is called a **basis** of V if it satisfies the conditions:

1. A is a linear independent system of vectors.

2. $\text{Lin}(A) = V$

✴ Examples A.28.

1. The vectors $e_1 = (1, 0, \ldots, 0)$, $e_2 = (0, 1, \ldots, 0)$, $\ldots$, $e_n = (0, 0, \ldots, 1)$ forms a basis of the vector space $\mathbb{R}^n$.

2. The vectors $v_0 = 1$, $v_1 = x$, $v_2 = x^2$, $\ldots$, $v_n = x^n$ forms a basis of the vector space

$$K_n[x] = \{a_0 + a_1 x + a_2 x^2 + \cdots + a_m x^m \ : \ m \le n, \ a_i \in K\}$$

of polynomials of degree $\leq n$ over a field K.

3. The vectors $v_0 = 1$, $v_1 = x$, $v_2 = x^2$, ..., $v_n = x^n$, ..., forms a basis of the vector space

$$K[x] = \{a_0 + a_1 x + a_2 x^2 + \cdots + a_m x^m \ : \ a_i \in K\}$$

of all polynomials over a field K.

The next theorem gives an equivalent definition of a basis of a vector space.

✳ Theorem A.29.

For a system of vectors $B = \{v_1, v_2, \ldots, v_n\}$ of a vector space V over a field K the following conditions are equivalent:

1. B is a basis.

2. B is a maximal linearly independent system of vectors.

3. Each vector $x \in V$ can be uniquely written in the form:

$$x = \alpha_1 v_1 + \alpha_2 v_2 + \cdots + \alpha_n v_n,$$

where $\alpha_i \in K$ for all i.

In the theory of vector spaces, the following results are important.

✳ Theorem A.30 (Steinitz).

Each vector space has a basis.

✳ Theorem A.31.

1. If a basis of a vector space V consists of $n \geq 1$ vectors, then each other basis also consists of n vectors.

2. If a basis of a vector space V is infinite, then each other basis is also infinite.

✳ Definition A.32.

If a vector space V has a finite basis, then V is called **finite dimensional**. The number of vectors in a basis of a finite dimensional vector space V is called the **dimension** of V and is denoted by $\dim V$.

If a vector space V has an infinite basis, then V is called **infinite dimensional** and we write $\dim V = \infty$.

✳ Examples A.33.

1. $\dim \mathbb{R}^n = n$.

2. $\dim \mathbb{R}_n[x] = n + 1$.

3. $\dim \mathbb{R}[x] = \infty$.

4. $\dim M_{n \times m}(\mathbb{R}) = n \times m$.

✳ **Theorem A.34.**

If the dimension of a vector space V is equal to n then a set $S = \{v_1, v_2, \ldots, v_n\} \subseteq V$ of linear independent vectors is a basis of V.

✳ **Theorem A.35.**

If U is a vector subspace of a vector space V then

$$\dim V = \dim U + \dim(V/U).$$

Index

T - #0138 - 071024 - C0 - 234/156/22 - PB - 9780367609085 - Gloss Lamination